Parallel Processor Systems, Technologies, and Applications

Parallel Processor Systems, Technologies, and Applications

Edited by

L. C. Hobbs, Hobbs Associates, Inc.
D. J. Theis, Hobbs Associates, Inc.
Joel Trimble, Office of Naval Research
Harold Titus, Naval Postgraduate School
Ivar Highberg, Naval Weapons Center

Spartan Books

New York • Washington

Library of Congress Catalog Card Number 77-118150
International Standard Book Number 0-87671-161-1

Sole distributor in Great Britain, the British Commonwealth, and the Continent of Europe:

Macmillan & Company, Ltd.
4 Little Essex Street
London, W.C. 2

Printed in the United States of America

Contents

PREFACE

Increased application requirements, new computer technologies, and new concepts in system organization have led to increasing interest in more highly parallel computer organizations.

A considerable amount of work has been done in this country during the past ten years on different types of parallel computer organizations. At the same time, many problems have arisen, particularly in military systems, which are not efficiently or effectively handled by conventional computer organizations. Since much of this work has been sponsored by the Navy and other parts of the Department of Defense, and since a number of Navy problems can benefit from the application of parallel processing techniques, it seemed appropriate that the Navy sponsor a symposium to bring together those with problems and applications in which parallel processing is a potential advantage and those specialists from different computer disciplines working on or interested in parallel processors. The symposium on "Parallel Processor Systems, Technologies, and Applications" was held at the Naval Postgraduate School in Monterey, California on June 25, 26, 27, 1969, sponsored by the Office of Naval Research, the Naval Postgraduate School, the Naval Weapons Center, and Hobbs Associates, Inc. The major objective of this symposium was the interchange of information and discussion of problems among those who may use parallel processor systems, those who may design or program such systems, and those who are working on basic technologies that may be used in such systems.

Much of the work to date on parallel processor organizations has been directed toward the ability to solve very large problems that cannot be handled with conventional machine organization approaches within the circuit speeds permitted by the existing state-of-the-art. Smaller scale parallel processor systems are also attractive for problems which are inherently parallel in nature. Effective utilization of LSI semiconductor technology, which places importance on minimizing interconnections between arrays and on the repetitive use of very large logic arrays, offers a further reason for considering parallel organizations. More exchange of information between users with specific applications that can benefit from parallel processing techniques and those working on system organization concepts, batch-fabricated hardware technologies (e.g., LSI), and software techniques is needed.

This symposium provided a forum for this exchange of information by bringing together active workers and interested participants from system, device, software, and application disciplines.

In an attempt to maximize the exchange of information between active workers and other interested persons in different disciplines, the format of this symposium was somewhat different from that usually encountered in computer conferences. Most of the speakers prepared longer papers which are published in these Proceedings, but each speaker was requested to limit his oral presentation to a 20 minute summary of the more important points of his paper in order to permit more time for discussion of the topics raised by the papers in each session and related topics of interest to the audience. A panel of experts in the area covered by each session was included in the session in addition to the formal papers to facilitate these discussions. Copies of the complete papers for each session were furnished to the panelists in advance, with the request that they read the papers prior to the symposium and be prepared to discuss topics raised by the papers as well as any related subjects which they considered of importance and interest. The panelists and authors in each session also discussed any questions raised by the audience. The exchange of information and ideas was stimulating and a lively discussion resulted in contrast to the usual one-way presentation in which an active speaker presents ideas to a passive audience.

Some of the work on parallel system organizations frequently appears to be a solution in search of a problem. On the other hand, there are many problems in military systems searching for solution which may be afforded by some form of parallel organization. This symposium attempted to bring together problems and solutions hoping to better match some problems with potential solutions. Hence, the first paper in the Proceedings presents an introduction and survey of parallel processors, and the next five papers are devoted to a discussion of several specific applications for parallel processing–primarily those involving military radar and sonar systems.

Following the discussion of applications of parallel processing systems, the remaining papers discuss the means of implementing such systems in three different categories–device technology and utilization, system and machine organization, and programming and software techniques. The next three papers treat two different technologies which are applicable to the implementation of parallel systems–large scale integrated circuit arrays and plated wire associative techniques. The next six papers discuss five different approaches to parallel machine organization and the analysis and synthesis of control mechanisms for such systems.

Finally, papers 16 through 19 consider the programming and software techniques and problems involved in such systems. The identification of inherent parallelism in problems and means of taking advantage of this in parallel systems is given particular consideration. Other problems in programming parallel systems are also considered.

The sponsors and organizers of this symposium focused attention both on applications and implementation for parallel systems and on the interrelation and trade-offs between device technologies, system organizations, and software. It is hoped that these Proceedings, in expanding the availability of this material beyond those who were able to attend the symposium, will lead to further consideration and understanding of these applications, interrelationships, and trade-offs.

L. C. Hobbs
D. J. Theis
Joel Trimble
Dr. Harold Titus
Dr. Ivar Highberg

INTRODUCTION AND SUMMARY

CHAPTER 1
SURVEY OF PARALLEL PROCESSOR APPROACHES AND TECHNIQUES

L. C. Hobbs and D. J. Theis
Hobbs Associates, Inc.

INTRODUCTION

In this survey of various parallel processor design approaches, the term "parallel processing" is interpreted in the broadest sense to include any type of computer or system organization in which multiple operations are performed in parallel. This includes types of organizations commonly covered by the terms "multicomputer systems," "multiprocessing systems," "parallel processors," "array processors," and "associative processors." Other types of organizations are also considered that meet the criteria of performing multiple operations in parallel.

Several reasons can be cited for justifying interest in parallel processor organizations. These are considered here in two categories. The first category includes functional reasons, the second includes hardware and economic reasons. Even if application or functional requirements cannot justify the use of parallel processing for a specific type of problem, the economics of new hardware technologies may still justify a parallel organization.

FUNCTIONAL REASONS FOR PARALLEL PROCESSING

Several reasons can be cited for justifying research and development work on parallel processor systems and techniques from the standpoint of functional needs. These include:

1) very large processing and computation problems
2) problems with inherent parallelism in either data streams or processing operations
3) multiprogramming and simultaneous use by multiusers
4) reliability and graceful degradation.

Large Problems

The history of the computer field has indicated that at any given point in time some users have very large processing and computation problems which place requirements on a computer system that exceed those which can be implemented with conventional computer organization approaches using state-of-the-art circuit and memory speeds available at that point in time. In spite of the continual efforts and achievements in increasing the speeds of logic circuits, memories, and input/output equipment, requirements for processing and computation capabilities have increased at a somewhat comparable rate. As computer performance increases, there is a continuing pressure to solve problems for which solutions have been impossible in the past. In a sense, computer users are as creative as computer designers.

If the magnitude of these large processing and computation problems results from a need for processing a number of different sets of data within a given time, one approach to a solution is to use different processing units to work simultaneously on different data sets rather than sequentially forcing different data sets through the same processor in time sequence. Examples of such problems include the global weather problem, nuclear physics problems, large hydrodynamic problems, and others in which an array or mesh of data points are processed.

Problems with Inherent Parallelism

There is another class of problems for which speed and performance requirements may not justify a parallel organization but in which the data processing involved is inherently parallel. Such inherent parallelism may result from the presence of several data streams which can be processed in parallel, and subsequently under the control of a single instruction stream. Another type of inherent parallelism results from the requirement to perform several relatively independent processing operations on different sets of data (i.e., the data in one operation does not depend upon that in another) between two points in the program. In other words, the program may progress to one point and then branch into several processing paths which can be performed in parallel and simultaneously before joining again into a common processing path. Such parallel computations are normally performed in a conventional computer organization by forcing them into sequential operations. When such parallel computation paths are relatively infrequent, this is usually the better approach. However, if such parallel computations are quite frequent in the program, the overall program may be executed more efficiently (from an equipment and economic standpoint) as well as in a shorter elapsed time by the use of some type of parallel organization.

Parallel data streams are typically found in phased array radar, sonar and radar signal processing, and pattern recognition problems. The correlation of radar returns with specific target tracks in track-while-scan radar systems is a good example of parallel computations.

Multiprogramming and Multiuser Systems

Increasingly, the performance capabilities of modern, large, high-speed computers are being effectively utilized by multiprogramming techniques in which several programs are in different stages of execution in the computer at the same time. Multiprogramming usually makes more effective use of the computer's capabilities since the high-speed portions of the computer can be used by one program at times when they are not used by another. For example, if one program is awaiting the completion of a disk access, another program can be using the high-speed arithmetic unit, and another can be initiating input/output operations or data transfers.

Another strong trend in current systems design is that toward multiuser systems. In such systems, a number of users may be operating on line with the system simultaneously (e.g., time-sharing systems), again in an attempt to more effectively use the high speed computation capabilities of the machine. In a multiuser system, one user may be utilizing the high-speed arithmetic unit for internal computation, while another user is accessing disks or performing an input/output operation. In this sense, a multiuser system is a special case of the multiprogramming system. In multiprogramming and multiuser systems, the systems software must keep track of the execution of the different programs and make sure that one program does not interfere with another, while at the same time assuring that the high speed portions of the machine are effectively utilized. Such software systems impose an appreciable systems overhead in terms of both execution time and core storage requirements. These requirements on the systems software can be reduced in a multiprocessor system—particularly if separate control units can keep track of each program while sharing the processing and input/output capabilities with other control units which are executing other programs.

Reliability and Graceful Degradation

The failure of any major unit (e.g., the arithmetic unit, the control unit, the memory, etc.) in a single processor system effectively puts the complete system out of operation until the malfunction is corrected. However, in a parallel processor organization, a failure in one unit need not interfere with the continuing operation of other parallel units. Hence, a parallel organization can increase the system reliability—particularly if some redundancy is employed. Even without adding redundancy, a parallel organization can permit graceful degradation of the system by dropping off less critical operations in the event of the malfunction of a part of the system, while permitting the more critical operations to continue in those parallel units which are still operable. For example, in a three computer multicomputer system, the major functions of the system can continue in the event of a malfunction in one computer by transferring the critical systems operations to the two remaining computers, while temporarily ceasing to perform desirable but less critical operations until the third computer is again operable. In a highly parallel organization, the failure of a single unit would decrease the total system capability by a very small percentage and would have an almost negligible effect on the necessary functions in the overall system. This is particularly true if the system is designed such that it can automatically detect a malfunction in one unit and reconfigure the problem to perform the critical functions on the remaining units.

While graceful degradation is desirable in many commercial applications (e.g., process control, time-sharing systems, etc.), it is essential in any military system where the results of complete system failure for even a short period of time could be catastrophic. The importance of this aspect of parallel processor techniques is further enhanced in military systems since they face the loss of part of the system hardware not only from normal equipment and hardware failures, but also from the results of enemy action such as shell or bomb damage. In the latter case, however, it is important to note that this advantage of a parallel organization may be largely nullified unless the components of the parallel system are distributed physically as well as conceptually.

HARDWARE AND ECONOMIC REASONS FOR PARALLEL PROCESSING

Even if parallel processing systems could not be justified on the basis of problem or application requirements, there are still technology reasons for utilizing parallel organizations in the future rather than more conventional machine organizations. While conventional computer organizations have been married quite effectively to existing hardware technology over a period of years, this will not automatically be the case in the future. Some of the new developments and technologies, such as large-scale integrated circuit arrays, LSI semiconductor memories, nondestructive-read-out plated wire memories, optical techniques, etc., may be used more effectively in some form of parallel organization.

One specific example of this is found in the problem of repetitive use of large semiconductor logic arrays. If semiconductor technology progresses a few years in the future to the point that it is possible to successfully interconnect several thousand logic elements on a single silicon chip, a relatively sophisticated computer could be implemented with 20 to 30 chips of this type. With present conventional machine organization approaches, each chip would tend to be unique so that the number of types of chips required would approach the total number of chips in the machine. It is very difficult to partition present computer organizations into large segments and yet permit a given segment to be used in more than one place in the system. A highly parallel organization or a functional organization provides an approach to the repetitive use of large arrays. In a parallel system implemented with LSI arrays, a few chips may be unique but a large percentage of the chips could be identical, thus providing repetitive use with the consequent design and manufacturing economies. Similarly parallel organizations may permit the number of interconnections required between LSI chips to be significantly reduced. This is extremely important since interconnections require pins or lead on the chips which may easily be the limiting factor on the size of the array that can be fabricated on a single chip.

Another example of the effect of changing technology on the desirability or feasibility of parallel organizations is afforded by recent work on semiconductor memories. The use of small distributed memories is now feasible since LSI memory costs tend to be linear with size. Hence, it is no longer economically necessary to design around a large central memory. The total internal memory capacity can be distributed over a number of smaller memory modules which can be accessed in parallel.

Still another example is provided by the way in which the nondestructive-read-out capability of plated wire and LSI memories permit associative techniques which facilitate the design of associative processors.

SCOPE OF PARALLEL PROCESSING AND TYPES OF SYSTEM ORGANIZATIONS

In this paper, the term "parallel processing" is interpreted broadly to include any type of system organization in which multiple operations are accomplished simultaneously, or multiple hardware control or processing units are working simultaneously. This broad definition permits consideration of widely different types of organizations and systems including multicomputer systems, multiprocessors, associative processors, array or network processors, and functionally partitioned systems.

Classification or categorization of parallel processing approaches is difficult at this time because the field is far from mature, and overlap is almost inevitable between any types or categories that may be selected. Several attempts have been made to delineate clear cut classification criteria, but a number of different parallel organizations that have been investigated to date could reasonably be placed in two or more categories in such classification schemes. In 1966, M. J. Flynn[1] set forth the following classes of high speed computers:

1) single instruction stream, single data stream (SISD)
2) single instruction stream, multidata stream (SIMD)
3) Multiinstruction stream, single data stream (MISD)
4) Multiinstruction, multidata stream (MIMD).

The first of these is a conventional system organization, but the last three can be considered as one approach to dividing parallel organizations into three broad categories, depending upon whether the instructions and/or data are handled one at a time or several at a time.

In 1964, J. C. Murtha and R. L. Beadles[2] set forth three broad classifications of parallel processors:

1) general purpose network computers
2) special purpose network computers characterized by global parallelism
3) nonglobal computers where each module is only semiindependent (i.e., locally parallel).

General purpose network computers were further divided into:

1) parallel networks with a common central control
2) parallel networks with many identical processor elements which can independently execute instructions.

Special purpose network computers were divided into:

1) pattern processors
2) associative processors.

"Nonglobal" was used as a catchall for other types of parallel organizations that do not fall into the first two categories. They set forth a criteria for differentiating between multiprocessors and highly parallel processor organizations–the highly parallel processor is one that has more than ten processors and a degree of integration between the processors that allows them to work on the same algorithm.

Another approach to classification is to consider those things which can be in parallel:

1) control functions or units
2) functional processing units
3) data streams.

Parallel control units can simultaneously provide independent instruction streams, either operating on parts of the same problem or on different problems. Parallel functional processing units, which may be identical or which may differ, can simultaneously operate on either a single or multiple data stream under control of a single or multiple control unit. Parallel data streams may be operated upon under the control of an identical instruction sequence performing the same operation on each data stream simultaneously, or by independent instruction sequences operating independently on each data stream. Although this approach to classification is relatively clear cut from the conceptual standpoint, many of the parallel processing systems that have been developed or proposed do not fall neatly into one or the other of these categories. In fact, some of them involve parallel control units, parallel functional processing units, and parallel data streams in the same system.

Hence, in this paper a different approach has been taken—that of grouping together the different parallel systems that have been developed or proposed and then assigning an appropriate classification title to each group. This is admittedly less academic but hopefully more practical. On this basis, four categories can be established:

1) multicomputers and multiprocessors
2) associative processors
3) network or array processors
4) functional organizations.

Table 1 shows the distribution of a number of systems into these four categories and cites a reference for each. The following paragraphs describe each category and present additional comments on the strengths and weaknesses peculiar to each design approach.

Multicomputers and Multiprocessors

Type I includes all multicomputers and multiprocessors except highly parallel machines (i.e., ones with more than ten processors and where the processors are integrated to allow them to work on the same algorithm). In early multicomputer systems, several complete computers were interconnected in a manner which facilitated the transfer of data and the assignment of processing tasks between them. This permitted them to work together on different parts of the same

Table 1
Categorization of Parallel Processor Types

Type I *Multicomputers and Multiprocessors*	Type II *Associative Processors*	Type III *Parallel Network of Array Processors*	Type IV *Functional Machines*
IBM 9020[3]	Rosin-Associative Cryogenic Computer[10]	Holland Machine[17]	IBM 360/91[30]
RW-400[4]	Librascope's Associative Parallel Processor[11]	Comfort's-Modified Holland Machine[18]	CDC 6600[31]
Pilot[5]	Lee-Intercommunicating Cells[12]	Gonzalez-Multilayer Iterative Circuit Computer ICC[19]	CDC 7600[32]
Intrinsic Multiprocessor IMP[6]	Davies–An Associative Processor[13]	Squire-Multidimensional Machine[20]	RCA LIMAC[33]
Burroughs D825[7]	Distributed Logic Memory DLM[14]	Vector Arithmetic Multiprocessor VAMP[21]	Bull Gamma 60[34]
Multics[8]	Hughes Associative Storing Processor[15]	Unger's Machine[22]	
Hughes 4118[9]	Goodyear's Associative Processor[16]	Solomon Machine[23]	
		Knapp-Iterative Array Computer[24]	
		Associative Logic for Highly Parallel Systems ALPS[25]	
		Conway's Machine[26]	
		ILLIAC IV[27]	
		Autonetic's Distributed Processor[28]	
		Litton's Block Oriented Computer, BOC[29]	

program or on different programs in an overall problem (e.g., air defense). The computers in the Navy's NTDS are one of the earliest and most successful examples of a multicomputer configuration.

In more recent systems, the use of several complete computers, each of which is capable of operating as a single unit, has given way to multiprocessors in which several processing units share common memory and common input/output equipment. Although each processor usually includes its own control unit, a higher level control unit may be used to control the transfer of data and to assign tasks and sequences of operations between the different processors. Frequently, one of the processors may itself serve this higher level control function. In these organizations, the processors may be identical, or some of them may be different to accomplish specific types of operations more efficiently (e.g., an arithmetic processor, an input/output processor, etc.). Frequently many of the processors are subordinate to the main central processor and their function is to keep the I/O peripherals busy.

Associative Processors

An associative processor utilizes associative memory techniques but with the inclusion of additional processing logic at each cell or word location to permit actual processing operations on each word of data. An assocative memory is a device capable of retrieving stored data by means of testing part or all of the contents of each word simultaneously (by hardware means) in order to find one or many desired words. An associative processor performs one operation on N operands simultaneously. In a typical associative processor system, a given bit position of each word is simultaneously read and some processing operation performed on it by the logic common to each word position. Successive bit positions of all words are sequentially passed through this logic at each word position.

The Hybrid Associative Processor is also included in this category. This type of organization uses a general purpose digital computer to control the associative processor. One distinction that should be clarified is the difference between Knapp's Iterative Array Computer[24] and Goodyear's Hybrid Associative Computer.[16] The Iterative Array Computer uses a CDC 160 computer incorporated integrally into a design of a 10 × 10 array of Processor Elements (PE's) which results in a Parallel Network configuration. The hybrid machine, however, uses a CDC 1604B to control the Goodyear 2048 word associative processor and, as such, the combination is called a Hybrid Associative Processor. The intent of the Goodyear system is to evaluate associative processors, not hybrid associative processors. It is only hybridized for the expediency of time and cost.

Davies Associative Processor[13] consists of many parallel networks, each with a very small memory and an associative memory with a considerable amount of arithmetic logic interleaved throughout the associative memory. Lee's Intercommunicating Cells[12] is a modular design with considerable distributed logic for performing retrieval testing and searching functions. This unit is more a memory than a full blown computer. In general, the organization and software for these types of machines have been complex and difficult, but the Goodyear Associative List Searcher seems relatively straightforward to program.

Network or Array Processors

Network or array processors involve a large number of processing units interconnected in some kind of network, frequently a matrix. Usually each processing element communicates with its immediate neighbor (above, below, to the right, and to the left) although provision is sometimes made for communicating with more remote units, but in a highly restrictive fashion, i.e., each processing element usually cannot communicate indiscriminately with any other element in the array. Some network or array processor systems have completely distributed all control functions among the elements of the array, but more recent approaches for this type of organization have tended to provide some central control in addition to the control within each processor element. Although perhaps not necessary from the hardware standpoint, a centralized control is almost essential to permit the programming of array processors—at least within the present state of the programming art.

The first efforts at this type of architecture were the Unger[22] and Solomon[23] machines. The intended applications were partial differential equations and pattern recognition data processing. In this type of machine, the Broadcast Register provides simultaneous access to the same piece of data by all processor elements (PE's) so redundant storage of data in each PE memory is not necessary. The actual programs are stored separately in a program memory; whereas, the operands and other essential references are stored in the PE memory. The Network Sequencer decodes all instructions. With new LSI technology providing higher speed circuitry in the PE's, the overall system will be limited by memory speed and instruction decoding time. Another problem in this type of design is the need for a more flexible means of data exchange between PE's instead of each PE being restricted to communicating with its four nearest neighbors. This problem is usually referred to as path building for establishing communication links to pass information from one PE to any other PE. Memory addressing seems to be very cumbersome in this type of architecture. As more experience was gained on this type of architecture, it became evident that much more capability was needed in each PE.

The Holland machine[17] approach was more theoretical than practical. The concept is challenging, but the realization problems are formidable to say the least, especially in terms of economics. This network of identical processors, which can execute instructions independently of one another, is seriously limited from implementing programs because of the lack of coordination between PE's. In addition, the hardware costs for building this structure are very high in view of the limited programming performance. The modified Holland machine[18] has a 4:1 hardware reduction over the Holland machine, and the instruction set is more versatile. Squire's multidimensional machine[20] seems well beyond practical economics, and the same might be said for the Iterative Circuit Computer[19] which is limited to a few special purpose applications related to geometric patterns. These machines continue to require improvement in their path building capabilities which accounts for considerable timing overhead and for extra circuitry.

The highly parallel network ILLIAC IV[27] approach appears very promising for scientific calculations involving a wide variety of sophisticated mathematical techniques including matrix multiplications, partial differential equations for modeling weather problems, linear programming, statistical communication algorithms, convolution and correlation equations, etc. A significant amount of research and development effort is being expended in developing software on the ILLIAC IV for solving most kinds of scientific computations.

Floating point arithmetic hardware is the cornerstone for any high-speed scientific computer, and presently this hardware is very expensive. Hopefully LSI technology will reduce the impact of these hardware costs. Integrated (monolithic) circuits are used throughout the ILLIAC IV. PE's constitute 90% of the hardware in the system excluding the peripheral equipment.

Associative Logic for Highly Parallel Systems, ALPS[25], presents a good approach to the traffic handling (scheduling) problem between many processors operating simultaneously with many semiindependent processing paths, but some kind of data flow analysis is needed to minimize the hardware requirements. Programming ALPS appears straightforward in the sense that a systems' programmer could learn it quickly. Conway's machine[26], which was most notable for presenting two parallel instructions, FORK and JOIN, closely resembles the ALPS machine. As it is true for most of these highly parallel processors, memory organization and data routing are problem areas where the hardware solutions yet are to be found.

Functional Organizations

A functional organization is one in which a number of functional modules are provided to permit performing different types of operations concurrently on a different data item within a single program or on different programs. The advent of LSI is giving large impetus to the interest in this type of organization because of the opportunity it offers for overcoming many of the problems in the use of LSI arrays. In one sense, this type of organization is an extension of the multiprocessor organization in which several different types of processors are used. The processors are smaller and are functionally organized. The further distinction is that each of the functional organizations in this type of machine may not have an internal programmable control unit. In other words, each of the functional units is hard-wired to perform a specific type of operation. The LIMAC[33] being developed at RCA Labs is a primary example of this type of organization.

Other Considerations

The evaluation of types of parallel processor configurations, both proposed and implemented, emphasizes several key considerations that can significantly favor a parallel processor configuration which would not be technically favored otherwise. Some of these are:

1) A hardware fabrication breakthrough may allow one parallel processor design approach to take precedence over all others.

2) One approach may be successful in a particular company because of its compatibility with that company's present generation of computers.
3) Software algorithms may be developed that provide high equipment utilization for one particular approach.
4) Government support of a particular approach may result in one parallel processor approach dominating regardless of whether it is technically superior or not.

References

1. M. J. Flynn, "Very High Speed Computing Systems," Proceedings IEEE, 54, No. 12 (December, 1966), pp. 1901-1909.
2. J. C. Murtha and R. L. Beadles, "Survey of the Highly Parallel Information Processing Systems," Prepared by Westinghouse Electric Corporation, Aerospace Division, ONR Report No. 4755, November, 1964.
3. G. R. Blakeney, *et al.*, "IBM 9020 Multiprocessing System," *IBM Systems Journal,* 6, No. 2 (1967), 80-94.
4. R. E. Porter, "The RW-400–A New Polymorphic Data System," *Datamation,* 6, No. 1 (January/February, 1960), 8-14.
5. A. L. Leiner *et al.*, "PILOT–A New Multiple Computer System," *Journal, ACM,* 6 (July, 1959), 315-35.
6. R. A. Aschenbrenner, M. J. Flynn, and G. A. Robinson, "Intrinsic Multi-Processing," Proceedings AFIPS, SJCC, 1967, pp. 81-86.
7. J. P. Anderson *et al.*, "D825–A Multiple Computer System for Command and Control," Proceedings AFIPS, FJCC, 1962, pp. 86-96.
8. F. J. Corbato and V. A. Vyssotsky, "Introduction and Overview of the Multics System," Proceedings AFIPS, FJCC, 1965, pp. 185-96.
9. HM-4118 Reference Manual, Hughes Aircraft Company, Hughs Fullerton, March, 1966, FR 66-11-64.
10. R. F. Rosin, "An Organization of an Associative Cryogenic Computer," Proceedings AFIPS, SJCC, 1962, pp. 203-207.
11. R. H. Fuller and R. M. Bird, "An Associative Parallel Processor with Application to Picture Processing," Proceedings AFIPS, FJCC, 1965, pp. 105-16.
12. C. Y. Lee, "Intercommunicating Cells, Basis for a Distributed Logic Computer," Proceedings AFIPS, FJCC, 1962, pp. 130-36.
13. P. M. Davies and R. G. Ewing, "An Associative Processor," Proceedings AFIPS, FJCC, 1964, pp. 147-58.
14. B. A. Crane and J. A. Githens, "Bulk Processing in Distributed Logic Memory (DLM)," *IEEE Transactions on Electronic Computers,* April, 1965, pp. 186-96.

15. D. A. Savitt, H. H. Love, and R. E. Troop, "ASP: A New Concept in Language and Machine Organization," Proceedings AFIPS, SJCC, 1967, pp. 87-102.
16. Russ G. Gall, "Hybrid Associative Computer Study," RADC-TR-65-445, Vol. 1, Final Report, July, 1966.
17. J. H. Holland, "A Universal Computer Capable of Executing an Arbitrary Number of Subprograms Simultaneously," Proceedings AFIPS, EJCC, 1959, pp. 108-13.
18. W. T. Comfort, "A Modified Holland Machine," Proceedings AFIPS, FJCC, 1963, pp. 481-88.
19. Rudolfo Gonzalez, "A Multi-Layer Iterative Circuit Computer (ICC)," *IEEE Transactions on Electronic Computers,* December, 1963, pp. 781-90.
20. J. S. Squire and S. M. Palais, "Programming and Design Considerations of a Highly Parallel Computer," Proceedings AFIPS, SJCC, 1963, pp. 395-400.
21. D. N. Senzig and R. V. Smith, "Computer Organization for Array Processing," Proceedings AFIPS, FJCC, 1965, pp. 117-29.
22. S. H. Unger, "A Computer Oriented Toward Special Problems," Proceedings IRE, October, 1958, p. 1744.
23. D. L. Slotnick, W. C. Borck, and R. C. McReynolds, "The Solomon Computer," Proceedings AFIPS, FJCC, 1962, pp. 97-107.
24. Morris A. Knapp, "Parallel Processing Computer System" Rome Air Development Center, Report No. RADC-TR-66-567, AD 803485, November, 1966.
25. R. R. Seeber and A. B. Lindquist, "Associative Logic for Highly Parallel Systems (ALPS)," Proceedings AFIPS, FJCC, 1963, pp. 489-93.
26. M. E. Conway, "A Multi-Processor System Design," Proceedings AFIPS, FJCC, 1963, pp. 139-46.
27. ILLIAC IV System Study Final Report, Burroughs Corporation, University of Illinois, Purchase Order No. 09852-B, December, 1966.
28. G. J. Burnett, L. J. Koczela, and R. A. Hokom, "A Distributed Processing System for General Purpose Computing," Proceedings AFIPS, FJCC, 1967, pp. 757-68.
29. J. O. Campeau, "The Block Oriented Computer," IEEE Computer Group Conference Digest, June 25-27, 1968, pp. 57-60.
30. D. W. Anderson, *et al.*, "Machine Philosophy on System 360 Model 91," *IBM Journal of Research and Development,* 2, No. 1 (January, 1967), pp. 8-25.
31. J. E. Thornton, "Parallel Operation in the Control Data 6600," Proceedings AFIPS, FJCC, 1964, Part II, pp. 33-40.
32. Control Data Corporation, "CDC 7600 Preliminary Reference Manual," St. Paul, Minnesota, 1968.
33. S. Y. Levy *et al.*, "System Utilization of Large-Scale Integration," *IEEE Transactions on Electronic Computers,* October, 1967, pp. 562-66.
34. R. Dreyfus, "System Design of the Gamma 60," 1958 Proceedings WJCC, pp. 130-33.

Bibliography

Alt, F. L., and Rubinoff, M. *Advances in Computers,* Vol. 7. New York: Academic Press, 1966.

Amarel, S. "Iterative Circuit Computers," *IRE Transactions on Electronic Computers,* ED-9 (September, 1960), pp. 384-85.

Amdahl, G. M. "New Concepts in Computing System Design," Proceedings IRE, 50 (May, 1962), 1073-77.

Amdahl, G. M. "Validity of the Single Processor Approach to Achieving Large Scale Computing Capabilities," Proceedings AFIPS, SJCC, 1967, pp. 483-85.

Anderson, J. P. "Better Processing Through Better Architecture," *Datamation,* August, 1967, pp. 37-41.

"Associative Memory Computer System Description and Selected Naval Applications," AD 466-313, Computer Command and Control Company, Washington, D. C., April, 1965.

Ball, J. R. "On the Use of the Solomon Parallel-Processing Computer," Proceedings AFIPS, FJCC, 1962.

Barnes, G. H. *et al.* "The ILLIAC IV Computer,"*IEEE Transactions on Electronic Computers,* C-17, No. 8 (August, 1968), 746-57.

Beelitz, H. R. *et al.* "System Architecture for Large-Scale Integration," Proceedings AFIPS, FJCC, 1967, pp. 185-200.

Bucholz, W. *Planning a Computer System.* New York: McGraw-Hill, 1962.

Chapin, N. "Logical Design To Improve Software Debugging–A Proposal," *Computers and Automation,* February, 1966, pp. 22-24.

Critchlow, A. J. "Generalized Multi-processing and Multi-programming Systems," Proceedings AFIPS, FJCC, 1963, pp. 107-25.

Eddey, E. E. "Associative Processors and Their Application to Radar Systems," Goodyear Aerospace Report No. GER-12394, November 29, 1965.

Estrin, G., and Turn, R. "Automatic Assignment of Computations in a Variable Structure Computer System," *IEEE Transactions on Electronic Computers,* EC-12, No. 6 (December, 1963), 755-73.

Estrin, G., *et al.* "Parallel Processing in a Restructurable Computer System," *IEEE Transactions on Electronic Computers,* EC-12, No. 6 (December, 1963), 747-54.

Flores, I. "Hardware Features That Aid Programming," *Computer Design,* January, 1967, pp. 48-54.

Flynn, M. J., and Amdahl, G. M. "Engineering Aspects of Large High Speed Computer Design," *Microelectronics and Large Systems.* Washington, D. C.: Spartan Books, 1965, pp. 77-95.

Gibson, D. H. "Considerations in Block-Oriented Systems Design," Proceedings AFIPS, SJCC, 1967, pp. 75-80.

Gonzalez, R., Gunderson, D. C., and Timmons, J. A. "Associative Techniques for Control Functions in a Multi-Processor Simulation Investigation," Honeywell Systems and Research Center, Technical Report No. RADC-TR-67-500, November, 1967.

Gregory, J., and McReynolds, R. "The Solomon Computer," *IEEE Transactions on Electronic Computers,* EC-12, No. 6 (December, 1963), 774-81.

Hanlon, A. G. "Content-Addressable and Associative Memory Systems-A SURVEY," *IEEE Transactions on Electronic Computers,* August, 1966, pp. 509-21.

Hennie, F. C. *Iterative Arrays of Logical Circuits.* New York: J. Wiley and Sons, Inc., 1961.

Hobbs, L. C. "Effects of Large Arrays on Machine Organization and Hardware/Software Trade-Offs," Proceedings AFIPS, FJCC, 1966.

Holland, J. H. "Iterative Circuit Computers: Characterization and Resume of Advantages and Disadvantages," *Microelectronics and Large Systems.* Washington, D. C.: Spartan Books, 1965, pp. 171-78.

Hudson, D. M. "The Applications and Implications of Large-Scale Integration," *Computer Design,* June, 1968, pp. 38-48.

Huttenhoff, J. H., and Shively, R. R. "Arithmetic Unit of a Computing Element in a Global Highly Parallel Computer," IEEE Computer Group Conference Digest, June 25-27, 1968, pp. 54-56.

Katz, J. H. "Simulation of a Multi-Processor Computer System," Proceedings AFIPS, SJCC, 1966, pp. 127-39.

Knight K. E. "Changes in Computer Performance," *Datamation,* September, 1966, pp. 40-54.

Lass, S. E. "A Fourth-Generation Computer Organization," Proceedings AFIPS, SJCC, 1966, pp. 435-41.

Lehman, M. "A Survey of Problems and Preliminary Results Concerning Parallel Processing and Parallel Processors," Proceedings IEEE, 54, No. 12 (December, 1966), pp. 1889-1901.

Levy, S. "The Last Decade of Computer Development," *Computer Design,* May, 1968, pp. 72-81.

Liptay, J. S. "Structural Aspects of System 360 Model 85," *IBM Systems Journal,* 7, No. 1 (1968), 2-29.

Lynch, J. T. "The Burroughs B8500," *Datamation,* August, 1965, pp. 49-50.

McCluskey, E. J. "Iterative Combinatorial Switching Networks–General Design Considerations," *IRE Transactions on Electronic Computers,* ED-7 (December, 1958), pp. 285-88.

McCormick, B. H. "The Illinois Pattern Recognition Computer," 1963 ACM National Conference, Denver, Colorado August 27, 1963.

McCormick, B. H. "The Illinois Pattern Recognition Computer–ILLIAC III," *IEEE Transactions on Electronic Computers,* 12, No. 6 (December, 1963), 791-813.

McLaughlin, R. A. "The CDC 7600," *Datamation,* January, 1969, pp. 61-64.

Ossanna, J. F., Mikus, J. E., and Dursten S. D. "Communication and Input-Output Switching in a Multiple Computing System," Proceedings AFIPS, FJCC, 1965, pp. 231-41.

Petritz, R. "Technological Foundations and Future Directions of Large-Scale Integrated Electronics," Proceedings AFIPS, FJCC, 1966, pp. 65-87.

Pirtle, M. "Intercommunication of Processors and Memory," Proceedings AFIPS,

FJCC, 1967, pp. 621-33.

Rhea, J. "Industry Moving Into Fourth Generation," *Aerospace Technology,* June 17, 1968, pp. 25-28.

Rohrbacher, D. L. "Advanced Computer Organization Study," Vol. 1 and Vol. 2, AD 631-870; Goodyear Aerospace Corporation, April, 1966.

Russell B., and Estrin, G. "An Evaluation of the Effectiveness of Parallel Processing," 1963 IEEE Pacific Computer Conference, pp. 201-20.

Schwartz, J. "Large Parallel Computers," *Journal ACM,* January, 1966, pp. 25-32.

Senzig, D. N. "Observations on High-Performance Machines," Proceedings AFIPS, FJCC, 1967, pp. 791-99.

"The Sigma 7 From SDS," *Datamation,* March, 1966, pp. 53-57.

Slotnick, D. L. "Parallel and Concurrent Computer Systems," IFIPS Proceedings, 1965.

Slotnick, D. L. "Unconventional Systems," Proceedings AFIPS, SJCC, 1967, pp. 477-81.

Squire, J. S., and Palais, S. M. "Physical and Logical Design of a Highly Parallel Computer," Information Systems Laboratory Technical Report 04794-2T, University of Michigan, October, 1962.

Stanga, D. C. "UNIVAC 1108 Multi-Processor System," Proceedings AFIPS, SJCC, 1967, pp. 67-74.

Unger, H. S. "Pattern Detection and Recognition," Proceedings IRE, October, 1958.

Wald, B. "Utilization of a Multi-Processor in Command and Control," Proceedings IEEE, December, 1966, pp. 1885-88.

Walter, D. J., *et al.* "Setting Characteristics for Fourth Generation Computer Systems," *Computer Design,* 7, No. 10 (October, 1968), 48-55.

Yang, C. C., and Yau, S. S. "A Cutpoint Cellular Associative Memory," *IEEE Transactions on Electronic Computers,* EC-15 (August, 1966), pp. 522-29.

APPLICATIONS FOR PARALLEL PROCESSING

CHAPTER 2

APPLICATION OF ILLIAC IV TO URBAN DEFENSE RADAR PROBLEM

Morris A. Knapp, Gary M. Ackins and John Thomas

University of Illinois
Urbana, Illinois

Abstract

The large phased array radar system required for urban defense application places severe time requirements on a data processing subsystem. This report deals with the applicability of the ILLIAC IV array processors for the handling of these data processing requirements.

This report will describe the particular application, the phased array radar, and its capabilities that were assumed for this study. The basic functions that require data processing interfacing for this application are (1) radar beam forming and control, (2) scan, (3) designation, (4) tracking, and (5) discrimination and interception. These functions with the exception of point 5, are described in detail, and the solution of each major function on the ILLIAC IV structure is shown along with flow charts and timing estimates.

INTRODUCTION

Phased array radars cannot operate effectively without close coupling with data processing equipment, while at the same time placing extremely critical time requirements upon the data processor to obtain

---

The authors wish to thank the following from RCA's Missile and Surface Radar Division for their assistance in describing the radar system, the basic data processing requirements, and some of the particular techniques of solution: H. Halpern, S. Gross, and J. Courtney.

maximum performance and efficiency for the complete system. More flexibility is obtained in the entire system when all the functions are performed under program control and not hard wired control. As more signal processing is performed by digital computers under program control, the overall radar system will be more flexible to changing environment and applications.

The phased array radars have more flexibility than the older mechanically steered radars; however, they require much more data processing equipment to control them. Their pointing direction can be switched rapidly, and they can collect data from many items in parallel. Therefore, the raw collected data rates are much higher than other radar systems.

When applying large phased array radars to urban defense situations, the data processing time for this real time application cannot be handled by conventional "general purpose" computers. The data processing rates greatly exceed the capabilities of these machines. The phased array radar is a parallel beast (not only in construction, but also in performance of function). Therefore, it is only sensible to use an array data processor which is capable of handling the data rates. The ILLIAC IV is the first large computer capable of performing many array operations, and is capable of handling the data rates generated by these large radar systems.

This report will describe how several of the major functions of a large phase array radar in an urban defense situation can be solved by a single quadrant of the ILLIAC IV computer. These major functions are (1) radar beam forming and control, (2) scan for objects, (3) confirmation, (4) cloud track, (5) designation, and (6) object track. Only the basic capabilities of ILLIAC IV that are required for handling these functions will be described. Further information regarding the ILLIAC IV computer can be obtained from University of Illinois' reports or "The ILLIAC IV Computer," published in the August 1968 *Proceedings of the IEEE.*

RADAR DESCRIPTION

The phased array radar system assumed for this report is capable of scanning 10^{11} resolution elements/second which may have as many as 10^9 apparent targets from which 3×10^2 to 3×10^3 should be tracked as possible real targets for further discrimination. (The 10^{11} resolution elements examined/second is a maximum rate in the high resolution mode. It is made up of approximately 3×10^4 angle cells/second,

3×10^5 range cells/angle cell, and 10 doppler cells/range cell.) This situation is possible for an urban defense against enemy attack by ICBM's and/or satellite warheads with heavy decoy capability.

In this specific radar system, the transmitter and receiver arrays are not identical; they are separate arrays. The transmitter array can transmit the full power in a single beam, or be split into independent arrays with corresponding increased beam width and reduced power. The number of transmit beams is selected in powers of 2—a system that has a maximum of 4 beams was used. Therefore, 1, 2, or 4 beams may be transmitted simultaneously.

The receiving array will consist of clusters of elements. Each cluster can perform independently, and each cluster can receive 3^2 beams simultaneously. The receiver has 20 clusters for scanning, 2 for designation, and about 5 for tracking and other purposes. There are approximately 3×10^4 possible beam positions that have to be covered by the receiver array.

Both the transmitter array and the receiver array have to be coupled to the controlling computer. The transmitter requires a lot of data to control the phase shifters and time delay setting for steering the transmitted beam into the correct direction. The receiver also requires some control data, and then sends the received information to the digital computer for analysis and processing. Figure 1 illustrates the basic radar system diagram.

The radar system has to operate in the following modes (1) scan, (2) confirmation, (3) cloud track, (4) designation, and (5) object track. The process flow is shown in Fig. 2. However, since there are many different receivers and since the transmitter is capable of several beams simultaneously, several modes may be in progress simultaneously. The process of determining *when, where,* and *type* of energy pulse to be transmitted and the corresponding type of receiver and position of that receiver is called the Beam Scheduling Task. It is a special function of the array processor, and requires the decision from many factors. In the performance of this scheduling task, the computer would ideally consider the following:

1) Receive and transmit cannot occur simultaneously.
2) A certain limit exists as to the amount of energy available in any period of time, and pulses can be delayed for a certain maximum time.
3) A certain minimum scan rate must be maintained or targets will not be detected.

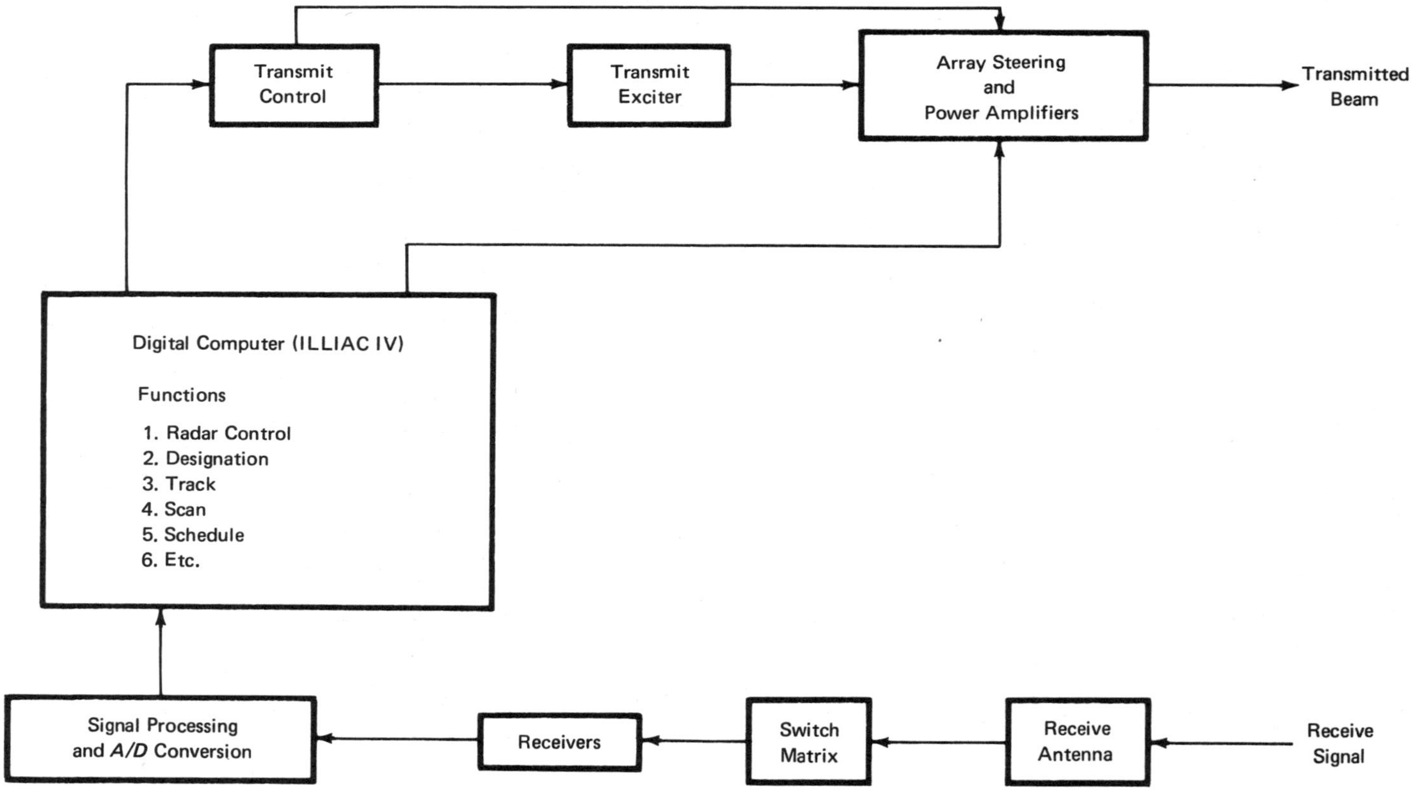

Fig. 1. Radar System Diagram

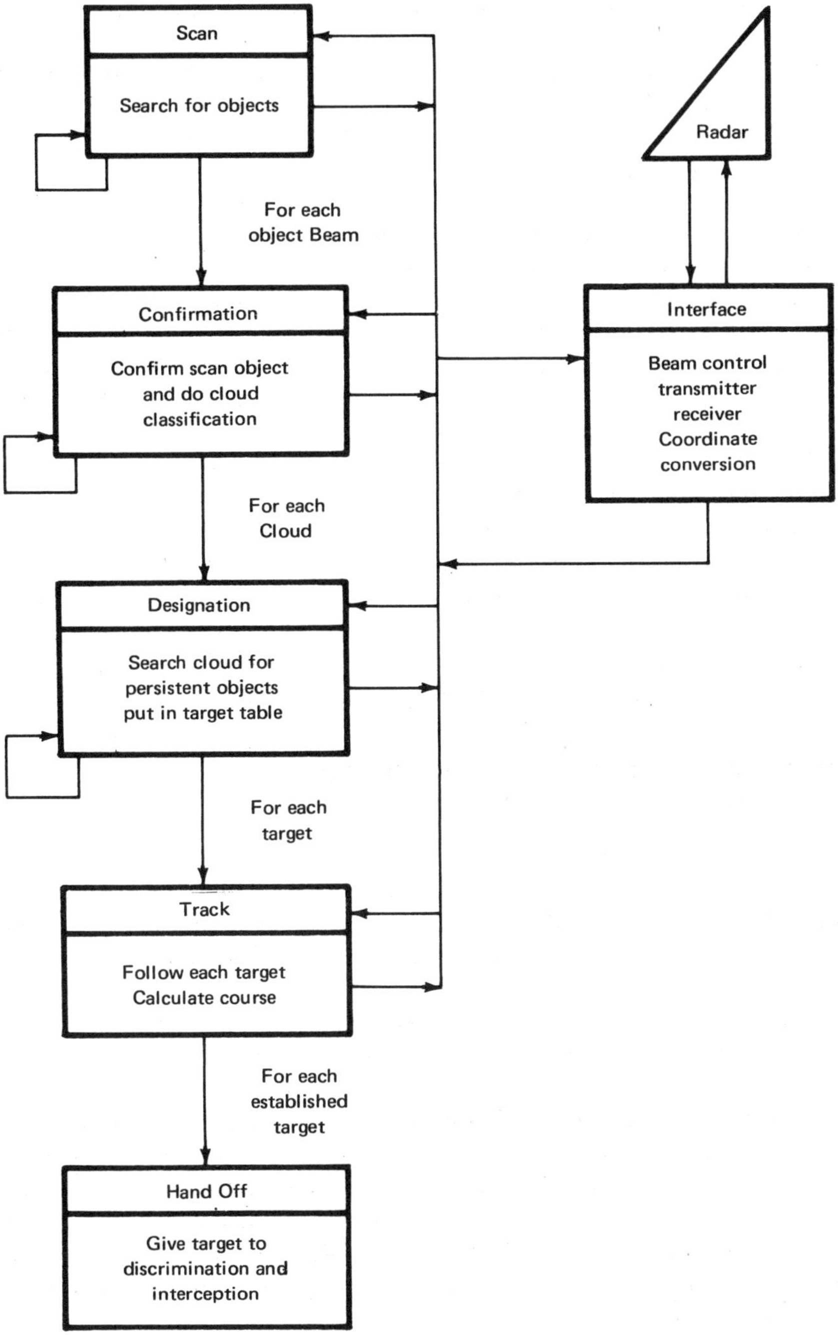

Fig. 2. Processing Flow for Radar Data

4) A pulse will have a selectable beam width, direction, wave form, and power.
5) Each transmission has a purpose which can be assigned a cost or priority basis.

The time between scans is determined by the requirements for detecting a target of minimum radar cross section with a certain cumulative probability of detection (P_D) and a certain probability of being a false target (P_f). These characteristics are relatively invariant in the way the power is divided in the transmitter or how the receivers are simmered, since the power aperture scanned product is the determining item.

The rate at which transmission occurs in the radar system is 5×10^2/second. During each 2-millisecond period, as many as 5 separate transmissions (each transmission may use all 4 quadrants, or each transmission may have 2 separate pulses, each using 2 quadrants, or each transmission may have 4 separate pulses, 1 in each quadrant) are spaced 8 microseconds apart. The remainder of the 2-millisecond period is used for receiving the returning signals. Therefore, both the transmitter and receiver have to be scheduled and steered at this rate.

TRANSMITTER AND RECEIVER CONTROL AND COORDINATE CONVERSION

The computer is required to handle the conversion of radar coordinates (azimuth, elevation, and range) to earth Cartesian coordinates and conversion in the opposite direction. It must also compute all the steering commands for the transmitter and receiver. For each transmission, many phase and time delay tapers have to be computed and sent to the transmitter to direct the transmitted energy beam into the correct direction. Figure 3 depicts the transmitter from the viewpoint of steering command requirements. These phase and time delay tapers, which are required for each transmission, can be computed for several transmissions in advance and sent to the transmitter to set up each beam to be transmitted. As given before, the basic frame time for the sequence of transmitter and receiver operations is 2 milliseconds (several beams are transmitted and received in this period). Therefore, many of the other data processing operations must be performed and completed within this cycle.

The beam steering for the transmitter, as shown in Fig. 3, requires 10-bit commands for the "fine" steering (5 bits for time delay and 5

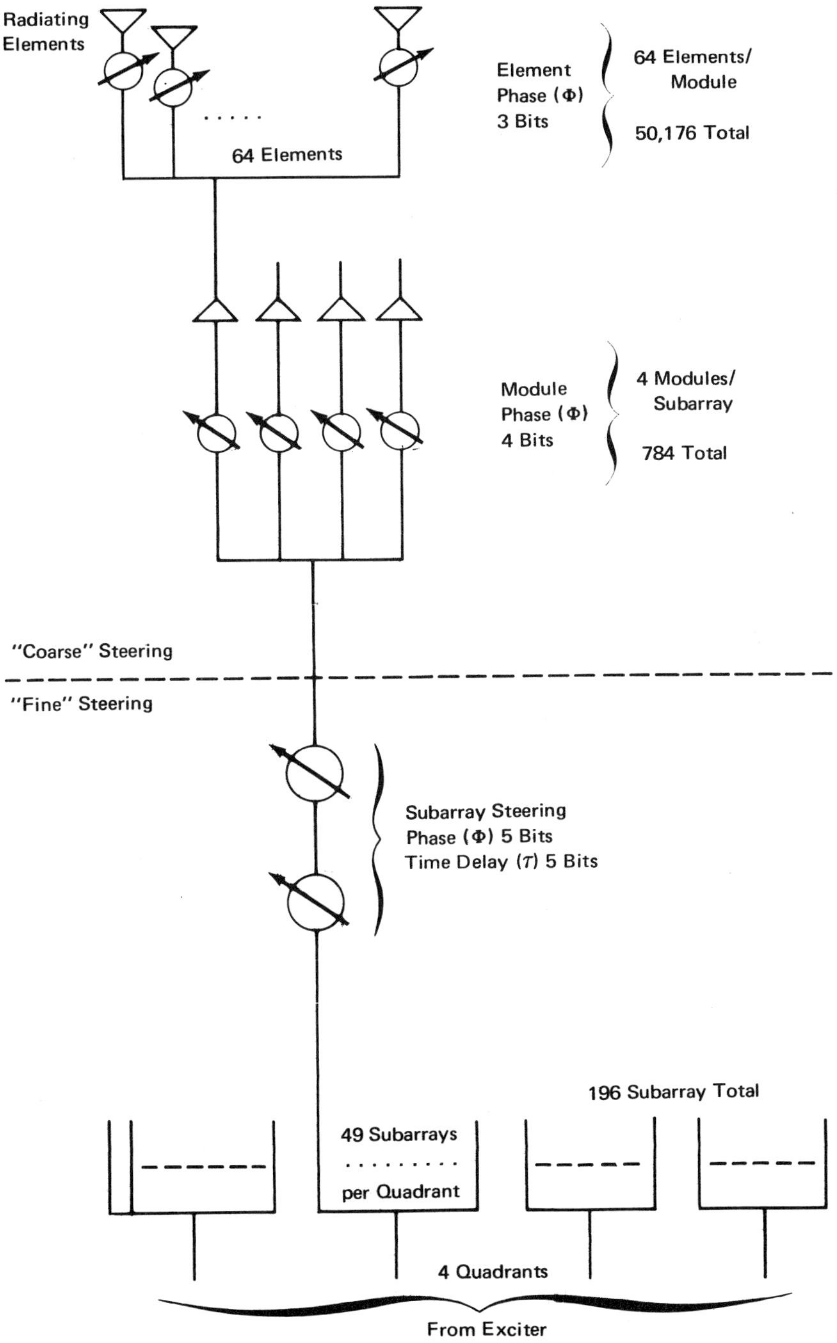

Fig. 3. Transmitter Array

bits for phase) for each of the 196 subarray units. Therefore, the fine steering transmissions in each 2-millisecond frame require 480,000 10-bit commands per second. In the coarse steering area, the same commands are used for many elements. Althogether 64 separate commands are needed per quadrant per transmission, requiring 640,000 7-bit commands per second. The total for both the fine and coarse steering of the transmitter is 1,130,000 commands or 9,380,000 bits per second. However, if storage is not provided at the transmitter to sum up the commands for each of the 5 transmissions in the 2 milliseconds, the computer must supply 452 commands on less than an 8-microsecond base for each transmission.

Calculation of Steering Commands

The terms *coarse* and *fine* as used above and in Fig. 3 are perhaps a misnomer. The big difference between the two is that *fine* steering is applied to each individual subarray (in a quadrant or a full array, depending on how many simultaneous beams are formed), and *coarse* steering is applied to the adjustment of the elements within each subarray with respect to the center of the subarray. The procedure used to calculate the fine steering commands is outlined as follows:

1) Calculate $\Delta\tau$'s time delay required between adjacent subarrays along x and y axis of the array.
2) Add $\Delta\tau$ to determine total delay (τ) required for all subarrays.
3) Round τ's to 5 bits for control of timing delays.
4) Use frequency (wavelength) of carrier to connect residue delays (after round off) to phase, and round again to 5 bits for control of phase steering.

The procedure requires conversion of spherical radar coordinates (azimuth and elevation) to array coordinates. Three values of $\Delta\tau$ are needed– $\Delta\tau_{\alpha 1}$, $\Delta\tau_{\alpha 2}$, and $\Delta\tau_{\beta}$. Subscript α refers to the x axis of the array and subscript β to the Y axis. Two values of $\Delta\tau_{\alpha}$ are needed, since structural elements of the antenna cause the columns of subarrays to be grouped in pairs. Relative delays are calculated from:

$$\Delta\tau = \frac{D}{c} \cos\theta$$

D is the spacing of the subarrays in meters, as shown in Fig. 4

c = speed of light = 3 (10^8) meters/sec.
θ = pointing angle of the beam with respect to the array axis.
Calculation of $\cos\theta$ depends on whether $\Delta\tau_\alpha$ or $\Delta\tau_\beta$ is desired.

For $\Delta\tau_\alpha$:

$$\cos\theta = \cos E \sin(A_o - A).$$

For $\Delta\tau_\beta$:

$$\cos\theta = \sin E \cos E_o - \cos E \cos(A_o - A) \sin E_o.$$

Here, E = elevation of beam
E_o = elevation of array
A = azimuth of beam
A_o = azimuth of array (will vary with each installation)

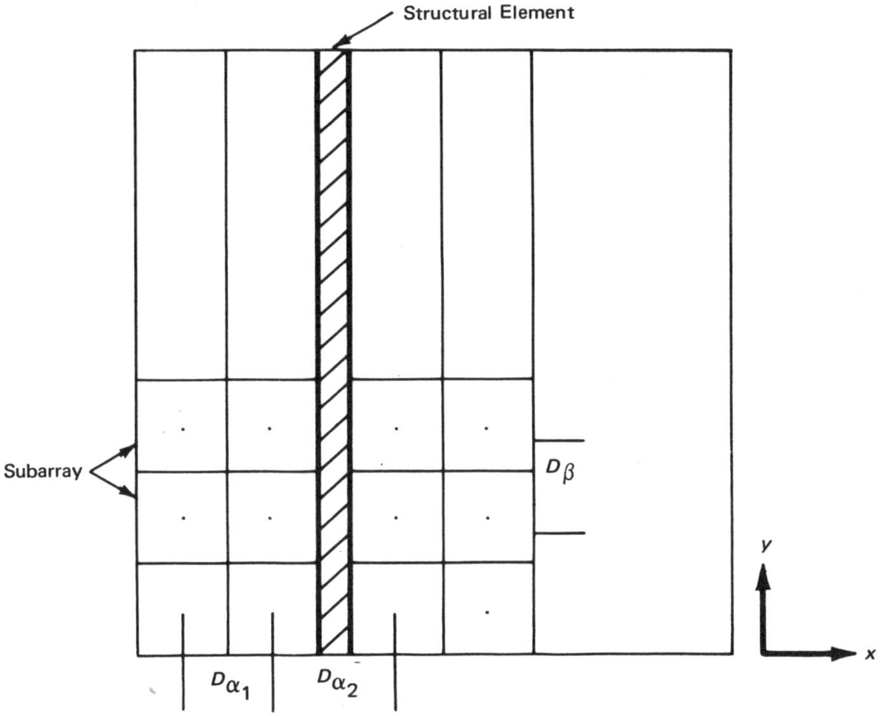

Fig. 4. Spacing of Subarray

The value of τ for a given subarray is the total of all the $\Delta\tau_\alpha$ and $\Delta\tau_\beta$ between the subarray and the reference subarray. In order that τ be positive, the reference subarray must be chosen in the bottom row when $E > E_o$, in the top when $E < E_o$, in the left column (facing outward) when $A > A_o$, and in the right column when $A < A_o$. The $\Delta\tau$'s which must be summed are illustrated in Fig. 5.

Each value of τ is computed to 12 bits and then rounded to 5 bits, as explained above, and the residue converted to phase by the relationship

$$\psi = 360\ (\delta\tau)\ f_o ,$$

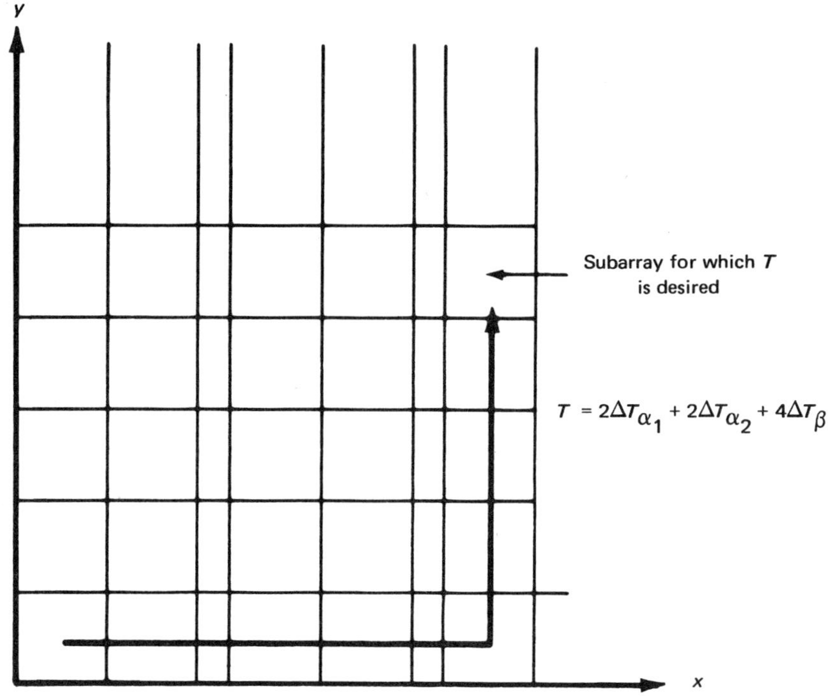

Fig. 5. Method of Calculating Delay for Steering

where ψ is the phase in degrees,
δ is the delay residue after round off,
f_o is the center (carrier) frequency in hertz.*

ψ is computed to 7 bits, and then rounded to 5 bits.

For coarse steering, phase differences ($\Delta\phi$) are calculated between adjacent elements of a module in the same way in which $\Delta\tau$'s are

*The carrier frequency data is supplied as an 8-bit word to the TGG from the receiver subsystem.

calculated for subarrays

$$\Delta\phi_{\alpha,\beta} = 360 f_o \frac{d_{\alpha,\beta}}{C} \cos\theta \, .$$

Here d_α and d_β are the distances between array elements along the X and Y axis, and $\cos\theta$ has the same value used for fine steering. The total phase shift for a given element is found in summing $\Delta\phi_\alpha$'s and $\Delta\phi_\beta$'s from a reference element in a similar manner to calculating τ in the fine steering case. The phase effect for each module is the same as for certain of the elements. It is automatically found because all the ϕ values are computed. More specifically, the phase effects for the 4 modules comprising a subarray are as follows:

1) 0
2) $8\Delta\phi_\alpha$
3) $8\Delta\phi_\beta$
4) $8\Delta\phi_\alpha + 8\Delta_\beta$

These phase calculations are carried out to 7 bits and then rounded to 3 bits for element steering and 4 bits for module steering.

The operation of the transmitter is continually checked on a regular cycle (one out of each one hundred frames) by a diagnostic check out and monitoring (CAM) procedure. The scheduling of the function will be a part of the overall system schedules and interleaved with the other functions. Samples of forward and reflected power are taken at critical points in the transmitter section by directional couplers. The sample signals are compared along with the exciter output in a phase detector. In case of failure of a power amplifier, a spare can be switched in. Amplifier failures and phase errors beyond the range of the trimmers are reported to the operator, and a record of the failure is automatically made along with the time of day.

Coordinate Conversion

The conversion from R, A, E coordinates to X, Y, Z with respect to the transmitter are given by the following equations:

$$\Delta \sin A = K_A \epsilon_A$$
$$\Delta \sin E = K_E \epsilon_E$$

where

E = elevation subscript

A = azimuth subscript

ϵ = monopulse error signal (4 bits)

K = proportionality factor

E_o and A_o = measured elevation and azimuth from angle (9 bits)

E_N and A_N = actual angles to targets (13 bits)

R = range (21 bits)

$$\sin E_N = \sin E_o + \Delta \sin E$$
$$\sin A_N = \sin A_o + \Delta \sin A$$
$$\Delta \cos A = \tan A_o \, \Delta \sin A$$
$$\Delta \cos E = \tan E_o \, \Delta \sin A$$
$$\cos A_N = \cos A_o - \Delta \cos A$$
$$\cos E_N = \cos E_o - \Delta \cos E.$$

Now solve for $X_T \; Y_T \; Z_T$ when

$$X_T = R \cos E_N \sin A_N$$
$$Y_T = R \cos E_N \cos A_N$$
$$Z_T = R \sin E_N \,.$$

The conversion from X, Y, Z coordinates to radar R, A, E coordinates are given by the following equations:

$$R_t = (X^2 + Y^2)^{1/2}$$

$$A = \sin^{-1} \frac{Y}{R_t}$$

$$R = (Z^2 + R_t^2)^{1/2}$$

$$E = \sin^{-1} \frac{Z}{R} \quad .$$

BRIEF DESCRIPTION OF ILLIAC IV SYSTEM

ILLIAC IV is a fast (240-nanosecond cycle time), large (.5 X 10^6 words of 64-bit thin film memory), highly array (256 processors in 4 autonomous group) computer being built by Burroughs Corporation for the University of Illinois. The system is an array of 256 coupled computers (in 4 autonomous groups of 64 each) driven by instructions from a common control unit. Each of the 256 processing elements (PE's) has 2,048 words of 64-bit memory with a 240-nanosecond cycle time. Each PE is capable of 64-bit floating point multiplication in 400 nanoseconds and addition in 240 nanoseconds. The PE's also have 32-bit capability where each quadrant could be considered as 128 parallel operating units. There is, also, an 8-bit fixed point capability in each PE. The PE instruction set is similar to that of a conventional computer, but with several exceptions. First, the PE's are capable of communicating data to form neighboring PE's by means of routing instructions. Secondly, the PE's are able to set their own mode registers to effectively enable or disable themselves.

Instructions are decoded, and control signals are sent to the PE array from the control unit (CU). Some instructions are executed directly in the CU. For example, operand addresses may be indexed once in the CU and again separately in each PE. In addition to instructions, the CU may send data to each PE by way of the broadcast instruction. This instruction may take an operand from a CU local data buffer (buffer of 64 words of 64 bits each is loaded from PE memory), and send it to each enabled PE.

The complete ILLIAC IV system for the University of Illinois includes a B6500 computer to perform input/output operations and compilation, and contains an operating system to control ILLIAC IV which is depicted in Fig. 6.* Also, there is a 10^9-bit head per track disk file with a 40-millisecond rotation speed and an effective transfer rate of 10^9 bits/second. This radar application has been programmed for a single quadrant of 64 PE's, and would require some type of front end computer similar to the B6500, for which the software is presently being developed.

COMPUTATION OF STEERING COMMANDS IN ILLIAC IV

The procedure, flow charts, and detail codes for generating the transmitter steering commands have been completed for ILLIAC IV. There exists an I/O path consisting of 1,024 bits/word in the ILLIAC IV that can transfer data at a rate of 4 per microsecond. The steering

*The real time data link, at present, consists of a 1,024-bit path which connects to 16 PE's in the 64 bit mode. This path could be expanded to all 64 PE's if the data rates require that band width.

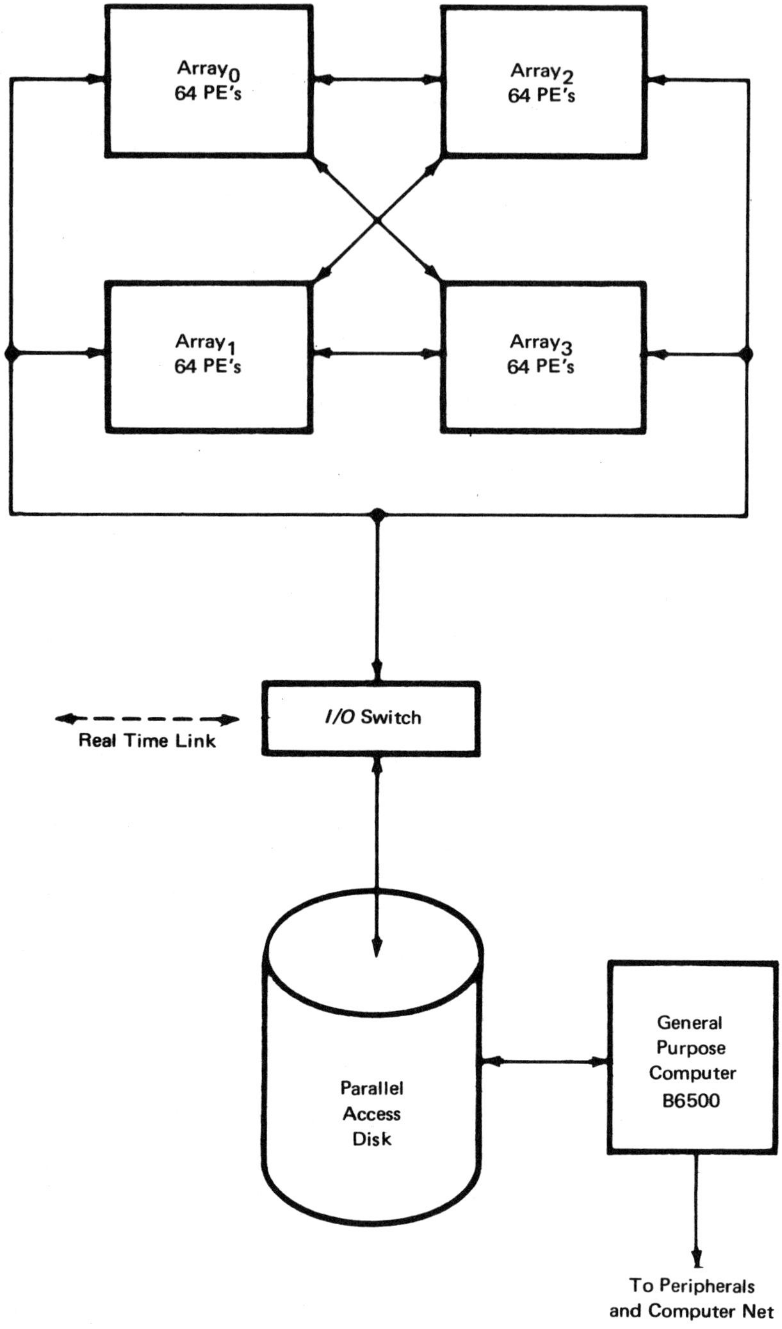

Fig. 6. ILLIAC IV System

commands for each transmission are packed into 4 I/O words. Therefore, with the transfer rate of ILLIAC IV, external storage at the transmitter is not required except for the commands to control a single transmission. Following is a detailed description of how the commands are computed and packed for transfer to the transmitter.

Given an E and A of where you want to look, this scheme computes the steering commands needed by the transmitter to send a signal to that position consisting of 5-bit time delay (T) for each of the 49 subarrays. It also consists of 5-bit phase delay (Ψ) or these subarrays, 4-bit phase delays (M) for each of the 4 modules per subarray, and 3-bit phase delays (φ) for each of the 64 radiating elements per module.

Since small word length is involved, the machine operates in 32-bit mode; so, in effect, the scheme consists of 128 32-bit PE's operating in parallel. Actually, only 120 E's and A's are treated at a time, taking care of 6 bursts or 12 msec worth of pulses.

Initially, each 32-bit PE is assumed to contain a 9-bit E specifying elevation of the desired beam position and a 9-bit A specifying azimuth. All PE's operate in parallel to compute:

$$\cos\alpha = \cos E \sin(A_O - A)$$

$$\cos\beta = \sin E \cdot \sin E_O - \cos E \cdot \cos(A_O - A) \cdot \sin E_O$$

$$\Delta T\alpha_1 = (DX_1/c) \cdot \cos\alpha$$

$$\Delta T\alpha_2 = (DX_2/c) \cdot \cos\alpha$$

$$\Delta T_\beta = (Dy/c) \cdot \cos\beta$$

$$\Delta\varphi_\alpha = (dx/c) \cdot \cos\alpha \cdot 360 \cdot F$$

$$\Delta\varphi_\beta = (dy/c) \cdot \cos\beta \cdot 360 \cdot F$$

$$M_1 = O,\ M_2 = 8\Delta\varphi_\alpha,\ M_3 = 8\Delta\varphi_\beta,\ M_4 = 8\Delta\alpha + 8\Delta\varphi_\beta .$$

The symbols have the following meanings:

A_0 = broadside azimuth

E_0 = broadside elevation

DX_1 = physical separation of subarrays in X direction

DX_2 = physical separation of subarrays in X direction

Dy = physical separation of subarrays in y direction
c = speed of light
dx = physical separation of radiating elements in X direction
dy = physical separation of radiating elements in y direction
F = frequency of pulse
M_i = ith module phase delay.

The ΔT's are the relative time differences between subarrays in a block of 49 where $\Delta T\alpha_1$ is the time difference between two subarrays separated by a distance of DX_1. $\Delta T\alpha_2$ is the time difference between two subarrays separated by DX_2, and $\Delta T\beta$ is the time difference between those separated by a distance of Dy. Two possible separations in the X direction, DX_1 and DX_2, of adjacent subarrays are possible because of structural elements in the subarray block.

Consider a corner of the block of 49 subarrays,

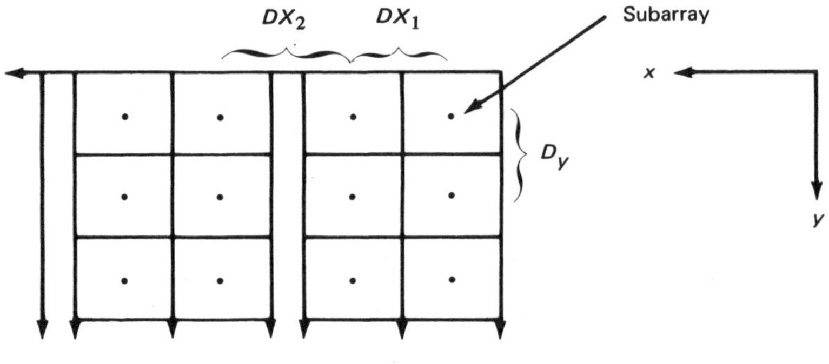

Structural Elements

After all PE's have computed these relative time and phase delays, the next job is to sum the relative phase and time delays with respect to the reference element to get absolute delays.

The first job is for all PE's to compare their $\bar{E}$ and A with broadside E and A (E_O and A_O) to determine which of the 4 corner subarrays in the block of 49 is the reference element, and which of the 4 corner radiating elements in a block of 64 is the reference element. After this is done, it is only necessary to sum the $\Delta T\alpha_1$'s, $\Delta T\alpha_2$'s, and $\Delta T\beta$ between each of the 49 subarrays and the reference element. The same summation is made for the 64 elements. For each of the 49 subarrays per E and A, the following must be computed

$$T = A \cdot (\Delta T\alpha_1) + B \cdot (\Delta T\alpha_2) + C \cdot (\Delta T\beta) \ .$$

For example, suppose the reference subarray turns out to be subarray #7 in the upper right hand corner of the block of 49,

1	2	3	4	5	6	7
8	9	10	11	12	13	14
		17				21

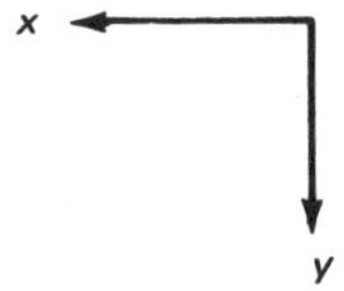

then, to compute T for subarray #6, $A = 1$, $B = 0$, $C = 0$. For subarray #12, $A = 1$, $B = 1$, $C = 1$; for subarray #17, $A = 2$, $B = 2$, $C = 2$.

A similar scheme is used for summing relative $\Delta\varphi$'s to get absolute phase difference φ for the 64 radiating elements. The $\Delta\varphi$'s are the relative phase differences between adjacent radiating elements in a block of 64. $\Delta\varphi\alpha$ is the relative phase delay between two elements separated by a distance of dx (i.e., adjacent elements in the x direction), and $\Delta\varphi\beta$ is the delay between elements in the y direction.

The T is computed to 12 bits for each of the 49 subarrays per E and A. The first 5 bits are the time delays for the subarrays and the last 7 bits are converted to the subarray phase delay (Ψ) by

$$\Psi = 360 \cdot F \cdot \text{(last 7 bits from } T\text{) and then rounded to 5 bits} \ . \qquad (1)$$

Since 49 T's, 49 Ψ's and 64 Φ's must be computed for each E and A, the parallelism of the machine is used to make these computations. At this point, all 120 PE's have computed $\Delta T\alpha_1$, $\Delta T\alpha_2$, $\Delta T\beta$, $\Delta\phi\alpha$, and $\Delta\phi\beta$ for their E and A. Using the $\Delta\phi$'s just computed, all PE's compute the 4-module phase delays, M_i $1 \leq i \leq 4$,using eq. (1) and round to 4 bits each. These four 4-bit words are then packed into one 16-bit word M.

To compute absolute T's for the 49 subarrays corresponding to the E and A in PE #1, the ΔT's computed from E and A in PE #1 are broadcast to a prescribed set of 49 PE's. The ΔT's computed in PE #2 are broadcast to a second set of 49 PE's. All 98 PE's that received this information calculate

$$T = A(\Delta T\alpha_1) + B(\Delta T\alpha_2) + C(\Delta T\beta) \ .$$

The coefficients A, B, and C in each PE are determined by the previous decision as to where the reference subarray lies for the E and A that these T's are for.

The first set of 49 PE's has now computed 12-bit T's for the 49 subarrays that were for the E and A that were originally in PE #1. The second set of 49 PE's has computed 12-bit T's for the 49 subarrays that were for the E and A originally in PE #2.

All 98 PE's take the last 7 bits of the T that they have just computed and use it to compute Ψ from eq. (1), and then round to 5 bits. The first 5 bits of T previously computed is the 5 bit T for the subarray delays.

The next job is to pack the 98 5-bit T's and 5-bit Ψ's that exist in 98 different PE's. The first 48 T's from both blocks of 49 are packed into PE's #0–15. Both of the 49th T's are packed into PE #65. ψ's are packed similarly so they exist in PE's #32–47 and PE #66.

The next two sets of 49 T's and ψ's corresponding to the E's and A's in PE #3 and #4 are now computed in the manner described above. Then the first 48 T's from both blocks of 49 are packed into PE's #16–31, and both of the 49th T's are packed into PE #65 along with the previous two 49th T's. ψ's are packed similarly into PE's #48–63 and PE #66. At the end of this, the four sets of 49 T's for the four different E's and A's lie in a common memory location in PE's #0–32.

The four 16-bit M's containing the module phase delay commands are then broadcast to this memory location in PE's #84, #85, #86, and #87 respectively.

The next procedure to calculate the 64 element absolute phase delays (Φ) for the same 4 E's and A's. The procedure is similar to that above because two sets are computed and packed, and then the other two sets are computed and packed. In particular, the $\Delta\Phi$'s from the first 2 E's and A's are broadcast to two specific sets of 64 PE's, and all 128 PE's calculate

$$\Phi = E(\Delta\Phi\alpha) + F(\Delta\Phi\beta) \ ,$$

using the $\Delta\Phi$'s they received on the broadcast. These 3-bit Φ's are then packed into location A's PE's #96–111. The procedure repeats using $\Delta\Phi$'s from the second 2 E's and A's, and these Φ's are packed in PE's #112–127. All 4 sets of 64 Φ's now exist in the common memory location across PE's #96–127.

The 4 complete sets of steering commands for the 4 E's and A's are now arranged in a memory location A across PE's as follows:

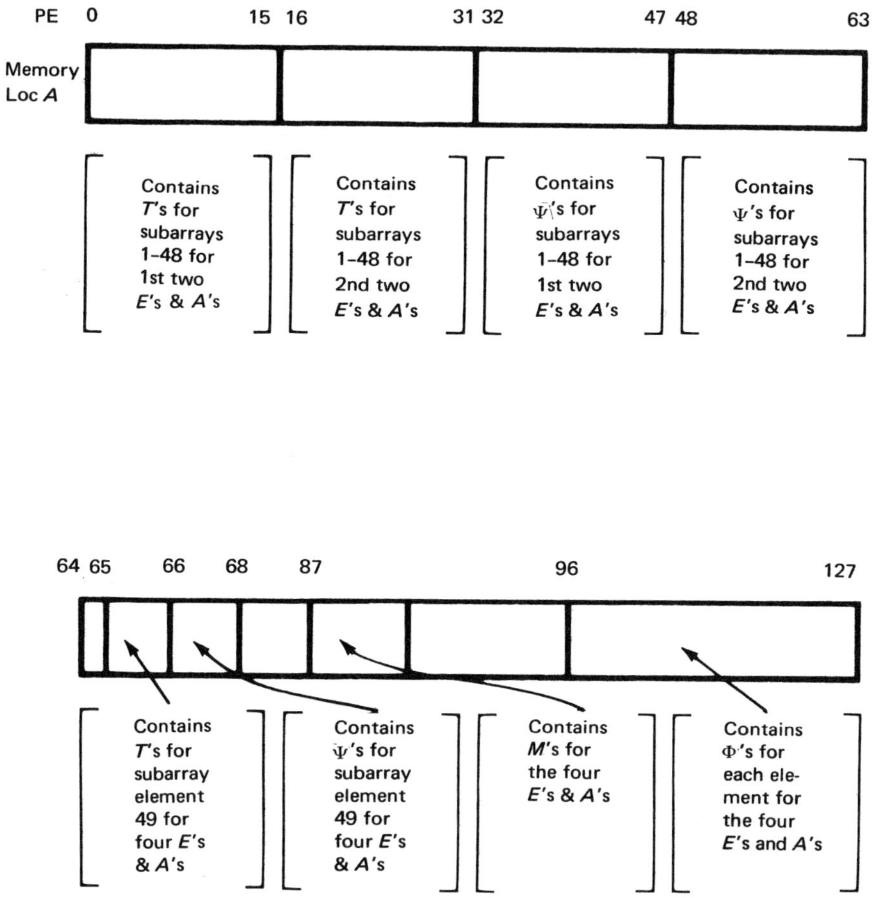

ILLIAC IV has an I/O path of 1,024 bits/word. Since they can be sent with 4 I/O words to the transmitter, the steering commands for the 4 quadrants of one transmission pulse are packed.

After the scheme has completed the above computation and packing of the steering commands for 4 E's and A's, it begins the computation and packing of T's, Ψ's, M's, and Φ's for the next four E's and A's. It continues in this manner until all 120 E's and A's have been processed.

Timing

Computation	*Time*
$\cos \alpha$	10 μsec
$\cos \beta$	11 μsec
ΔT's	2 μsec
$\Delta\varphi$'s	3.5 μsec
M_i's	2 μsec
comparison of E with E_O and A with A_O	1 μsec
pack M_1, M_2, M_3, M_4 into 16-bit words	1.5 μsec
	31 μsec

The above includes everything down to point I in the flow chart (Fig. 7). Since all PE's are operating in parallel, this is the time it takes to compute the above information for 120 different E's and A's.

The total time for all fetches of needed data for broadcasts and for all broadcasts of this data, such as M_1's, ΔT's, $\Delta\varphi$'s, and F, for the computation of 120 steering commands is 750 μsecs. The computation of 2 sets of 49 T's done in parallel takes 2 μsecs. Since 120 sets must be computed, the total time for the calculation of T's from ΔT's is 120 μsecs. The calculation of 120 sets of Ψ's requires 72 μsecs. The calculation of 120 sets of φ's requires 77 μsecs. To pack 4 sets of T's, Ψ's, and φ's takes 23 μsecs. Since 120 sets must be done, all of the packing of T's, Ψ's, and φ's requires 690 $m$sec. The total time for the computation of 120 sets of steering commands for 120 different E's and A's required by this scheme is ≈ 1.7 $m$sec. The scheme requires ≈ 60 memory locations.

SCAN/CONFIRMATION MODES AND BEAM SCHEDULING

The normal operation for the radar will be a continuous scan of the volume of space that can be covered by the radar system. The purpose of the scan operation is the continuous evaluation of the surveillance data to determine if any objects are present. The scan operation is carried out with a narrow-band signal with a range resolution (10 bits)

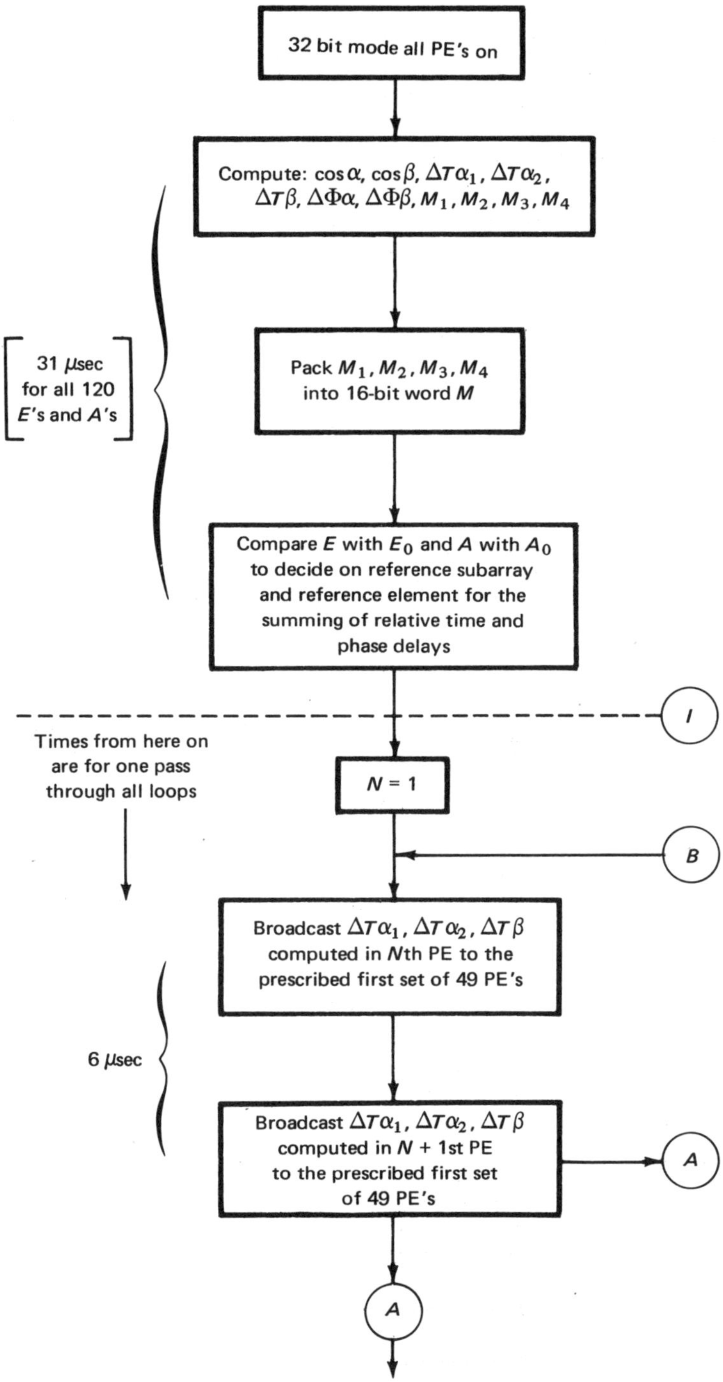

Fig. 7. Steering Commands Flow Chart

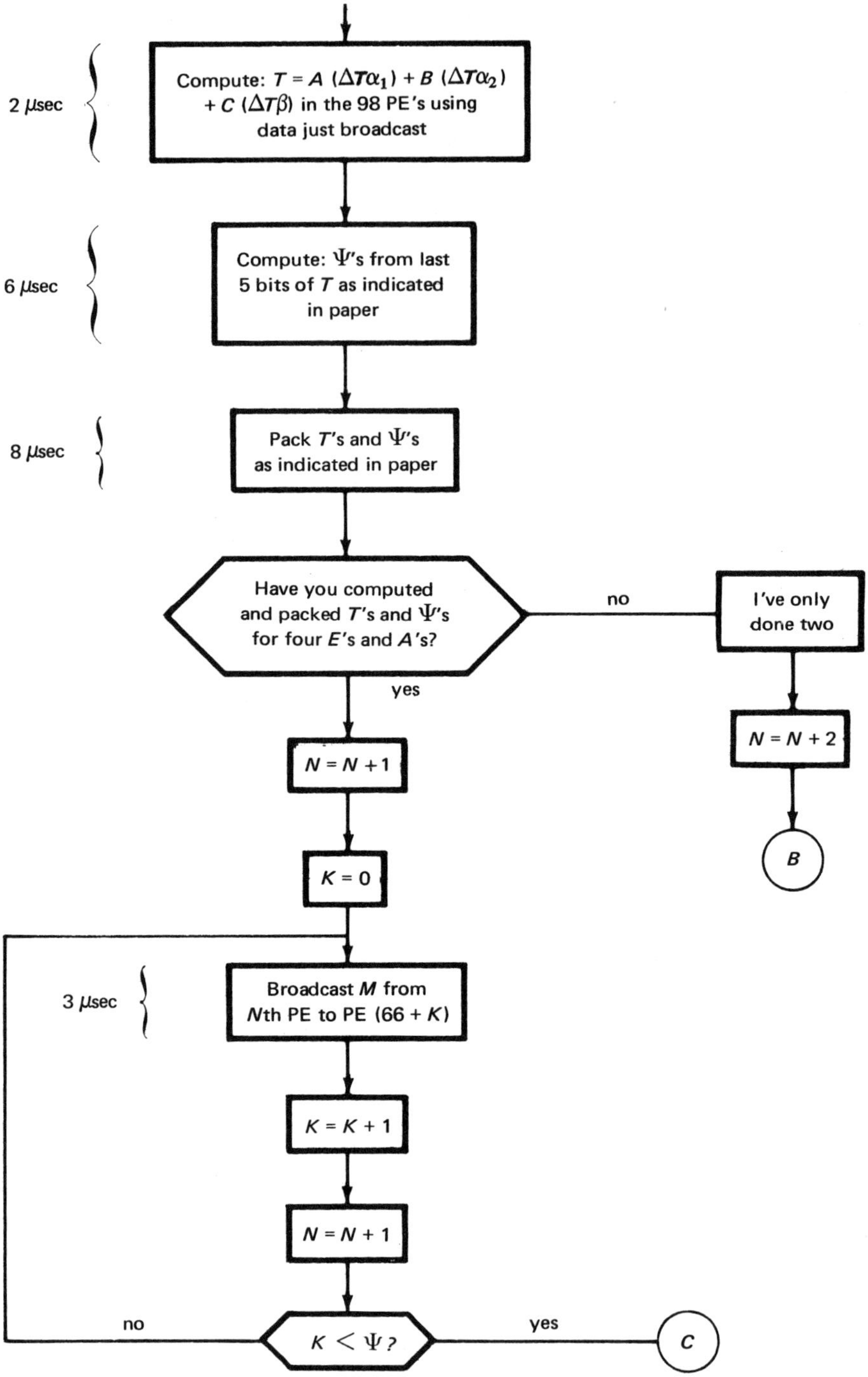

Fig. 7. Steering Commands Flow Chart (cont'd)

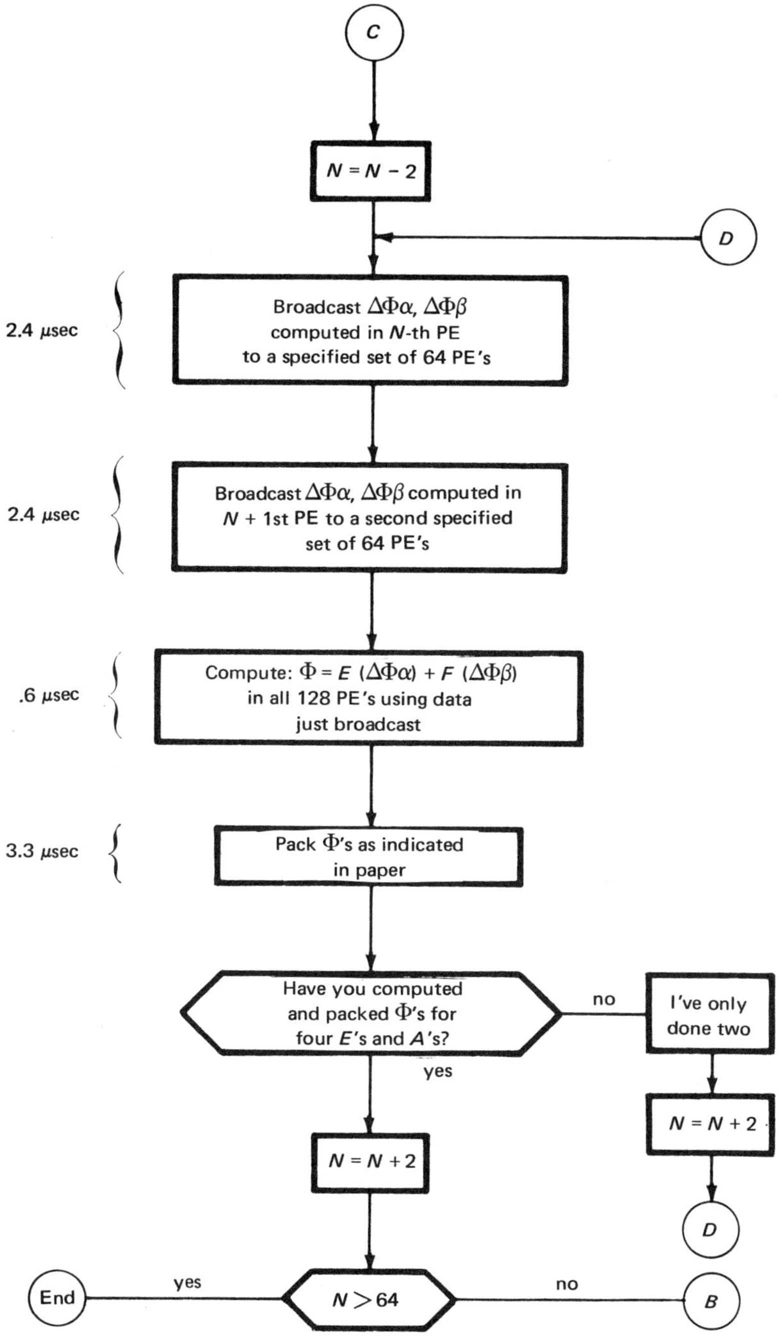

Fig. 7. Steering Commands Flow Chart (cont'd)

of approximately 0.4 nautical miles. If a return above a certain threshold is apparent on a scan receiver, then the volume of space will be assigned to confirmation for a higher resolution look. This scan receiver is made adaptive to certain parameters; that is, the threshold of the receiver is made a function of the noise level, and if the noise level rises above a certain threshold, which indicates jamming, the frequency of operation is changed. In the scan operation, no count of the number of objects is performed. Only the knowledge of the presence of an object will then schedule that volume for confirmation mode. The flow chart for the scan/confirmation modes is given in Fig. 8.

The confirmation mode uses two different radar pulses: the first is a high-resolution pulse (400 MH_z chip, 5 μsec long), and the second is a burst pulse. The first pulse (high resolution) is used to obtain more accurate range resolution (21 bits or approximately 1.5 feet resolution) and to obtain a count of the number of objects in this particular angular sector of space. This is needed (as shown by the flow chart) to determine how these objects are to be classified for further handling. The second confirmation pulse is used to establish the range rate. This information is also used in the classification of the items for further processing.

The amount of data and the amount of processing are very low for the scan and confirmation modes, except for where the data obtained in the confirmation mode will be used for further processing. Therefore, most of this part of the overall radar processing will probably be handled along with the beam scheduling in a controlling computer similar to the B6500 of the ILLIAC IV; however, it could be performed in the ILLIAC machine with extremely minor loading onto the overall processing. The interplay of the different modes for the beam scheduling problem is illustrated by Fig. 9. This again seems to require only a small amount of data processing capability, and has not been programmed in detail.

DESIGNATION MODE AND ILLIAC IV PROCESSING

The designation mode is a process that is applied to clouds of objects to reduce the number of objects for tracking. It is a simple time filtering of the cloud so only objects of interest are assigned to the track mode. The designation processing consists of comparing returns received from a cloud on .1 second and 1 second intervals for equal positions relative to the center of the cloud. The returns that meet this

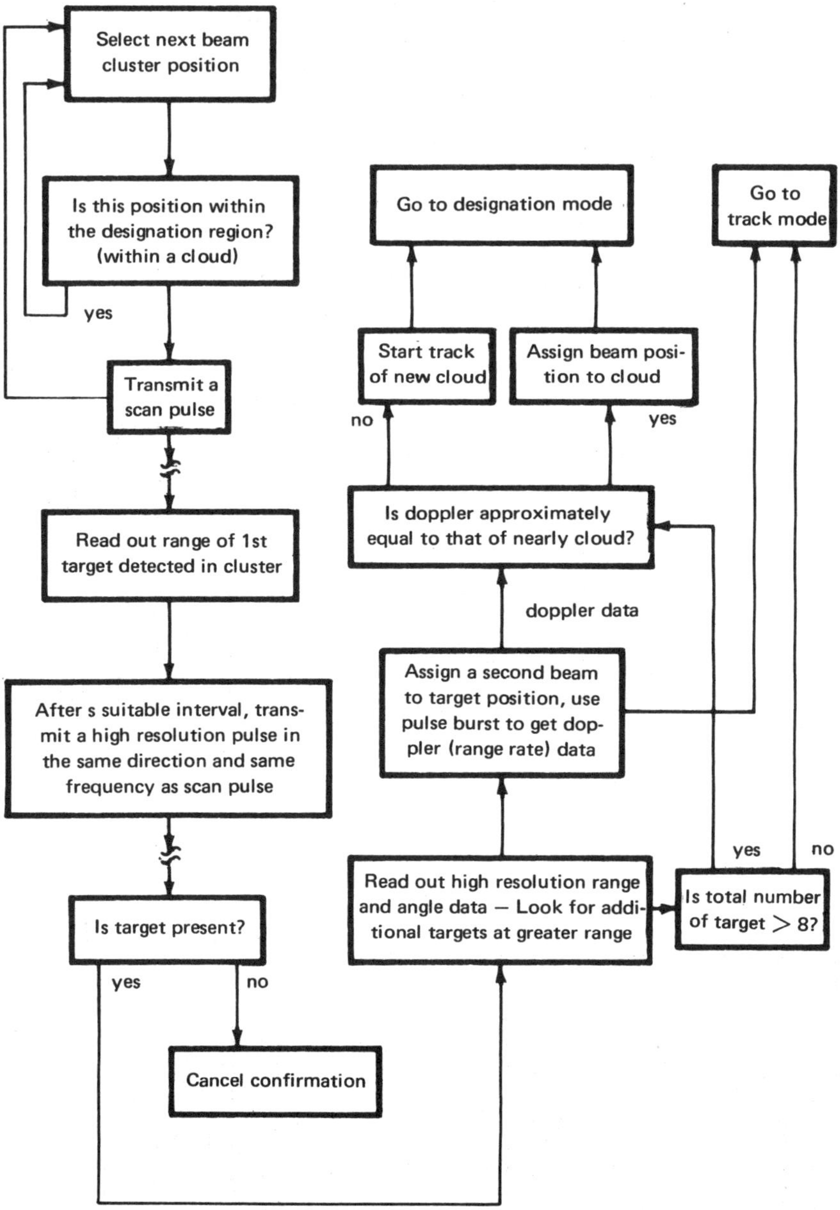

Fig. 8. Flow Chart of Scan/Confirmation

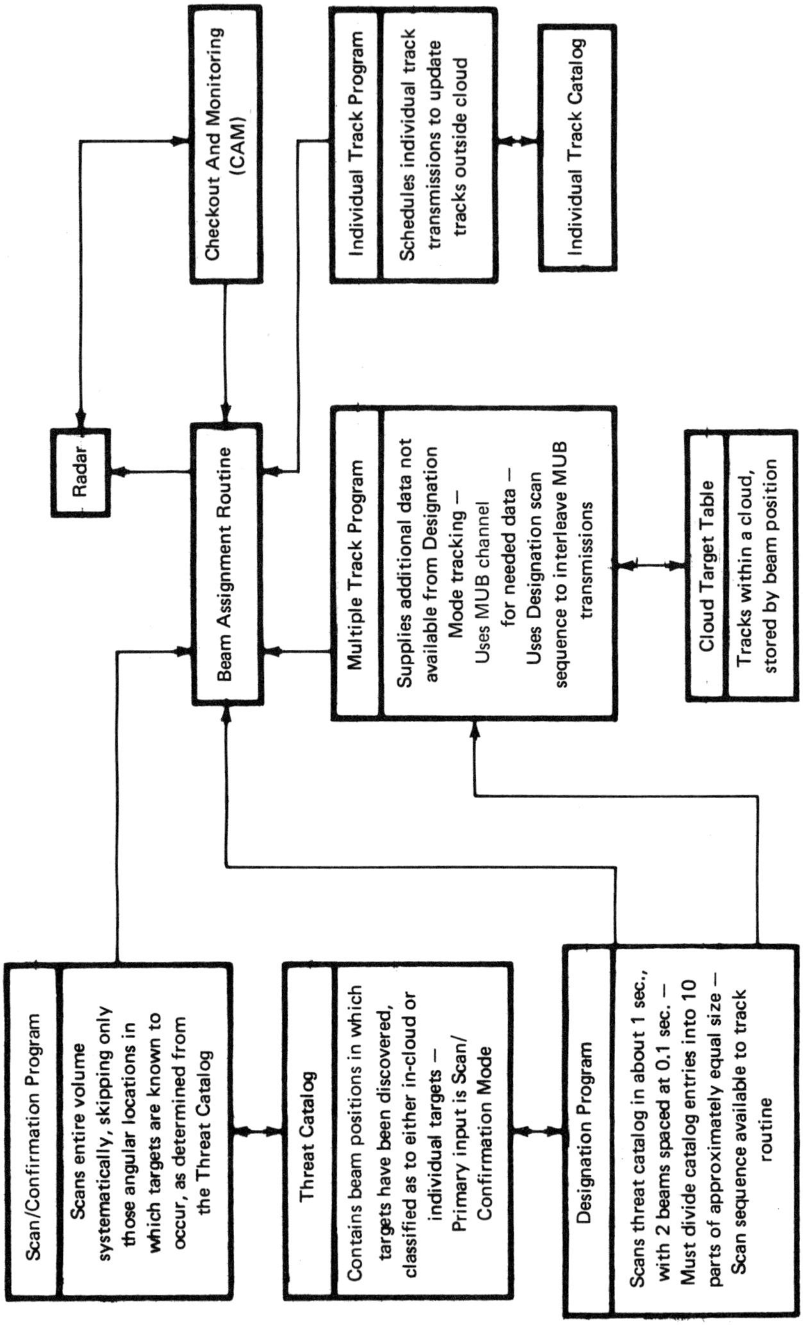

Fig. 9. Relationship of Operating Mode Programs to Beam Scheduling

time correlation criteria are designated as potential targets and are assigned to a track mode. The cloud (assume 30,000 items) is broken into 10 approximately equal segments. Each segment is swept with leading and trailing beams spaced .1 second apart. A correlation is performed on the returns between these two beams, and then the matched items are compared with the items that were present during the sweep of the previous second. This processing is illustrated in Fig. 10.

The radar returns from the receivers are grouped according to the transmitted beam from which they occurred. Each group of returns consists of a beam position address (E,A), a range (R), azimuth and

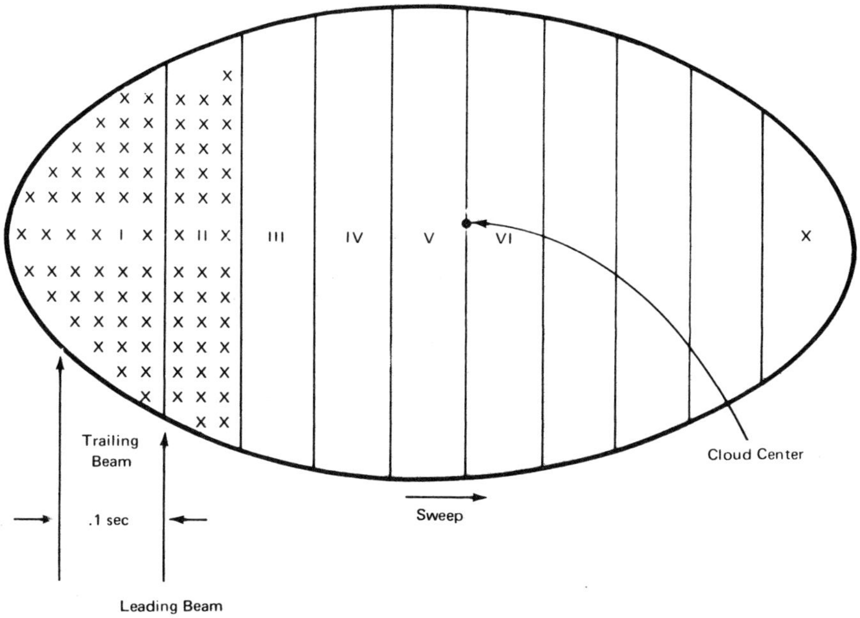

Section I

The leading beam has completed the sweep, and the *x, y, z* of each object in relation to cloud center has been computed and stored in the hash memory area for Section I.

The trailing beam is beginning to sweep for a .1 second correlation. This data will be correlated with that stored from the leading beam.

Section II

The leading beam is beginning to sweep.

Fig. 10. Cloud Designation Processing

elevation ($\epsilon_E \epsilon_A$) correction errors, and time associated with each beam. In ILLIAC IV, the beam position (E, A) is used by the control unit, while the $R, \epsilon_A, \epsilon_E$ for each object is distributed throughout the PE's in the 32-bit mode. Presently, the input/output path to ILLIAC IV is 1,024 bits or 32 PE's in the 32-bit mode. The mapping of this input data into ILLIAC IV is shown by Fig. 11. This figure also shows how the data is handled when 128 parallel PE's perform the coordinate conversion to Cartesian coordinates. After the conversion to Cartesian

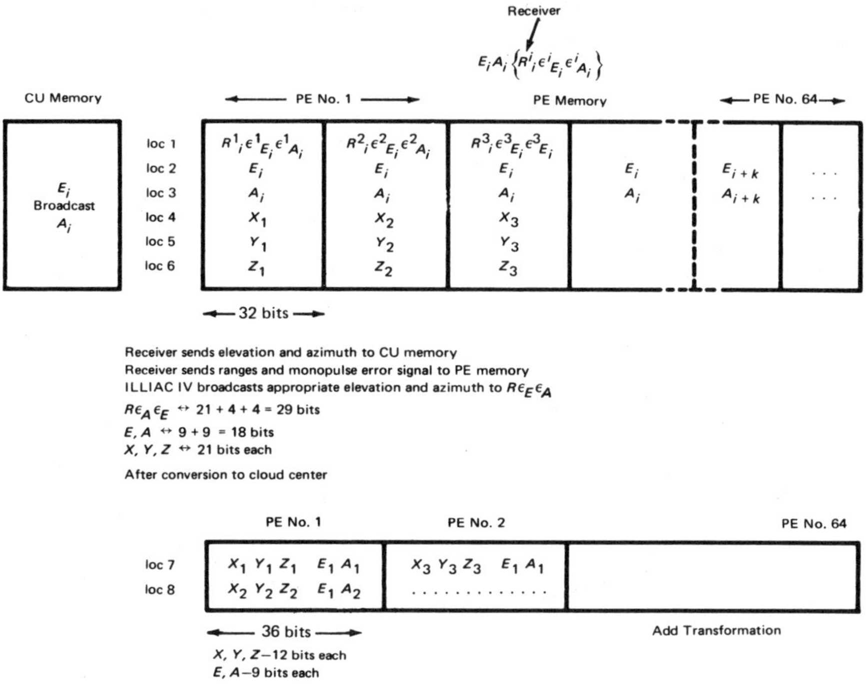

Fig. 11. Input of Data for Designation

coordinates, the x, y, z for each object in the cloud is obtained with respect to cloud center.* For this value (36 bits), the ILLIAC IV is used in the 64-bit mode and only 64 PE's. The correlation will be performed on this 36-bit number, which represents the coordinates of the object

*The cloud center is found by computing the average of the different coordinate values for the entire cloud. This cloud center is tracked, and the position is updated in time to perform the 1 second correlation.

with respect to cloud center. Depending on the radar accuracy, it is undetermined if all 36 bits will be used for an exact match. At present, the program assumes all 36 bits can be used; however, a mask can be used in the correlation if the radar accuracy does validate an exact match on all 36 bits.

The designation mode processing is illustrated by the flow chart given in Fig. 12. Each correlation (.1 second and 1 second) will give a reduction by a factor of 1 to 10; therefore, a total reduction factor will be 10 to 100 to the cloud. The cloud is broken into 10 equal segments;

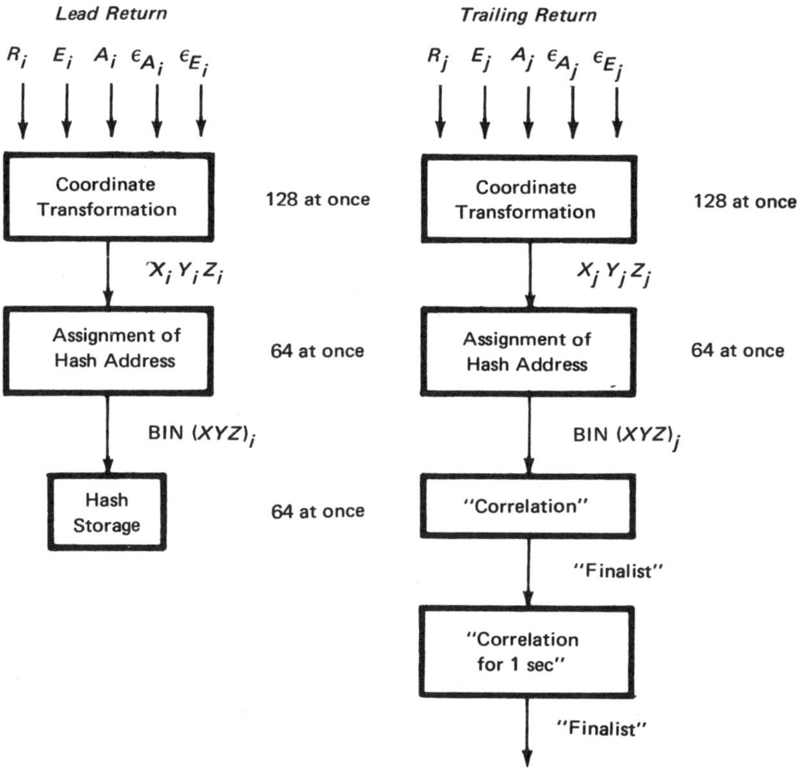

Since $E_i, A_i, R_i, \epsilon_{A_i}, \epsilon_{E_i}$ each contain $<$ 32 bits, the quadrant may be used as if it were 128 PE's

After Coordinate Transformation, the (relative) X, Y, Z coordinates are formed into 36-bits words, and the quadrant is used as 64 PE's

Finalists which qualify for .1 sec are compared with the finalists from the 1 sec before

Finalists which qualify for both the .1 second and 1 second are sent to track mode

Fig. 12. Basic Flow Chart for Designation

therefore, a 3,000 by 3,000 object correlation has to be performed each .1 second, and the finalists compared with the object which persisted for the previous second. A hash processing technique was used to reduce the number of correlations for each .1 second in ILLIAC IV from the $(3{,}000)^2$ to 3,000. This same correlation process can be used for the 1 second correlation.

THE MECHANICS OF HASH PROCESSING

Each of the PE's now contains (say in their *A* reg.) a 64-bit word consisting of the 36 bits (x, y, z) which determine the origin of the return in space (see Fig. 11, loc 7).

Now, using these 36 bits the following procedures are done:

1) Assignment of hash address	"Pull out" a smaller number of these bits and use the binary number these bits represent to form an address in which to store the original 36 bits.
2) Hash Storage or Correlation	Perform the actual storage (in the case of a leading beam return) or compare each 36-bit word to all lead returns previously stored in that address (in the case of a trailing beam return).

Below is a pictorial idea of what is done to obtain the address.

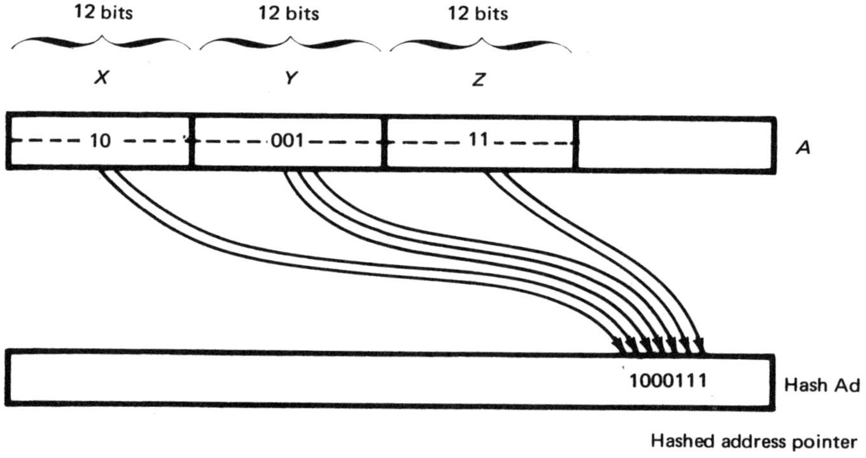

STORAGE BY HASH TECHNIQUE IN ILLIAC IV

There is only one minor "problem" that occurs in hash storage; otherwise, the procedure is quite straightforward.

Suppose the first 64 returns have been received, and 7 bits which determine up to 128 bins have been pulled. Now each return occurs once in each PE and is stored by that PE in one of the 128 bins.

Pictorially, it is:

	PE 1	PE 2	PE 3		PE K		PE 64
loc 1	Ret 1						
loc 2			Ret 3				
loc 3		Ret 2					
⋮							Ret 64
loc n					Ret K		

Now, the next 64 returns are received, and it is desired that they be stored in the appropriate location (bin).

The Problem is: WHAT if PE K receives a second return which wants to go into loc n? Well, the answer is simply that it must be stored in some loc n in another PE. (Say $k + 1$, if it is empty.) This is the "problem" alluded to above.

Now, it may be convenient (or necessary) to provide more than one location per PE for a particular hash address (bin). This situation will arise when "bursts" of targets must be processed. The program has been developed so that, in general, it can handle bins of "depth" n. In this case, when the expected value of returns per hash address is 24, n is equal to 1. When the problem of wanting to store two returns in the same memory location arises, the shifting to the neighboring PE is accomplished before more returns are handled.

CORRELATION OF TRAILING RETURNS

After all the leading returns from a strip of cloud have been processed (i.e., hash stored), and the returns from the trail beam on the same strip are being received, the following is done:

1) Repeat for trail returns the exact procedure of:
   a) "Coordinate Transformation"
   b) "Assignment of Hash Address"
2) At this point, when 64 trail returns are in the accumulators of 64 PE's the following is done:
   a) Feed the trail returns and their associated Hash Ad to the ILLIAC IV CU
   b) Put the Hash Ad in Car1 Reg.
   c) Broadcast the return to all 64 PE A Registers
   d) Subtract the *initial* return which is stored in the loc referred to by Car1
   e) If any result is zero, send the *trail* return (which is in the CU) to "Track Finalists" for the 1 second correlation
3) The correlation of the "finalist" items on this .1 second return with the "finalist" items on the previous second of the cloud sweep needs to be performed:
   a) The procedure is the same as the .1 second correlation
   b) The objects which match the 1 second correlation are sent to the track mode as potential targets to track.

The objects which are to be sent to the track mode must be converted back to the x, y, z coordinates of the earth center. Also, the associated E and A and time must be sent along to the track mode.

Timing

The time needed for preliminary coordinate transformation and conversion to cloud center coordinates is 15 μsec for 64 objects. Assume a 30,000 item cloud, therefore, 3,000 objects for .1 second correlation for both the leading and trailing beams. This gives a total for .1 second correlation of 1.5 m sec.

The needed amount of time for obtaining the hash code for the storage of the leading beam returns and the hash code for the trailing beam to perform the correlations is 5 μsec for 64 objects, giving a total of .5 m sec.

The storage of the leading beam returns requires an average time of 4 μsec per 64 objects. The total storage time for all 64 objects is .2 m sec.

The correlation of trailing returns with leading returns requires 1.75 μsec per object. This is a total time of 5.4 m sec.

The total time for the .1 second correlation of 3,000 objects is ($\approx$ 8 m sec.).

The 1 second correlation will be performed in the same way ∴ it adds 1 m sec. per .1 sec.

The total time for designation is 9 m sec./.1 sec.

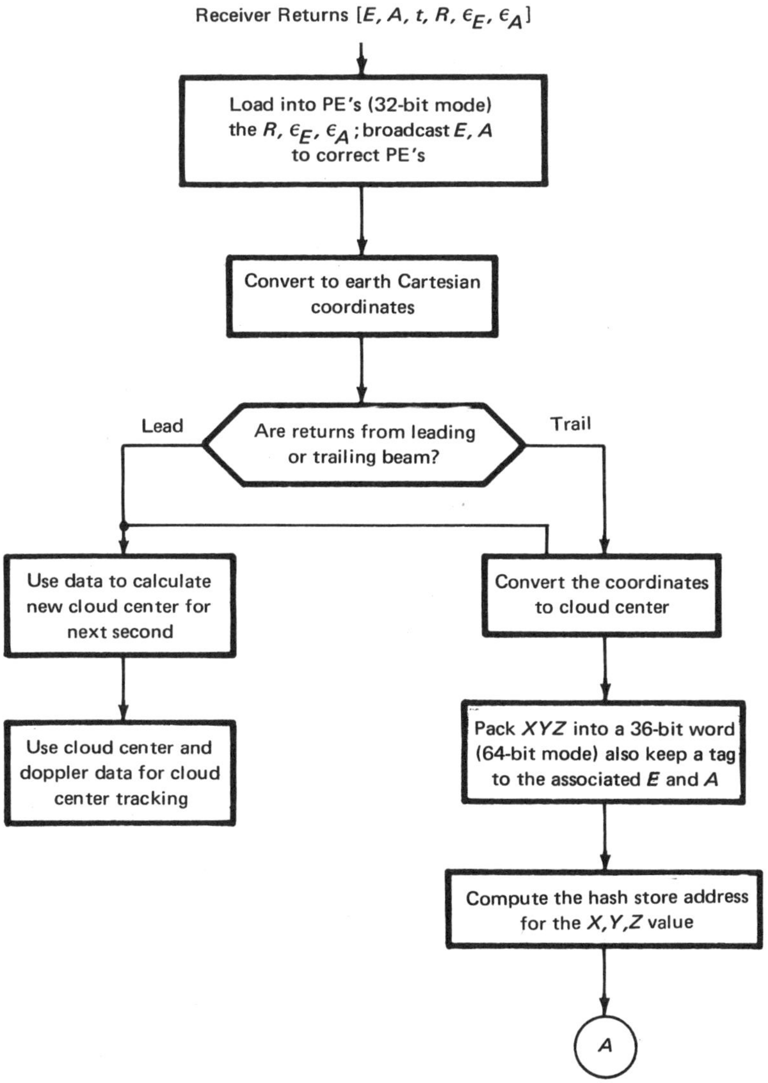

Fig. 13. Flow Charts for Designation

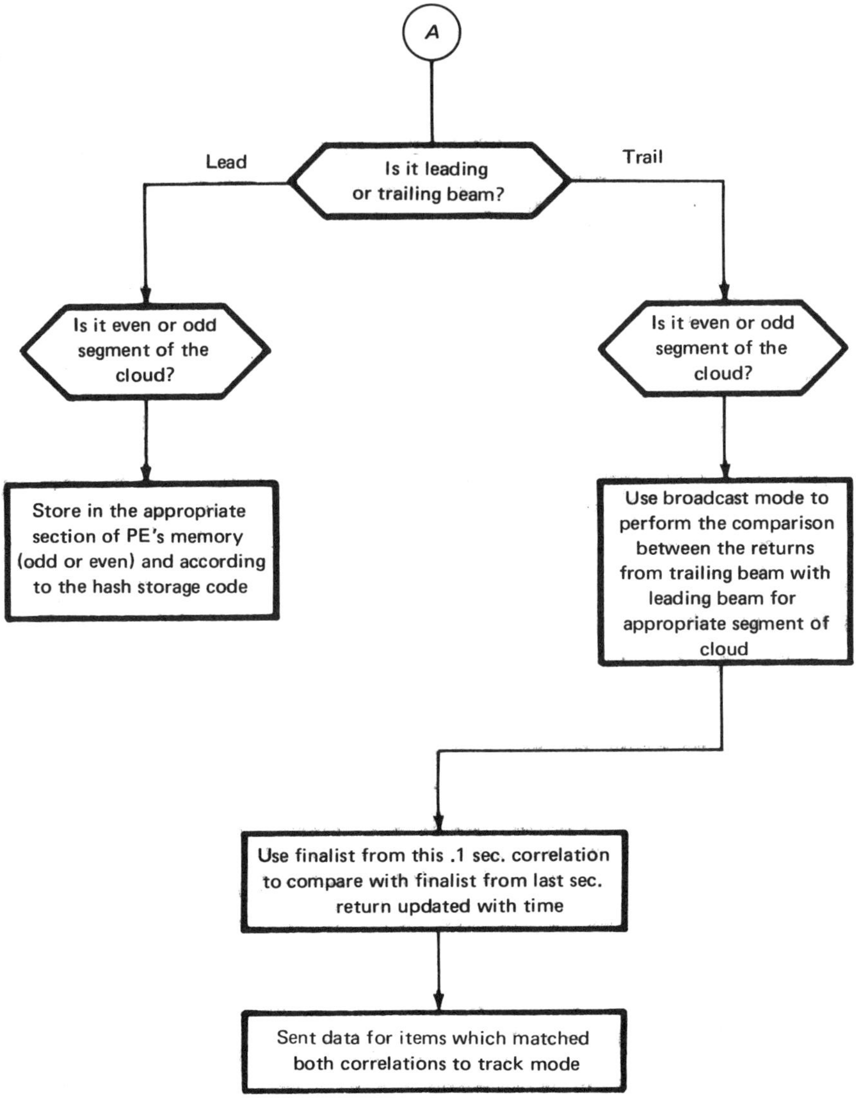

Fig. 13. Flow Charts for Designation (cont'd)

MEMORY REQUIREMENTS

The storage of the leading beam requires 128 memory locations. Memory area is needed for the even and odd .1 sec. correlation and for the 1 sec. correlation. The total memory requirement is 374 locations.

TRACK MODE

The tracking of objects involves the filtering of radar data obtained at discrete intervals, and requires periodic estimates of various continuous parameters such as target position, velocity, and acceleration. In general, there will be one output sample for each input sample. The output sample will represent an estimate of the input data, its integral, or a nth derivative. This estimate will be for some specified time before or after the current sample.

Much work has been published relating to this problem, but no general survey as to the best approach depending upon the assumed environment has been performed. The approach taken and given in the following few paragraphs is based upon material found in Monroe,* which seems to fairly well reflect the basic state of the art. If conditions require more complex techniques, they can be handled.

The smoothing equations assumed for this situation were least-square curve fitting. It involved the fitting of a polynomial to the input data samples. They were considered for only the dimension of the dependent variable; although the data given is in three dimensions, it is resolved and filtered one dimension at a time.

LEAST-SQUARE FIT OF POLYNOMIAL AND DATA

The samples are called X_i, and X_i is the sample taken at time t_i, and $i = 1, 2, \ldots, N$. The object is to determine a polynomial in t which provides a least-square fit to the N given points. This method also allows us to assign a weight, W_i, to each point to take into account any estimate of the quality of the data.

For the case of a first-degree polynomial, a curve of the form

$$y = a + bt$$

is to be fitted to the data, such that the expression

$$\sum_{i=1}^{N} W_i(a + bt_i - X_i)^2 \tag{2}$$

---

*Alfred J. Monroe, *Digital Processing for Sampled Data Systems,* John Wiley and Sons, 1962.

is minimized.

If the x_i represents readings of the target position, a will be the estimate of target position at $t = 0$, and b will be the estimate of its velocity.

The expression (see [2]) is minimum when

$$a = \frac{1}{D} \{(\Sigma W_i t_i^2) \Sigma W_i X_i - (\Sigma W_i t_i) \Sigma W_i t_i x_i\}$$

and

$$b = \frac{1}{D} \{(\Sigma W_i) \Sigma W_i t_i X_i - (\Sigma W_i t_i) \Sigma W_i X_i\} ,$$

where

$$D = (\Sigma W_i)(\Sigma W_i t_i^2) - (\Sigma W_i t_i^2) .$$

If all the weights are assumed to be uniform, all $W_i = 1$.

For quadratic smoothing, the polynomial is

$$y = a + bt + ct^2 .$$

Here $2c$ is the estimate of acceleration, if the X_i represent position, and b is the velocity at $t = 0$. The least-squares criterion is satisfied when a, b, and c are given by the set of simultaneous equations

$$\begin{cases} a\Sigma W_i + b\Sigma W_i t_i + c\Sigma W_i t_i^2 = \Sigma W_i X_i \\ a\Sigma W_i t_i + b\Sigma W_i t_i^2 + c\Sigma W_i t_i^3 + \Sigma W_i t_i X_i \\ a\Sigma W_i t_i^2 + b\Sigma W_i t_i^3 + c\Sigma W_i t_i^4 = \Sigma W_i t_i^2 X_i \end{cases}$$

When acceleration is a variable, for example, at least a third-degree polynomial would be required.

If the general form for a polynomial of degree q is written

$$b + b_1 t + b_2 t^2 + \ldots + b_q t^q ,$$

the coefficients b_j for a least-squares fit may be found by solving the following set of q + 1 linear simultaneous equations:

$$\begin{cases} b_0\Sigma W_i + b_1\Sigma W_i t_i + b_2\Sigma W_i t_i^2 + \ldots + b_q\Sigma W_i t_i^q = \Sigma W_i X_i \\ b_0\Sigma W_i t_i + b_1\Sigma W_i t_i^2 + b_2\Sigma W_i t_i^3 + \ldots + b_q\Sigma W_i t_i^{q+1} = \Sigma W_i t_i X_i \\ b_0\Sigma W_i t_i^2 + b_1\Sigma W_i t_i^3 + b_2\Sigma W_i t_i^4 + \ldots + b_q\Sigma W_i t_i^{q+2} = \Sigma W_i t_i^2 X_i \\ \cdot \\ \cdot \\ \cdot \\ b_0\Sigma W_i t_i^q + b_1\Sigma W_i t_i^{q+1} + b_2\Sigma W_i t_i^{q+2} + \ldots + b_q\Sigma W_i t_i^{2q} = \Sigma W_i t_i^q X_i \end{cases} \quad (3)$$

As before, the range over the summation signs is $i = 1$ to $i = N$, where N is the number of points taken, and $N \geq q + 1$.

USE OF ILLIAC IV FOR TRACK MODE

Data input to the track mode consists of each PE receiving all information returned for one target. If the target has come from the designation mode, this information consists of:

1)	E and A	(9 bits each specifying beam position)
2)	x, y, z	(21 bits each specifying earth centered coordinates of the target)
3)	$\dot{x}, \dot{y}, \dot{z}$	(10 bits each specifying velocity of cloud)
4)	t	(time pulse was transmitted).

If the target is from a MUB or individual track receiver, the PE receives:

1)	R	(21 bit range)
2)	E and A	(9 bits each)
3)	$\Delta E, \Delta A$	(4 bits specifying monopulse).

After initial temporary storage of the above information, the first task is to compute earth centered x, y, z in those PE's that received MUB or individual track targets. The equations for the transformation were given before.

X, y, z are computed and rounded to 21 bits each. Now that everyone has *x, y, z* of the target, all PE's form a 63-bit word from the 21-bit *x, y, z*'s.

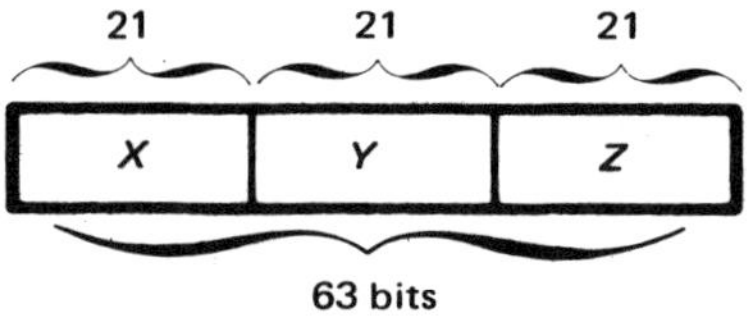

The coordinates of these words are the basis for the comparison of targets.

The next task for all PE's is to hash code an address from *E* and *A* to get an address that will tell where the targets are in memory that are to be compared to the new target. An address from *E* and *A* is hash coded, specifying beam position, and since the number of targets per beam position is limited because MUB receivers can only set 9 range gates per look, 4 to 5 different beam positions per memory location, each 64 PE's long, are gotten. The hash coding of *E* and *A* for an address, where the data from that beam position lies, consists of taking 6 bits to specify a PE memory location.

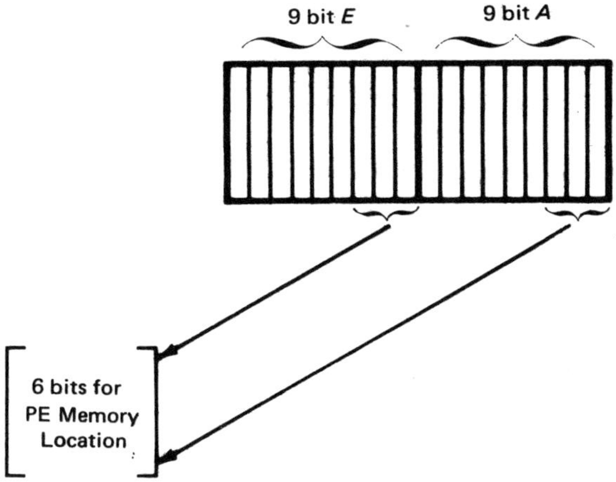

Since hash coding is expected to lead to sparse storage, the tables for each target—consisting of the last five returns on this target and their times and other necessary data for updating of tracks—cannot be stored in the hash coded area but must be stored somewhere down memory. A pointer is then necessary for each target in the hash coded area, telling it where its table is.

The above is accomplished by taking the 6 specified bits and tacking on a trailing zero to form the hash address. Hashing to every other memory location in a block, the predicted values of *X, Y, Z* are stored at these even locations, and the odd locations are reserved for the pointers. Therefore, if the predicted *X, Y, Z* for a particular target lie in memory location *A*, memory location $A + 1$ in that same PE will contain the address of the beginning of its associated 10 word table which also lies in that PE.

The comparison scheme works in this way: After each PE has formed its 63-bit *X*, *Y*, *Z*, and has hashed a 6-bit address, all PE's look at the contents of the hashed address which will either be empty or contain a predicted *XYZ* 63-bit word; in either case, all PE's compare that contents with their 63-bit *XYZ* words. If they are equal (i.e., the predicted *XYZ* agrees with the new target *XYZ*), the PE looks at the hashed address plus one for the pointer to the table of the data associated with that predicted *XYZ*, and stores the new *XYZ* and its time in the appropriate place in the table. Presently, this comparison is programmed to be performed in all 21 bits of *XYZ;* however, a mask can be used in the comparison if it is required. If no equality if found, no storage is made. After everyone has completed this, the *XYZ* word and the hash address are routed over one PE for the comparison scheme to begin again in that PE. This is done 63 times, so all PE's can try to find a comparison for a particular target, or until everything compares and is stored.

The next step is to check to see if there is anything stored in the overflow auxiliary hash coded area (i.e., if there are more than 64 targets associated with the beam positions hashed into one memory location). If there is, the hash coded address of all returns that have not compared yet are incremented to point to this auxiliary area. The above scheme for comparison is run to compare returns with stored returns in the auxiliary hash coded area. If after 64 attempts at comparisons (i.e., 63 routes), a target still does not compare, it is considered a new target and sent to the storage scheme. If no targets have overflowed the regular area, the auxiliary area is not checked before the decision of a new target is made.

The storage scheme now attempts to store the new targets. The problem is similar to the comparison scheme, only now the goal is to find "holes" in the hashed coded area rather than data that compare.

All PE's that contain data to be stored read the contents of their hashed address. If this address is empty, this return can be stored in this PE. The PE reads the contents of a counter LOC that tells it where it can start a new table in this PE. It enters a pointer to this table in the hashed address + 1 and stores the new return and data related to it in the new table.

If the PE found no "hole," it routes its information to its neighbor where the scheme runs again. This is done until all returns are stored, or until "holes" have been looked for in both the hash coded and auxiliary hash coded area. If a return does not store after 128 attempts (64 in the hash area and 64 in the auxiliary hash area), it is discarded.

The table associated with each target in the track mode consists of:

1) the last 5 *XYZ* words and their times
2) *E* and *A* of the beam
3) cloud velocity (if last return was from desig.)
4) a pointer to predicted *XYZ* in the hash coded area.

After the comparison and storage schemes, the next job is the updating of the data to get a new predicted *XYZ*. Each PE runs through its tables one at a time, applies the prediction scheme to the data in each table, and stores the predicted *XYZ* in the proper hash coded area address. For each table it encounters whose last entry was designation or individual track, it computes a predicted range of the target and forms a word in the table.

These words are then stored in the table for access when a MUB or individual track beam wants information about where it should next shoot and listen, and also at what range to set the range gates. These words are tagged to differentiate between those for MUB and individual track.

After all PE's have completed the above procedures for one table, they move on to the next. The scheme is repeated until all tables have been treated.

The tables are arranged as follows:

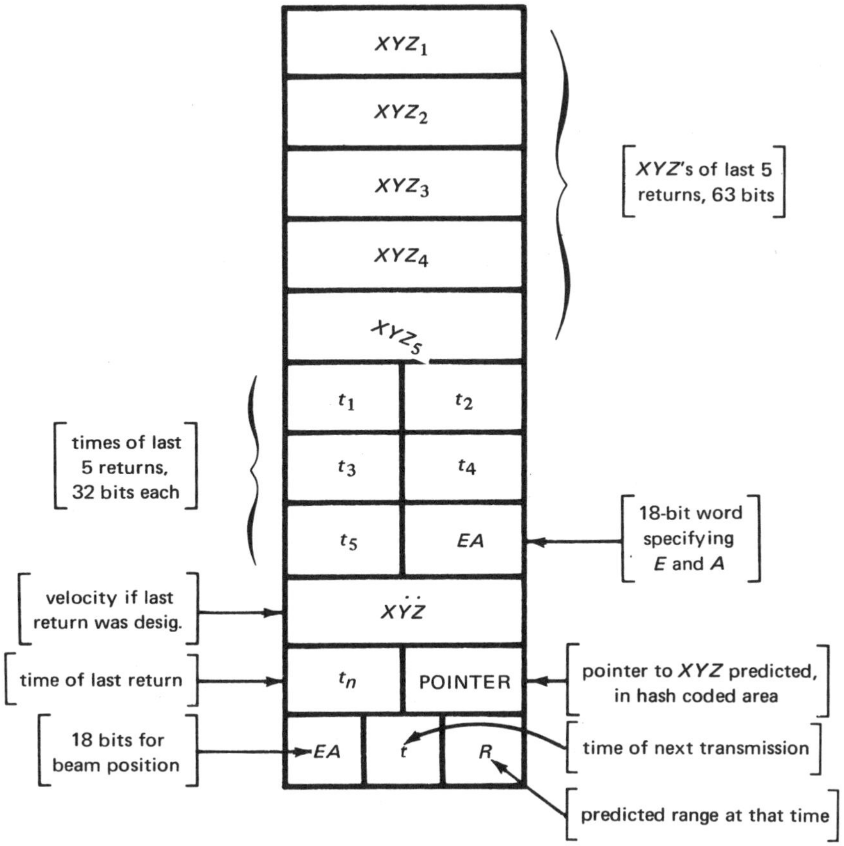

All predictions are now made and stored in the hash coded area for comparison with the next set of new returns. This track procedure is performed at least twice per second for each target.

Timing

Preliminary, coordinate conversion and hash coding requires 17 μsec for 64 targets—a total of 850 μsecs. This is down to point I in Flow Chart.

The comparison scheme requires 6.5 μsecs. Since it must be performed a maximum of 128 times (64 for comparison with 1st hash coded area, and 64 for comparisons with auxiliary hash coded area), the entire comparison can take at most 40 *m* sec. However, the overflow area will only have to be checked occasionally ∴ average 24 *m* sec.

The storage scheme requires 4.5 μsec to be performed once. Because it must be performed a maximum of 64 times when a new target appears, the total time required for the storage scheme is 2.8 *m* sec.

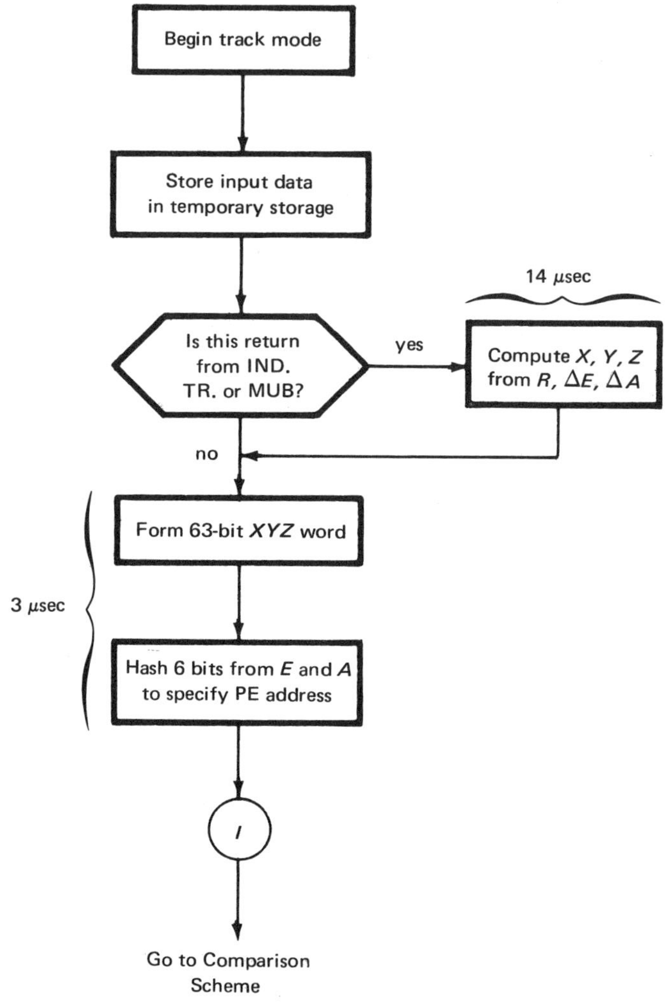

Fig. 14. Track Mode Flow Charts

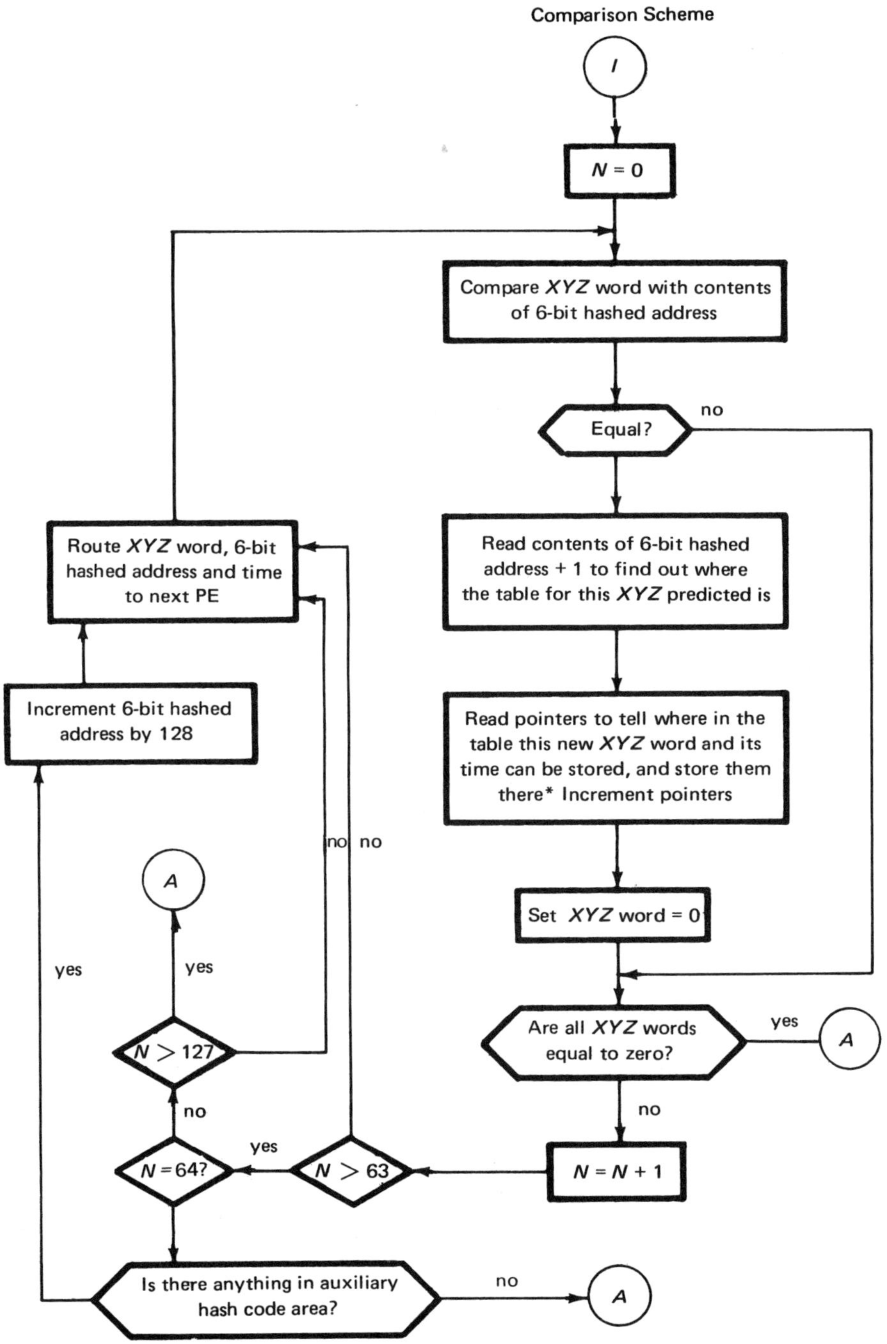

*These pointers point to the next empty location, if there are not already five returns in the table, or to where the oldest return is so the new return can replace it.

Fig. 14. Track Mode Flow Charts (cont'd)

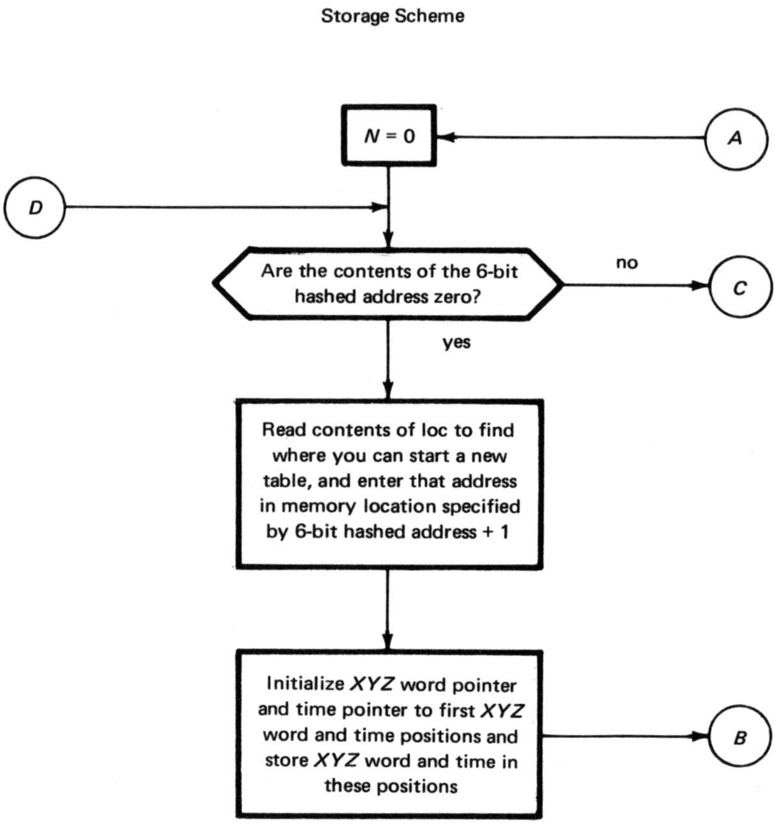

Fig. 14. Track Mode Flow Charts (cont'd)

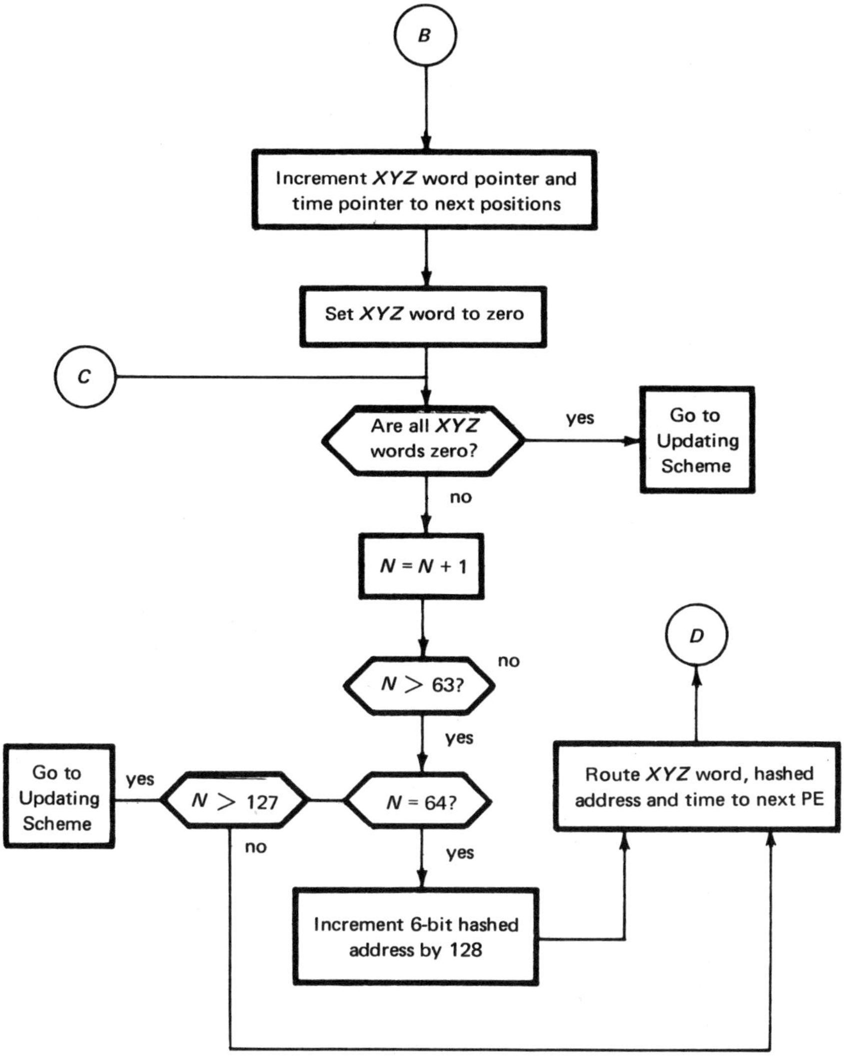

Fig. 14. Track Mode Flow Charts (cont'd)

The updating of target positions to get a predicted *XYZ* and also the forming of | *EA* | *t* | *R* | words, as explained in the paper, require 60 μsec per table. It is possible that a particular PE could have 128 tables if both its first hash coded and auxiliary areas are completely filled—although it seems extremely unlikely. If the maximum number of tables in a PE is N, the updating scheme requires $N \times 60$ μsecs. For the updating scheme, the greatest amount of time that could be consumed is when $N = 128$ which takes 8 *m* sec.

The absolutely longest time the track processing can take is approximately $\approx 35m$ sec.

SUMMARY OF TIME AND MEMORY REQUIREMENTS

The amount of time required by ILLIAC IV to handle the three major functions—(1) Steering Commands, (2) Designation, and (3) Tracking—has been given with three different time references. Therefore, it is appropriate to bring these to one common time of 1 second.

Steering Commands		144 milliseconds
Designation		90 milliseconds
Tracking		70 milliseconds
	TOTAL:	304 milliseconds/ 1 second

The memory requirements for the three major functions are as follows:

	Steering Commands		60 words
	Designation Area		374 words
Tracking	Hash Area		128 words
Tracking	Overflow Area		128 words
Tracking	History Area		825 words
		TOTAL:	1,515 words/ 2,048 words

More comments about what will take time and the available memory will be given in the following "Summary and Comment" section of this report.

Summary and Comment

This report has illustrated that a single quadrant of the ILLIAC IV computer is capable of handling the major functions of the phased array radar. These functions have been programmed for the ILLIAC IV and have been run on the ILLIAC IV simulator. As shown by this report, these three major functions (1) Steering Commands, (2) Designation, and (3) Tracking require less than half of the computer time. Several comments are in order to describe what other functions will require some of this available time, and what is needed for better reliability and what is available for the real time applications. These two factors are interrelated.

The tracking algorithm used in this report was least-square fitted to the last five report points of the track to predict the new position in time. More complex tracking algorithms (i.e., Kalman Filter, etc.) may be required for more accurate tracking of an object, especially when it starts re-entry through the atmosphere. These more complex tracking algorithms will require more computer time in the updating of the track history tables and predictions of target position. Also, if discrimination and interception are to be a function of this system, they will require some of the available computer time.

The computer diagnostics will have to be interleaved and run along with the other programs to check the availability of the computer hardware (the same as the diagnostic for testing the radar hardware has to schedule on a periodic basis). Techniques of recovery and the continuation of the system from a failure of a processing element is being investigated. The most important part is the ability to recover the track history of the objects which are stored in that particular processing element memory. A copy of the track history will be kept in external storage, i.e. in disk file, and when an element fails, this history will have to be obtained from the file. With the exception of this track history of important objects, the failure of a PE will not greatly affect the success of the system. Software techniques can be developed to work around failed PE's until they are replaced. Using the hash code technique for the storage of data on all objects does not require the operation of any particular PE. This aids in the ability to recover from a failure.

A separate program memory in the control unit, instead of the distribution of instructions through the PE's (as in the ILLIAC IV presently being built), would probably be the best approach to assure higher reliability against program self-destruction. This and other techniques for better system reliability are being investigated by Burrough's personnel.

ILLIAC IV will provide a large increase in computational capabilities which is required for many applications. This report illustrated how this system is capable of handling a large real-time application. It requires only a few minor changes in the computer system hardware to have fast input/output response (the *I/O* data path can even be expanded beyond the 1 *K* bits) and high availability.

CHAPTER 3

AN ASSOCIATIVE, HIGHLY–PARALLEL COMPUTER FOR RADAR DATA PROCESSING

J. A. Githens

Bell Telephone Laboratories, Incorporated
Whippany, New Jersey

Abstract

The organization of a global, highly parallel machine designed to provide a high performance facility for real-time track data processing in advanced radar systems is described. Intended as a supplement to conventional computers, the machine consists of an unstructured ensemble of any desired number of processing elements and a common control unit. In implementing a fully parallel concept, an individual processing element is assigned to each object in track, and it performs all of the data gathering, smoothing, prediction, and discrimination processing required on that track. The fully parallel approach allows that processing to be accomplished as highly parallel operations in which common instructions are executed simultaneously on all tracks, or any selected subset of tracks. Advantage, thus, can be taken of whatever inherent data parallelism exists in the application. The ensemble is content addressed and the input, output, and control operations use associative techniques to provide novel capabilities for data input and retrieval. The use of content addressing eliminates the artificalities and inefficiencies of location addressing, and makes the software independent of ensemble size and more fault tolerant.

Each processing element contains the registers and adder logic of a parallel arithmetic unit and a random-access memory. In keeping with the global nature of the machine, an element contains a minimum of control logic, the bulk of the control being concentrated in the common control unit. The elements are capable of executing a complete single address instruction repertoire permitting any desired arithmetic or logical operation. A major component of each element is a special purpose, concurrent input unit, called a *correlator,* made up of a number of associative registers. Specialized to make efficient the repetitive process of associating incoming radar reports with the stored track histories, this unit also provides programmable flexibility that permits any desired tracking algorithm and a variety of system configurations to be realized. The correlator, which is capable of concurrent and independent operation, is used to store predicted positions, gate limits, and input quantities freeing the arithmetic unit from the tedium of the comparisons inherent in the correlations. This allows the arithmetic unit to concentrate on processing preceding reports while the next set is input.

This paper describes the organization of the machine and its novel features. Emphasis is given to the design of the correlator and its role in tracking in a many target environment.

INTRODUCTION

As the role played by the digital computer in real-time control systems continues to increase, so does the need for greater computing rates. This need, created by improved sensors, wider band widths, increased traffic, and expanded functional requirements, exceeds the capabilities for today's machines. This paper describes the organization of a highly parallel computer designed to meet these needs in a variety of real-time systems. As a specific example of its application, its use in radar data processing is discussed in the following section. The system organization is given in the section "Highly Parallel Ensemble Organization," and the calculating elements that make up the machine are described in "Element Organization." The instruction repertoire and its novel features are discussed in the section "Instructional Repertoire." The section "Input-Output" explains the facilities provided for inputting radar data, and discusses their use in tracking. The final section "Summary" includes a comparison of this machine with other highly parallel computers.

THE RADAR PROBLEM

Phased array antennas provide the modern radar system with the ability to electronically point and shape one beam, or a multiplicity of beams, and to steer them at electronic speeds. This new versatility means that a single antenna can provide a variety of functions. To fully utilize these capabilities, direct computer control of the antenna functions seems a natural. The consequent close coupling between the data gathering and data processing functions is essential for high performance systems. The resulting data processing load with its stringent real-time requirements, however, is nontrivial. When combined with the load generated by applying such systems to difficult problems involving heavy traffic, the required processing rates often far exceed those possible with even the next generation of sequential computers.

The problem in radar data processing is not so much the complexity as it is the multiplicity of the calculations that must be performed. This is because the system must operate on a very large data base made up of information on all the echoes, ghosts, and noise the radar sees. Some systems, in the data gathering function, may have to correlate, store, and process the information on thousands, or even tens of thousands, of objects. A significant proportion of the computations must be performed on all objects at the sampling rate, which means that they must be completed within the pulse repetition period of the radar. This may be only 20 to 25 milliseconds in modern systems. The remaining calculations can be performed at lower rates, like once or twice per second.

The problem can be illustrated using the highly simplified program flow diagram of Fig. 1. In this diagram, rectangles represent blocks of instructions to be executed to accomplish some given function. Where rectangles are in parallel, the instructions to be executed on some classes of objects differ from those to be performed on other classes. The classes are determined by tests made at decision points indicated by diamonds in the diagram. Thus, in processing radar observations, for example, the first block represents operations to be performed on all objects. Different operations, however, may be more appropriate for objects above some given altitude than for those at lower altitudes, for example, and this situation is illustrated by blocks 3 and 4. The decision point (block 2) then tests the altitude to determine the class into which each object falls. In a similar way, the flow diagram proceeds in the dichotomous procedure characteristic of computer programs.

In a conventional sequential computer approach, one would take the data on an object and proceed through the diagram, examining the results at each decision point to determine the branch to be taken. Then the data on the next object would be obtained from memory by indexing, and the procedure repeated. To process n objects, using this approach, the procedure is repeated n times. In a real-time system these n repetitions must be made in the allotted time, which in the case of a

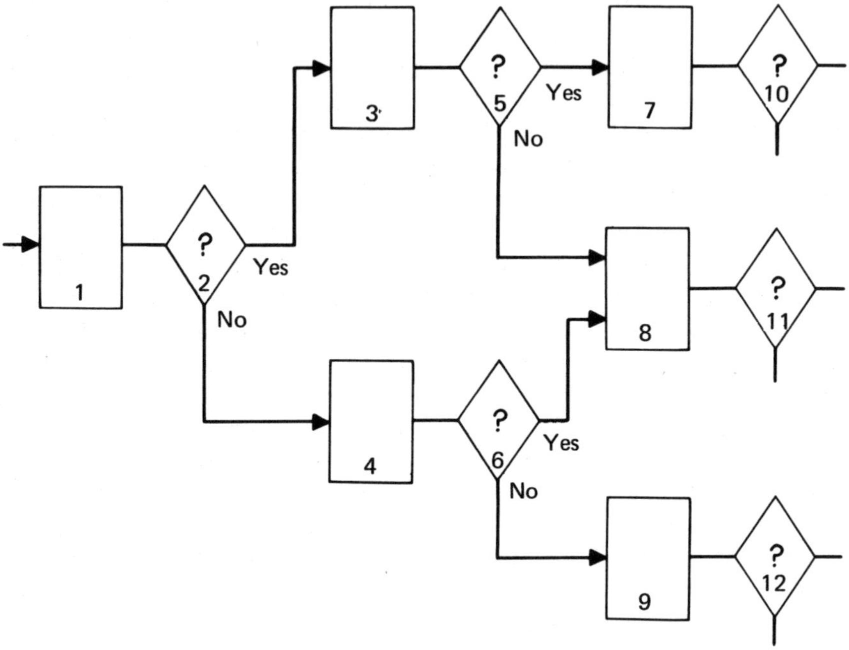

Fig. 1. Simplified Program Flow Diagram

radar system is the pulse repetition period. Thus, if a radar system is to have the capacity to handle 10,000 objects in a 25 millisecond pulse repetition period, only 2.5 microseconds on the average can be devoted to each object. This does not permit much processing with even the fastest sequential machine, its capability having been effectively dissipated by the combination of a large number of objects and high repetition rates.

The situation can be improved either by speeding up the machine or by using more machines—that is, by using parallelism. The former approach seems to offer opportunity for only limited improvement due to the current high state of solid-state logic circuit and memory development. A first step in the latter direction has been made in advanced third-generation multiprocessor systems where the simultaneous operation of several processing units and data channels yields degrees of parallelism as high as 15 to 20. The achievable parallelism is limited by memory access queuing and the problems of effectively controlling a multiplicity of processors.

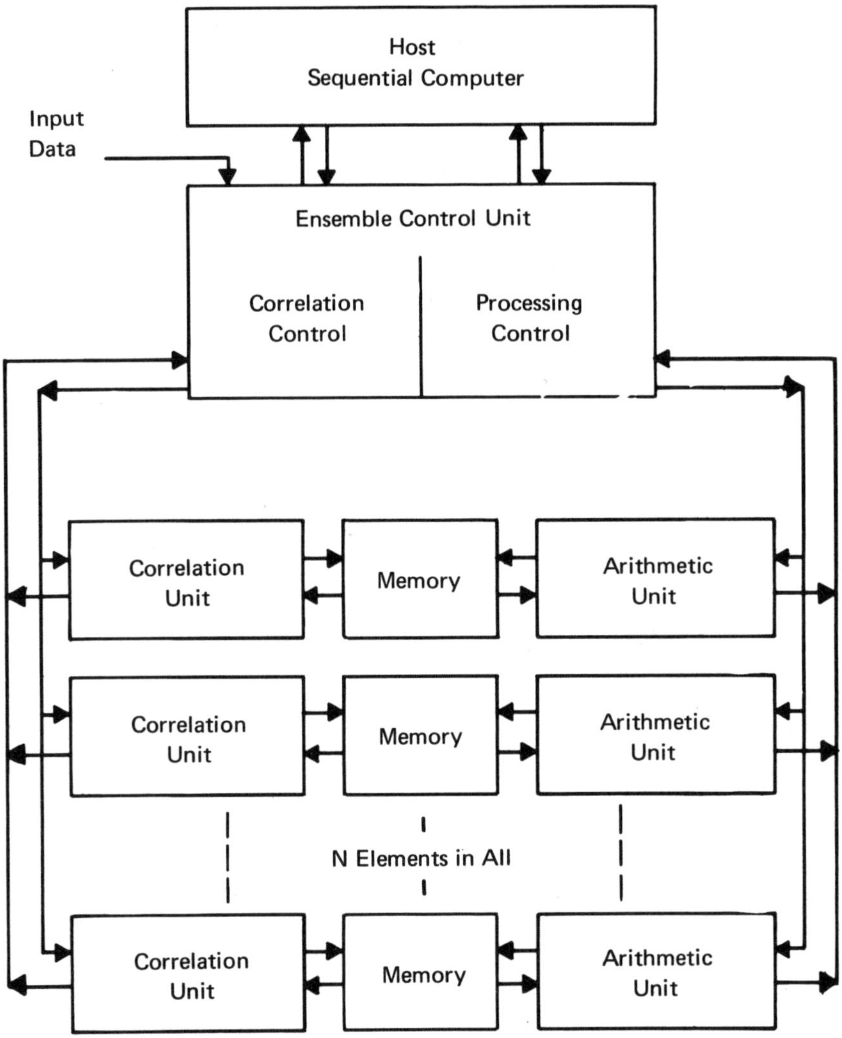

Fig. 2. Highly-Parallel Machine Organization

HIGHLY PARALLEL ENSEMBLE ORGANIZATION

This paper describes a much larger step in this direction. The machine, illustrated in Fig. 2, consists of an ensemble of processing elements operating under the control of an ensemble control unit. The ensemble is the arithmetic and operand storage organ of the machine. Program information is stored in the host sequential computer and instructions to be executed are supplied in the proper order.

The number of elements, n, is made large enough to equal or exceed the size of the data base that must be handled. This degree of parallelism permits each object being observed by the radar to be assigned to an individual processing element. The machine is then capable of operating on all objects simultaneously, and the processing time is no longer a function of the number of objects being observed. The problem of controlling this very large number of processors is solved by using the global control approach. Common input, output, and control buses go from the control unit to all elements. All elements receive the same inputs and the same control signals, and the ensemble as a whole performs one common operation at a time. Individual elements either participate or not, depending on their internal state. The operation can be illustrated by returning to the previous example using Fig. 1. In the global, highly parallel machine, the operations indicated by the first block are performed simultaneously on all objects. The altitude parameter would then be examined simultaneously in all processing elements. Those in which the altitude was below the specified level would be deactivated and the process indicated by block 3 would be performed in all active elements. These elements would then be deactivated and the others activated while the process indicated by block 4 was executed. It is instructive to think of this process as set manipulation. Block 1 is performed on the universe of all objects stored, and the branches of the diagram represent operations to be performed on various subsets defined by the decision tests. Regardless of the number of objects being processed, the program is executed by proceeding through the diagram, executing each branch once. Note that the efficiency of the machine varies in the process. In block 1, all elements are in operation, but in executing the branches, only that proportion of the machine corresponding to the size of the subset is active. The average effective processing rate, in sequential machine terms, is the actual rate multiplied by the average number of objects in a subset. This has been referred to as bulk processing[1], and for applications where the average set size is large, such as radar data processing, the effective rate can be very high. However, almost all programs will require some operations on sets having few members. Such program branches may be more efficiently processed in the sequential machine.

ELEMENT ORGANIZATION

Each of the identical elements takes the form of a small, parallel, digital computer. As shown in Fig. 2, each contains the familiar components of an arithmetic unit and a memory. In addition, each element contains a special purpose input unit, called a correlator, which will be discussed more fully in the section "Input-Output." The arithmetic unit and the correlator share the memory on a cycle-stealing basis with the correlator having priority. It can be seen that the resemblance to a conventional computer is quite strong except that one component, the control, is largely missing. In keeping with the global nature of the machine, the control circuitry in each element is kept to a minimum with most of the control being concentrated in the common control unit. All instruction decoding and the sequence control is performed by the common unit with microgating orders being transmitted over the common control leads to all elements. The small amount of control circuitry in each element is concerned with determining whether or not the element should participate in the current operation, as will be discussed later.

The arithmetic unit, then, consists of registers and the adder logic, but very little control circuitry. It performs floating-point operations with the following speeds: 1 sec. addition, 4 sec. multiply, and 20 sec. division times. These operations are executed on variable length fields. This allows the wide range of precisions characteristic of real-time applications to be efficiently handled. Furthermore, it allows the memory to be packed. Efficient utilization of storage capacity is of concern in this organization because only a modest memory capacity is provided in each element. This is possible since each element needs to store the information on only one object. A small memory capacity requirement allows the realization of very fast speeds in a physically small, low-power memory, which is entirely compatible and commensurate with the logic.

The memory size and the bit capacity of the registers are design variables that can be used to optimize the machine for particular applications. A memory size of 16,000 bits, organized as 512 32-bit locations, has been found to cover a broad range of radar applications. A register length determines the maximum field that can be directly processed. A length of 32 bits appears to be adequate for most real-time applications.

INSTRUCTION REPERTOIRE

The instruction repertoire of the machine provides a complete, though not elaborate, set of single-address arithmetic and logical operations.

These permit any desired processing to be performed by the elements on operands in their memories, or on common operands supplied through the ensemble control unit. The conventional aspects of the repertoire are many because the operations are performed by elements that are quite conventional in form. That is, operations such as *clear add, add, multiply,* or *and* can be performed with the only unusual feature being that they may be simultaneously executed by thousands of elements. However, the collection of elements as a whole, which is the ensemble, does have some unusual properties, and this results in some novel capabilities. To remove the artificialities and inefficiencies of location addressing, the ensemble as a whole is given the characteristics of a content-addressable, or associative memory. That is, each element appears to be a memory location at which the data file on a given object is stored. Access to this information is based on information content rather than on element location. The associative access is used in three functions (1) in determining which elements are to participate in each operation, (2) in distributing inputs to the ensemble, and (3) in obtaining outputs from the ensemble.

Each instruction to be executed by the ensemble is accompanied by an element activity specification. The activity specification details in an encoded way the attributes of the objects to be operated on by that particular instruction. The activity specification is compared in each element with the element's activity state which is stored in a tag register. The tag register contents can be altered by tests on the accumulator contents. Elements whose activity states match the specification, perform the instruction. Those whose states do not match, do nothing. Thus, as each instruction is executed, the ensemble is partitioned associatively into elements that participate and those that do not.

This type of operation can sometimes present a problem when the common input or output facilities are used. It is similar to what is often called the multiple response problem of associative memories. It arises on input or output operations when the activity specification causes multiple selections of elements. Since the common input and output paths are shared by all elements, this can result in inputs being accepted by many elements, or in many elements trying to output simultaneously over the common leads. For example, suppose the ranges of objects at altitudes between 40,000 and 45,000 feet and having velocities greater than 2,000 feet per second are desired. In general, more than one element contains data on objects meeting such a specification, and to read-out the ranges it is necessary to isolate the activity to one member of this subset at a time. This facility is provided by a set of instructions called *activate.* Activate instructions impose a special activity state on the ensemble causing the members of the specified set to be individually activated in a given sequence. Three versions of activate that result in different sequences are available. *Activate ascending* and *activate descending* activate the members of the specified set in ascending or descending order of any selected ordering variable. To continue the previous example, these instructions make it possible to read-out the ranges in ascending or descending order of the altitudes, ranges, or of any other desired parameter. They make use of the Lewin algorithm[2] for resolving multiple responses in associative memory. This very efficient algorithm makes it possible to retrieve an ordered list of m items from the array in $3m$-1 read operations in the general case of the ordering variable being different from the item being retrieved. These instructions also make determination of the element storing the maximum or minimum value of any selected variable a rapid and easy process.

The third form of this instruction is called *activate sequentially.* This instruction is used when the order of retrieval is not important. In this case, the elements storing members of the specified set are activated in the order that they happen to be stored in the ensemble, that is, by location.

In using activate instructions, the first execution selects the first member of the specified set in the desired order and leaves that element as the only active element in the array. In this state, any desired input or output operation may be performed. Arithmetic operations are not permitted; however, there is little motivation to perform such operations in this state because it would be very inefficient. Subsequent executions of the activate instruction cause the next member of the specified set to be selected. This process continues until the set is exhausted or the process is terminated by the control, returning the array to normal activity state.

An instruction is provided to test array activity. This produces a binary answer to the question whether there are any elements storing data on objects meeting a given specification. This could be used to determine branching in program control. However, this instruction finds little use in highly parallel programs. This is because in the types of problems to which these programs are best applied, it is generally safe to assume that there will be members of every set specified. Moreover, it is often more rapid to perform the branch than to test to see if there are any members. The result is that the problem of program branching in highly parallel machines is quite different than in sequential machines. One can afford to do program look ahead, secure in the knowledge that all program branches will be taken.

A related instruction that is more useful is called *count.* This instruction produces a number approximately equal to the number of elements storing data and meeting the activity specification. This number is determined without counting by encoding the output of an analog summing circuit which sums the precision current outputs produced by the active elements. The encoded number is accurate to ± 10% and is produced in less than 2 μsec. This provides a rapid means of determining the statistics of the objects being observed by the radar.

INPUT–OUTPUT

Inputs present a problem in any highly parallel machine with this degree of parallelism just due to the number that are to be expected. This is compounded by the real-time requirements in the radar application. Because the input process is basically sequential, utilizing common facilities, the real-time requirement places an upper bound on the number of objects that can be processed. To push that bound as high as possible, the input process is made concurrent so that the full pulse repetition period is available for inputting. As indicated in Fig. 2, the correlator section has input and control leads in common with all other correlators, but separate from those to the arithmetic section to permit concurrent application.

On the other hand, the number of outputs required is expected to be relatively small. This is a consequence of the concept of operation that uses the highly parallel ensemble as the sole storage site for the data on individual objects. That is, the array is not thought of as a parallel computation facility into which data is read, operated on in parallel, and then read-out to be stored elsewhere. That type of operation could quickly dissipate the capabilities with data shuffling. Rather, the data on an object is resident in the ensemble as long as it is being observed by the radar. It is manipulated in place. Detailed information on a small number of objects is output. On the great majority of objects only summary information and ensemble statistics will be output. The other output of consequence is the ordered lists required for radar control. Thus, the output is not made concurrent but interrupts and uses the arithmetic unit.

The principal input process in the radar application is one of associating incoming radar reports with the corresponding stored object history. That is, in the ensemble the input problem is one of directing incoming reports to the proper elements. For one important class of tracks this is a relatively simple matter. In "target-oriented" tracking, an individual radar transmission is scheduled for sampling the position of each object in track. An element storing such a track causes a radar order to be generated specifying the desired time of transmission, direction, predicted range, gate width, and wave form to be used. To this order, the element attaches a "correlation" number. When the radar has made the transmission and received the return, it delivers the encoded video tagged with the correlation number to the ensemble. Directing the return to the correct element then requires a simple, exact, and unique match on the correlation number. Unfortunately, in any system the number of tracks that can be handled as target-oriented tracks is severely limited by the available radar power and time. When the number of objects to be tracked exceeds this limit, resort must be made to "spatially oriented" tracking. In spatially oriented tracking, the volume to be covered is scanned by directing a radar beam, or a cluster of beams, sequentially to a raster of predetermined positions. The returns received in this manner are compared with the predicted positions of the stored tracks to determine which returns belong to which track. The process is complicated by the fact that (1) angle resolution is sacrificed in spatially oriented tracking, (2) an object can produce returns in several beam positions, and (3) as the number of objects increases so does the probability of crossing trajectories which can cause confusion. Tracking under these circumstances is complex enough that reliable and tested algorithms are not yet established, and an important requirement for the correlator is that it provide the flexibility and performance so that any desired algorithm can be implemented with efficiency.

To provide this facility, the correlator is realized as a number of associative storage registers. The number provided is a design variable with 4 as a minimum and 8, perhaps, as typical. As viewed by the processing unit, these registers appear as an extension of memory providing additional locations that can be loaded and read like the rest of the memory. However, these locations differ in that they are capable of operating concurrently with and independently of the rest of the memory and the processing unit. The operation is under control of the correlation control unit which has an independent set of input and control lines bused to all correlators. Each register contains equal-to-or-greater-than comparison logic which compares the register contents with the data on the common input lines from the correlation control. Thus, as viewed by the correlation control, the ensemble of correlation units appears to be an associative memory having a number of locations for each processing element.

To meet the flexibility requirement, the correlator registers are interconnected as a distributed logic memory in the manner of Lee and

Paul,[3] That is, the registers of the correlator are connected as a linear array with capabilities for communication between neighboring locations. In addition, each location is provided with additional bit storage capacity for control purposes. Forty-bit registers provide 8 positions for control tag storage in addition to 32 positions for data storage. It has been shown that with this form of associative memory, any desired arithmetic or logical operation can be performed on the contents of each location by a sequence of elementary comparison and store operations. In the correlator, this means that any given tracking-gate logic or tracking algorithm can be realized by microprogramming. Thus, for example, 4 registers can be programmed to provide a dual tracking gate that could be used to store the predicted range R_p and the gate limit values, $R_p + \Delta_2$ and $R_p - \Delta_1$. The ranges, R_I, of new reports are compared with these stored values. If, in a given report

$$R_p - \Delta_1 < R_I < R_p$$

the return is accepted in the leading gate with the input being automatically steered to the proper registers for storage. In one approach R_I can overwrite $R_p - \Delta_1$ closing up the gate. Similarly, if

$$R_p < R_I < R_p + \Delta_2$$

the return is accepted in the trailing gate. In this manner, it is possible to capture the nearest returns on either side of the predicted value. For lower grade tracks this degree of sophistication may not be warranted, and 4 registers could be programmed to provide 2 single gates, which stored the gate limits for 2 separate tracks. Single gates capture the return nearest the leading or trailing edge depending on how the gate limits are overwritten.

The correlation unit is capable of executing three distinct classes of operations. The comparison operations and the store operations in which input data are written into the correlation registers make up one class. These operations are completely concurrent and independent of the processing unit operations. It is anticipated that the bulk of the operations performed in handling the radar inputs will be of this class. The correlation registers provide a limited capacity for input data storage. Some inputs, such as information on extended targets, for example, will require more storage. For this purpose the correlation inputs are given direct access to the element memories. In these operations the correlation unit competes with the processor unit for memory access on a cycle-stealing basis with the correlator having priority. Finally, the correlator is capable of executing virtually the full arithmetic and logical repertoire of the processor, but for these operations it interrupts the processor and uses its arithmetic unit. It is expected that such operations make up only a small proportion of the operations performed in correlating the inputs.

With these facilities available, one could approach spatially oriented tracking in the following way. Suppose the returns from the ith beam position are ready to be input. The predicted positions of all tracks are updated in a highly parallel manner to the time of the ith transmission, and the gate limits are calculated. The appropriate values are stored in the correlator registers, freeing the arithmetic unit. The subset of predictions corresponding to the ith beam position are conditionally activated as are those of neighboring beam positions. The returns from the ith beam position are then input in sequence and the measured positions compared with the predicted positions. If there is a match with a predicted position in the ith beam position, the data for that return are input to the element holding the matching track. If there is no match in the ith beam, but one in a neighboring beam, note is made of that fact in the element holding the neighboring track. If there is no matching prediction at all, the data may be sent to an empty element to start a new object history depending on the initiation algorithm. And so on until the returns from the ith beam position had been input. At that point the processor unit would be briefly interrupted to make a new update so that the returns from beam position $i + 1$ could be input.

Consider now the problems that were mentioned above. Angle resolution is usually poor in spatially oriented tracking because the beam is often broadened to achieve coverage, and it can no longer be assumed that the target is near the beam axis. Frequently, no off-axis or beam splitting information at all is available, and the angle granularity corresponds to the beam position. In effect, a target is always reported at beam center, and there is no basis for predicting when the target will leave the beam. It is left to the data processing to detect the occurrence of beam crossing. This is accomplished by using the information on matches in neighboring beam positions. Thus, a sequence of misses in a given beam position accompanied by matches in a neighboring beam position indicates a beam crossing, and when this is detected the beam name in the element storing that track is changed to the name of the neighboring position.

It is worth noting here that in keeping with the associative nature of the machine, tracks are addressed by their attributes and, in general, one never knows which elements stores what track. Thus, when a new object history is to be initiated, the data are stored in the first available empty element. In general, the data on that track will remain in that element. As its status, grade, or position changes, the names of its attributes will be changed, but the inefficient and unproductive process of data transfer is avoided. Element location addresses are used only for diagnosis and maintenance.

The production of returns in several beam positions by a given object is handled by ignoring matches in neighboring beam positions when there is a match in the predicted beam position. Furthermore, when a new object is first detected in a given beam position, a set of gates is immediately predicted for that object so that as subsequent beam positions are processed, any other returns from the same object are ignored and do not cause multiple tracks. If the beam in which an object is initially detected does not correspond to its actual position, the beam crossing algorithm will soon correct the situation.

The best source of information for the detection of impending target crossing situations is the radar input. For this reason the correlation control contains facilities for preprocessing the returns as they are input. As each return is input, the range differences to the preceding and the succeeding are added to the report. This enables each element to determine range closing rates of the nearest tracks, and to anticipate crossings. When an impending crossing is detected, the strategy is to coast the tracks involved, ignoring the data that become garbled in this situation.

SUMMARY

The organization of a computer incorporating highly parallel, calculating arrays has been described. The combination of a high performance, sequential computer, and high capacity, highly parallel ensembles provides a powerful data processing system for application to real-time radar data processing. The ensembles give the system the capability of simultaneously manipulating the sets encountered in processing data on the large numbers of objects characteristic of this application. The ensemble inputs have been specialized to make efficient the scan-to-scan correlations that are an essential part of radar data processing and, on the output, ordered lists and summary information can be rapidly generated.

The highly parallel ensemble utilized is of the global type. As such, it belongs to the same family of machines as the SOLOMON[4] and ILLIAC IV[5] computers. There are, however, a number of differences. The most notable being that the ensemble described here is a special-purpose machine. While it is hoped that it will prove valuable in a spectrum of real-time control applications using radars or similar sensors, no attempt was made to provide a general-purpose facility. Similarly, this ensemble is not intended to stand alone, but is to be a part of a conventional machine, supplementing it where ensemble properties can be useful, and leaving to the sequential machine those operations that it does well, such as program control. As a consequence, the elements hold operands only, and store no program information. These elements have relatively modest memory capacity. They are intended to be used in a fully parallel manner with each element storing only one data group; i.e., one object. Thus, there is no need for sequential operation on different data groups and, hence, there is little need for local indexing.

An outstanding distinction is the lack of structure in the ensemble and the common data input. This, at first, is surprising, particularly in a system that may have parallel input channels (from multiple beams) and in which input time is a problem. To be useful, however, a connected ensemble must in some way model the process which, in this application, is three dimensional. In the face of the very difficult problems of realizing a three-dimensional array that could meet the practical constraints imposed by packaging, wiring, and maintainability considerations, it was decided to use the unstructured ensemble and the common input, and to rely on associative techniques to accomplish the relatively small amount of intercommunication needed in this class of problems. The unstructured ensemble has a number of other advantages. Packaging and wiring are simplified and, thereby, less expensive. The traffic capacity of the array is easily changed by adding or subtracting elements, and variable size ensembles made up from a pool of element modules becomes possible. Element location in the array has little meaning, and element failure merely results in a small decrease in array capacity.

Finally, it can be instructive to think of the ensemble as a special form of memory of the sequential computer. It is, after all, where the data on the objects are stored. The memory locations do, however, have the unusual property that arithmetic and logical operations can be performed on their contents in situ. Furthermore, any number of the locations, even all of them, can operate simultaneously. The content-addressable feature makes the input and output processes very efficient, and it allows the set manipulations (or vector operations as they are sometimes called) characteristic of parallel processing to be performed in a natural manner.

References

1. B. A. Crane and J. A. Githens, "Bulk Processing in Distributed Logic Memory," *IEEE Transactions on Electronic Computers,* EC-14; No. 2 (April, 1965), 186.
2. M. H. Lewin, "Retrieval of Ordered Lists from a Content-Addressed Memory," *RCA Review,* June, 1962, p. 215.
3. C.Y. Lee and M. C. Paull, "A Content-Addressable Distributed Logic Memory with Applications to Information Retrieval," *Proc. IEEE,* No. 6 (June, 1963), 924.
4. J. Gregory and R. McReynolds, "The SOLOMON Computer," *IEEE Transactions on Electronic Computers,* EC-12; No. 5 (December, 1963), 774.
5. D. L. Slolnick, "Unconventional Systems," *Proc. Spring Joint Computer Conference,* April 18-20, 1967, p. 477.

CHAPTER 4

PARALLEL PROCESSING FOR PHASED–ARRAY RADARS

J. E. Courtney and H. M. Halpern

Radio Corporation of America
Moorestown, New Jersey

INTRODUCTION

This paper reports some results of a study of the characteristics and requirements of a large radar computer system, with special emphasis on phased array radars and parallel processing computers. A survey of radar systems requirements was made with emphasis on the high density target environment problem. A detailed analysis of the various system tasks was performed, as they were related to digital processing of signals and data. A study of various parallel processing computer configurations was also performed, and a particular configuration was chosen for evaluation against the radar problem. This configuration was a modular array computer. A number of radar tasks were applied to this configuration, in order to determine its applicability under various conditions.

A major problem of parallel processors is to find a configuration that is both flexible and efficient. Flexibility is desired to enable the system to expand in an orderly manner when the task grows, so that the same concept may be applied to a wide variety of radar computer systems. Efficiency is used here in the sense that internal housekeeping and executive functions should be small as related to the total

The authors wish to acknowledge the contributions of R. Klein, and those of the RCA Defense Communications Systems Division who worked on the ILLIAC IV Project.

This work was supported by the U.S. Air Force, under Contract F30602-67-C-0211, Rome Air Development Center, Griffiss Air Force Base, New York.

computational capacity. The modular array system was found to be both efficient and flexible.

NATURE OF THE RADAR PROBLEM

A large phased array radar includes very many data processing tasks. These tasks include target search, target track, array steering, and beam scheduling. Beam scheduling alone is a considerable task; the radar may have multiple beams, which must be scheduled in accordance with target priorities and transmitter power constraints. In other words, the radar imposes a very large data processing load, which must be performed in real time.

Considerable flexibility is needed in the computer configuration, due to the manner in which large radar systems are evolved. A radar may take 3 to 5 years to develop, and have an operational life of another 10 years. Even after a system becomes operational, modifications are frequently made as the threat changes, and new technical developments become available. These facts have a bearing on the choice of special purpose versus general purpose computers. While a special purpose system may be more efficient for a specific computational task, changes in the system require changes in the hardware and are very expensive. On the other hand, the general purpose computer permits changes to be made by changing the software. While changes in the software are far from trivial, they do allow the system to respond to changing requirements. However, the general purpose processor must also be able to grow with the radar. Most changes required in the radar system put an ever greater load on the processor. Increasing the capacity of a conventional general purpose computer is very difficult, and it was for this reason that a modular configuration was chosen.

Some radar tasks are naturally parallel in nature. An example is the array-steering problem, in which settings must be computed for a large number of phase shifters or time delay units simultaneously. In other tasks the method of application of parallel processing is not so obvious, and some study is required in order to efficiently fit the problem to the computer.

SOME PARALLEL PROCESSOR CONFIGURATIONS

As mentioned earlier, one of the problems of parallel processors is efficiency. The efficiency of a computer becomes low when too much of the capacity of the computer is devoted to internal housekeeping. This includes such jobs as shifting data from point to point inside the computer, as it is needed in various areas, and assigning tasks to the various processors. If the proper configuration is not used, it is difficult to keep all the processors busy at one time. This is related to the problem mentioned in the previous section of fitting tasks to the computer. This task is greatly eased if the proper computer configuration is used.

The first parallel processor configuration to be considered is that of the multiprocessor. The multiprocessor configuration is one in which a number of processors and controllers are all interconnected so that any controller has access to any processor. The problem with the multiprocessor is communications. The amount of housekeeping required increases rapidly with the number of processors employed, so that a point is eventually reached, beyond which the addition of more processors does not result in an increase in overall capacity, due to the accompanying increase in housekeeping.

The second configuration is an array processor; this is one in which a single controller issues commands to a number of processors, and these all perform the same task simultaneously. The array processor avoids some of the housekeeping problems by using a single controller. However, the array processor tends to become limited in its capability as additional processors are added, since it is not able to keep all the processors working all the time. This is also related to the problem mentioned before of fitting the radar task to the computer. Many of the tasks that the radar must perform will not require the full capacity of the computer at any given time, and, as a result, the efficiency of the computer begins to decline as the number of processors grows.

A solution to some of these problems lies in a combination of the array processor and the multiprocessor. This is the modular array computer, the subject of this report.

THE MODULAR ARRAY COMPUTER

The modular array computer may contain from 1 to 16 modules, as shown in Fig. 1. Each module in turn may contain from 1 to 16 processors and 1 controller, arranged as in Fig. 2. Each module contains a Next Instruction Bus, a Data Bus, and an Input–Output Bus. Each processor has associated with it a 32-bit 4,096-word memory. The instructions for the controller are stored in the processor memories.

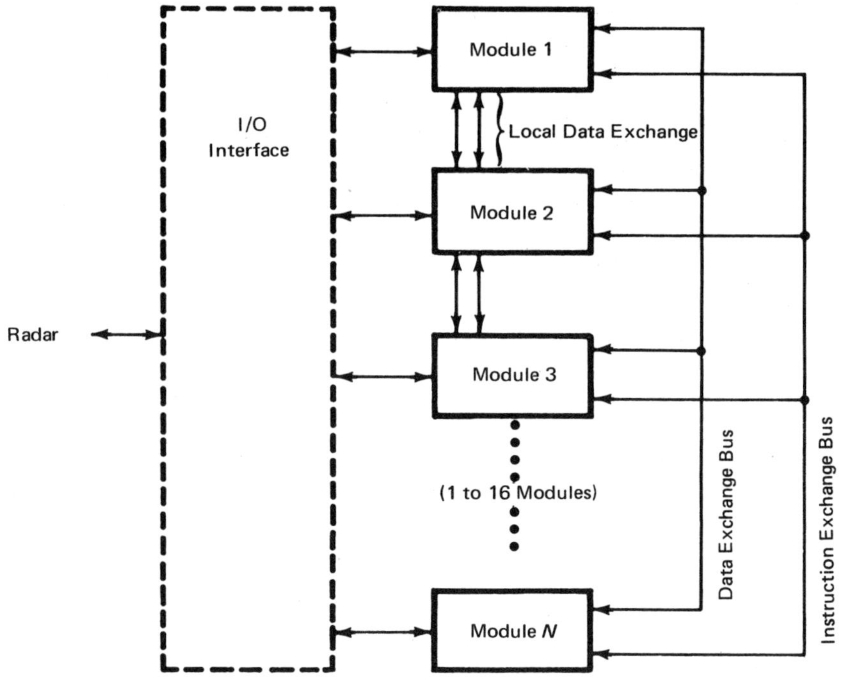

Fig. 1. Modular Array Computer Configuration

Communication is the key to success in parallel processing. Each module is interconnected to the others via a Data Exchange Bus and an Instruction Exchange Bus, but each processor also communicates directly with its immediate neighbors via the memory buffer registers. A processor communicates with its neighbors within the module by the East–West transfers shown in Fig. 2. In addition, each processor communicates with its counterpart in the two adjacent modules by means of the North–South transfer lines. Not only may data be transferred by this means, but data stored in the memory of one processor may be directly accessed by a neighboring arithmetic unit, without actually transferring the data to the neighbor's memory.

The resulting performance of this configuration may be illustrated qualitatively in Fig. 3. This figure shows the relative performance of the Multiprocessor, the Array Processor, and the Modular Array Processor. Here the effects of housekeeping and processor utilization are seen. By selecting the proper number of modules for the radar task at hand, most of the processors may be utilized most of the time, so that the efficiency of the modular array processor is kept high.

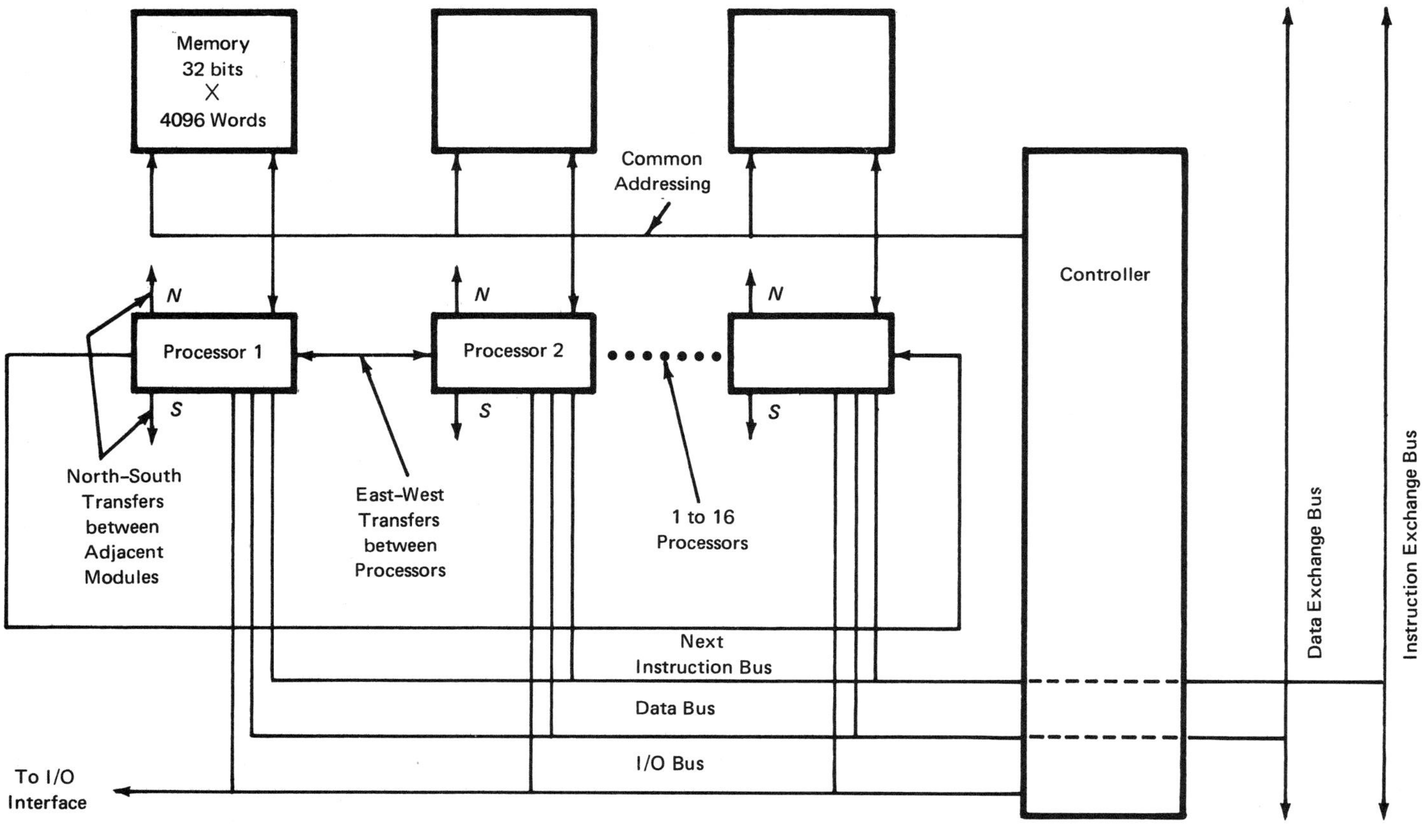

Fig. 2. Computer Module

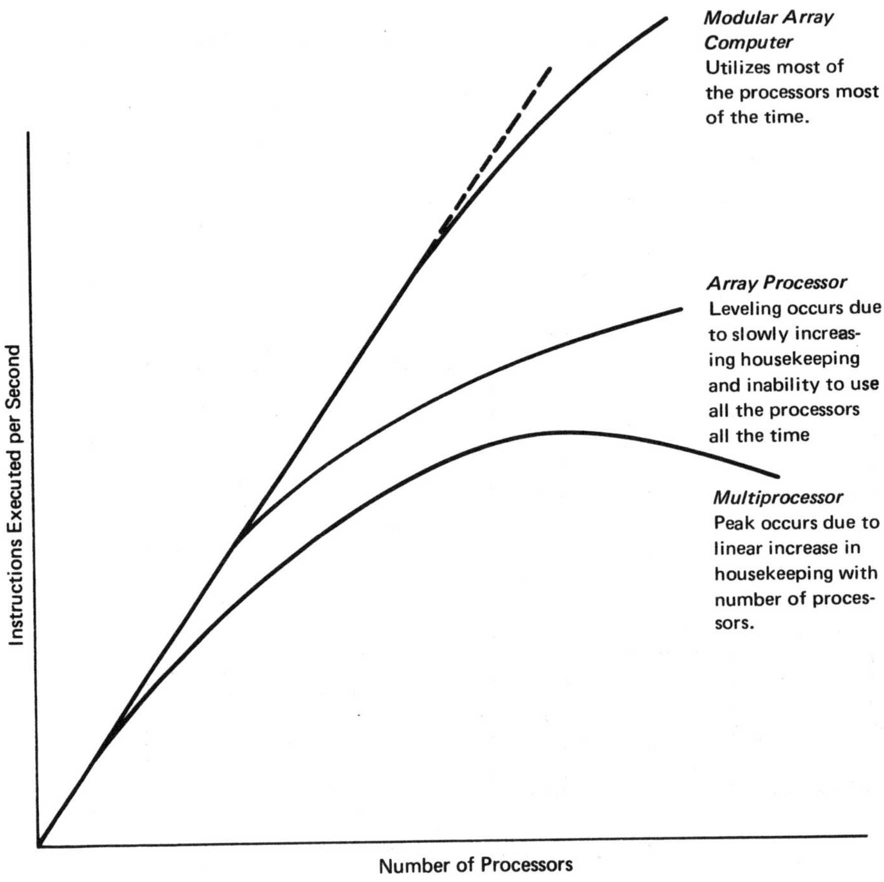

Fig. 3. Relative Performance of Computer Configurations

The processor which was assumed for the basic unit in each module has capabilities that might be expected to obtain in a system being designed today. That is, a memory cycle time of 200 nanoseconds, an add time (including the memory cycle) of 250 nanoseconds, and a multiply time of 650 nanoseconds. Extensive use of large scale integration (LSI) was assumed. This is important since the power of the computer may be rapidly dissipated if the individual components are so large that delays are encountered in transferring data from one component to another. Each processor or controller is to fit on a single platter 19" by 19."

EXAMPLES OF RADAR PROBLEMS

This section gives a description of some typical radar problems, and how the multiple array computer was applied to these problems.

Data Input

The first problem to be considered is that of data input. This problem is very important because no other processing may take place until the radar data has been loaded into the computer. The problem of loading radar data hinges on the special nature of this data. The output of a radar receiver is characterized by short bursts at very high rates, followed by relatively long quiescent periods, during which processing may take place. An example of the origin of one of these bursts of data is the contents of a range gate. The problem was aggravated in the particular case which was studied, since the radar had several parallel receivers, all operating independently, with more than one range gate in each receiver. We will now show how the particular structure employed in this computer is useful in loading radar data.

Referring back to Figs. 1 and 2, it should be remembered that each module has its own Input–Output Bus. This bus connects directly to the processors in each module. Since each processor within the module has the capability of loading a 32-bit word in one memory cycle, and there are 8 processors in each module, 256 bits may be loaded into each module during a single memory cycle.

Figure 4 shows the Input-Output interface for each module. Each radar receiver has an output of relatively few bits in each sample, and

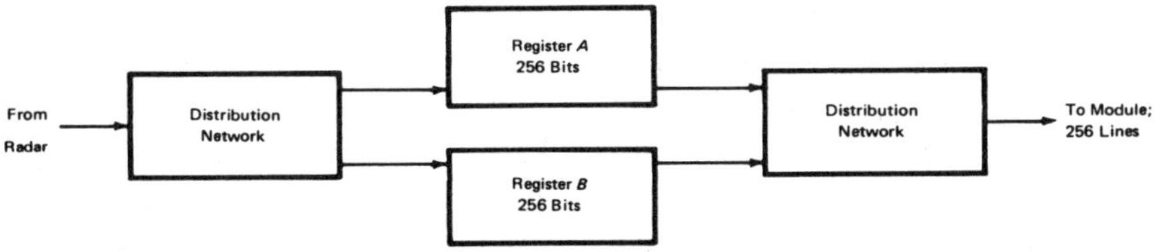

Fig. 4. I/O Interface for One Module

the buffer registers are used to collect enough of the short samples to make one large 256-bit word. Two buffer registers are used, so that one may be loading into the module while the other is accumulating data from the radar receivers. When enough data has been transmitted from the radar to fill one of the buffer registers, an interrupt signal is sent to the computer. Upon reception of the interrupt signal, computation is suspended and the contents of the buffer registers are transferred into the modules.

Figure 5 shows how the modules are used as buffers for each other during a high rate burst from the radar. For most of the processing problems, the data to be processed is needed in one module. However, due to the method of high speed loading, the data which has just been received is scattered through a number of modules. The sequence of operations shown in Fig. 5 is used to shift the data into the module in which the processing is to be done. This procedure makes use of the North–South transfer connections between modules. During each step of the sequence the data in each module is shifted to the module just below. That is, the data in Module 1 is transferred to Module 2, while the data which was in Module 2 goes to Module 3, and so on. During this first step 256 bits have been transferred from each module to the module just below. These steps are repeated until all the data required has been transferred into Module 4. (The Data Exchange Bus may also be used to transfer data between modules which are not adjacent.) The number of actual steps or memory cycles required to complete the transfer depends on how many words of data were contained originally in each module.

Computation of Phase Steering Commands

A second radar problem to be illustrated for application to the computer is the one of generating the steering commands for a phased array. This is an example of a problem with a large amount of parallelism; a single beam command will give rise to hundreds of phase shifter commands.

The geometry is illustrated in Fig. 6. Commands for positioning a beam are received in radar (spherical) coordinates, azimuth and elevation of the beam. Some radars have a variable carrier frequency, and if this is the case, the frequency to be used must be supplied with each beam order. Other parameters which enter the computation concern the structure and orientation of the array. The phase shifters

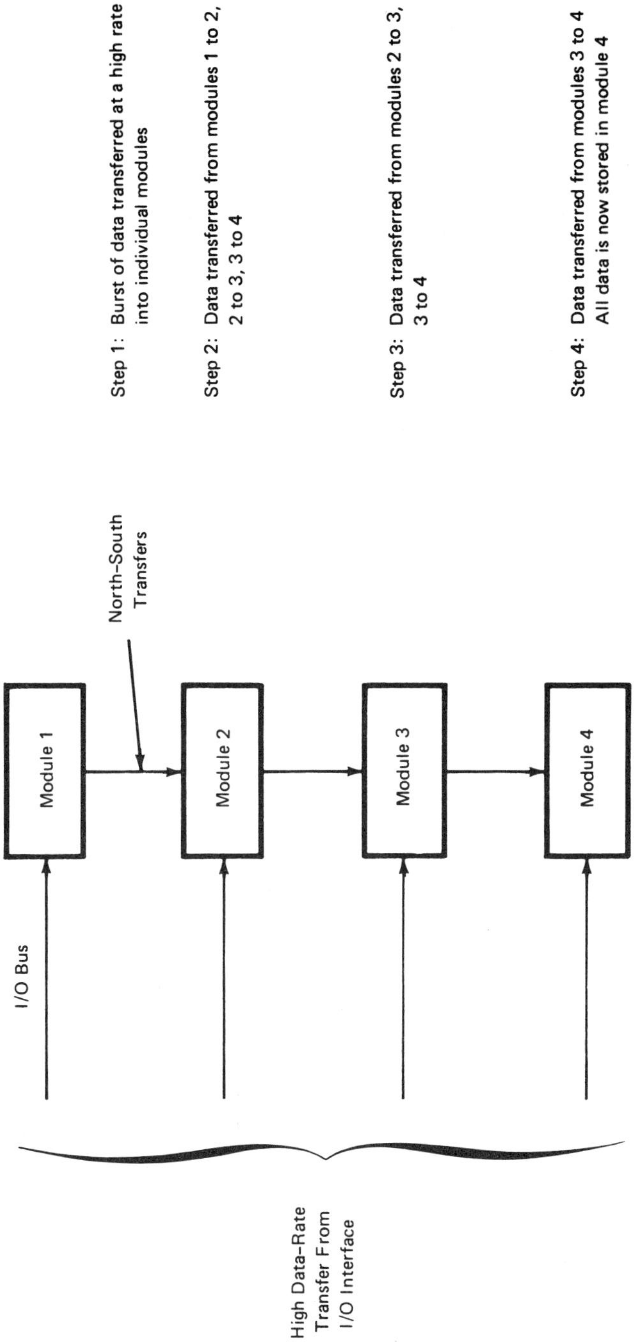

Fig. 5. Loading of Burst of Radar Data into One Module

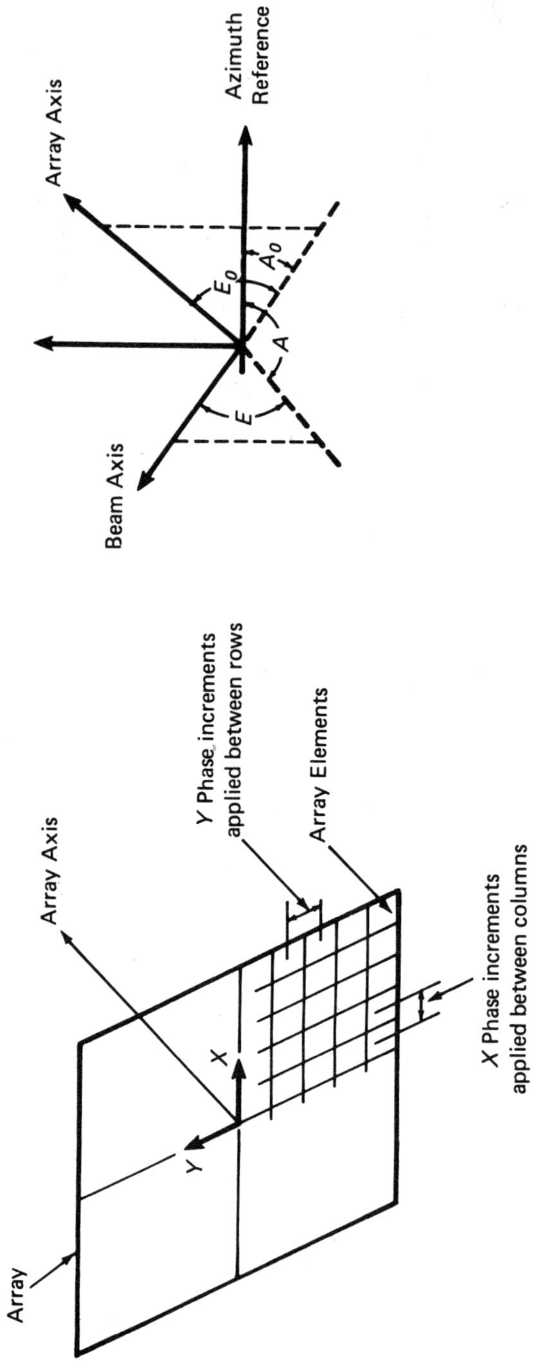

Fig. 6. Beam Steering Geometry

for the array being considered are arranged on rectangular (row and column) coordinates. The array also has a fixed azimuth and elevation, as shown in Fig. 6, which must be taken into account.

The command for each phase shifter is computed in two components, which are added together to obtain the setting for the phase shifter. The reason for this is that all the phase shifters in a row, say, will have the same "y" component, and the "x" components will be integer multiples of one phase increment. Hence the basic quantities to be computed are x and y phase increments, given by the following expressions:

x phase increment:

$$\phi_x = 360\, f D_x \left(\frac{1}{c}\right) \cos E \, \sin (A_o - A)$$

y phase increment:

$$\phi_y = 360 f D_y \left(\frac{1}{c}\right)\left(\sin E \cos E_o - \cos E \cos [A_o - A] \sin E_o\right),$$

where

- f is the transmitter carrier frequency
- D_x is the horizontal array element spacing
- D_y is the vertical array element spacing
- c is the velocity of light
- A is the beam azimuth
- E is the beam elevation
- A_o is the array azimuth, and
- E_o is the array elevation.

Not only is there parallelism in that many phase shifter commands are to be computed at once, but in the particular system which was studied, commands for up to 16 different beams are to be computed simultaneously. For each beam, a separate value of A, E, and f are given. The sequence of computations and the amount of parallelism for this problem is shown in Fig. 7. At each stage of the process the maximum number of processors which may be used simultaneously is given. The sequence shown in Fig. 7 is for a single module. Notice, however, that steps 1, 2, and 3 could be performed simultaneously using three adjacent modules, since the values of A, E, and f need be stored only in the middle module, the other two modules accessing the A, E, and f values directly from the middle memory.

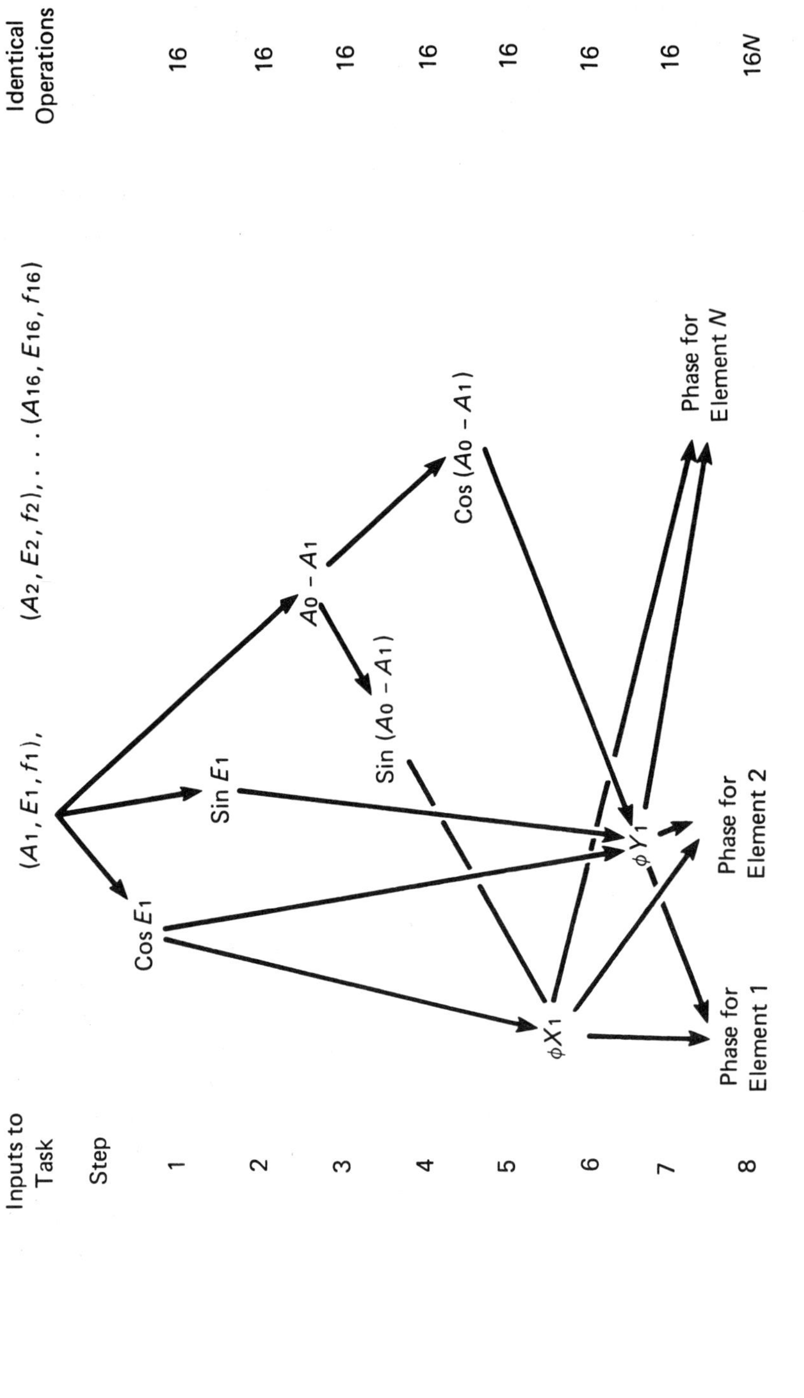

N = Number of Elements In Array

Fig. 7. Processing of Phase Steering Commands

Now, as an example of how the problem may be arranged to take maximum advantage of parallel processing, note that at some stages the cosine of an angle is needed, while at others the sine of the same angle is needed. The inclusion of a short addition step will allow the paralleling of the much longer sine or cosine computation, as shown in Fig. 8. If the processors are available to perform this parallel operation, a considerable amount of time may be saved by *adding* operations.

Problems of Target Data Storage and Handling

In illustrating the problems of target data storage and handling, it is helpful to consider the radar signal processing task as a sequence of filters. The earlier in the sequence, the more "targets" there are to be manipulated. After each step in the process, the number of targets is reduced and, consequently, more sophisticated filtering can be applied. Typical filtering operations require association of the targets resulting from successive transmissions to permit doppler filtering, drag estimation, scintillation tests, etc.

One can approach the association problem in two ways. The first is more commonly used but becomes extremely complicated for large numbers of targets. In this method, each target is assigned a fixed location in memory, containing position and coordinates. The target priority and expected position versus time is updated periodically in a schedular which groups the targets by beams and expected range, and orders radar pulses in accordance with the priority ranking. As the number of targets increases, the continuous shifting of order and establishment of range gates greatly increases the communication time between processors, particularly if the targets are distributed among several processors.

A second method of handling the target association task is to place the target data in memory by some associative scheme, so that the location in memory depends in some fashion upon the location of the target in space.* (All the target data need not be stored at this location, but a system of pointers may be used, with some of the data, such as target priorities, stored at some other location designated by the pointer.) A very simple method of doing this is to assign a separate memory location to each resolvable point in space. This of course is

*The location in space may be relative to the earth, or relative to some point within a large group of targets ("cloud-centered coordinates").

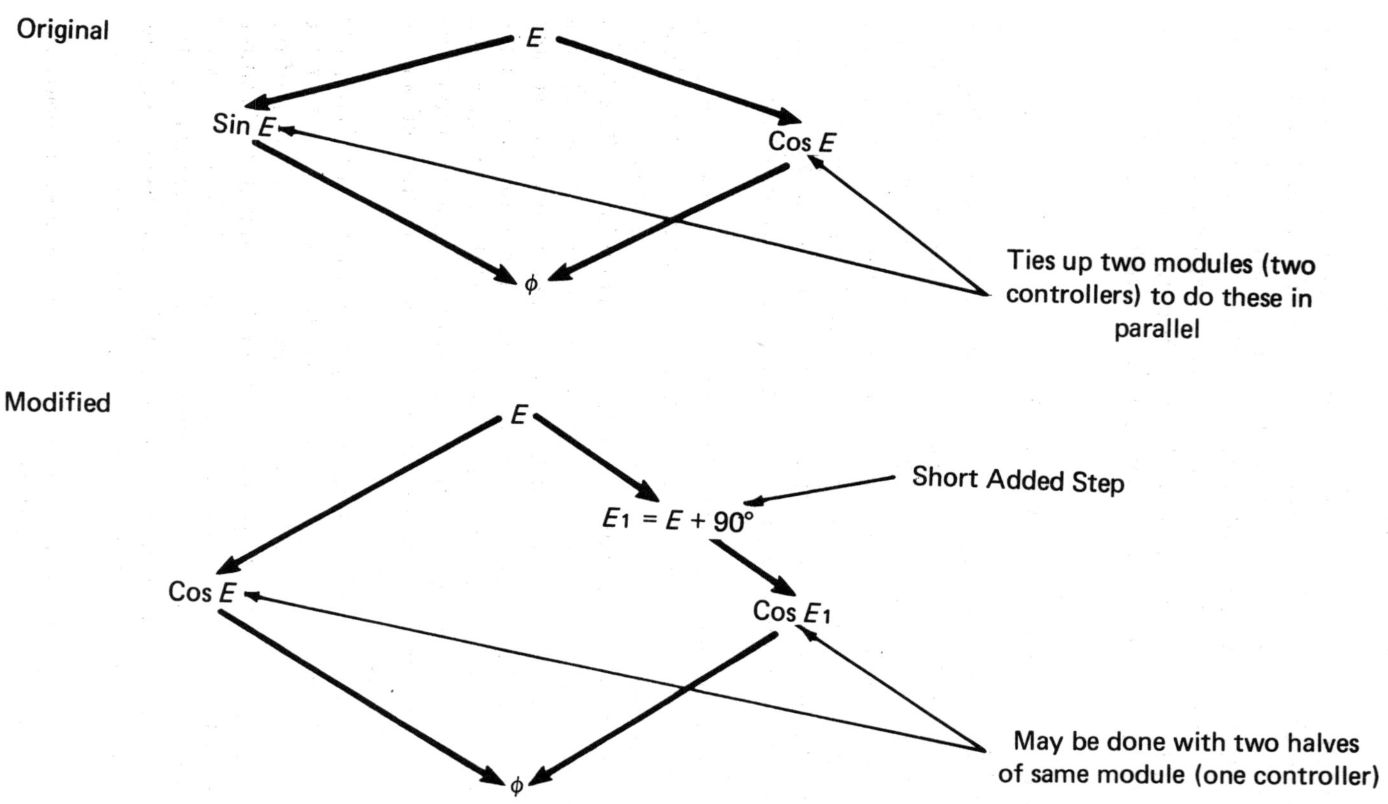

Fig. 8. Change of Program to Increase Parallelism

extremely uneconomical of memory. A more efficient approach is to establish bins in memory capable of holding many pointers. The pointers are placed in a bin by selecting the bin address through a random number computed from the target position. This method has proved to be very economical in computing time and memory requirements. Randomizing the address has a distinct advantage in that targets are more uniformly distributed among the processors, and a bunching of targets in space does not place an undue burden on any one processor.

Figure 9 illustrates the organization of one such memory used in a scintillation test. For this test two beams scan the same moving volume. The two beams maintain a fixed time interval, τ. A second pair of beams are swept over the volume about 10τ later. Targets that pass the first comparison (targets that persist for τ seconds) are associated with the targets that pass the second comparison. Only those targets that persist through the two successive examinations are passed by the scintillation filter.

The leading beams establish the contents of the memory bins. Shown in the figure is a 12-bit word for each of the x, y, z coordinates of the relative target position.* The intermediate bits of the coordinate words are used to generate a random address. As shown in Fig. 9, all the words in memory that are called out by the same address are lumped together as a superword. All the processors in a module have a common controller, and the controller manipulates data as superwords.

Each piece of target data that is stored is located in two superwords, one containing x and half of y, and the other containing the rest of y and z. For the scintillation filter studied, each bin consisted of 32 possible storage spaces, which required 8 superwords. A total of 256 storage bins were provided.

The system operates in this fashion: For the leading beam, a bin address is computed for each target, and the target coordinates placed in the next available storage space in the bin. For the trailing beam, a bin address is also computed for each incoming target. As each trailing beam target arrives, comparisons are made simultaneously by all processors going successively through superwords containing x coordinates until a match is found. If x and half of y compare, then the remaining superword is called for a comparison of the remaining y and z. A successful comparison is passed to the next store, to await the passage of the second pair of beams. A scintillating "target" that does

*Cloud-centered coordinates are assumed here.

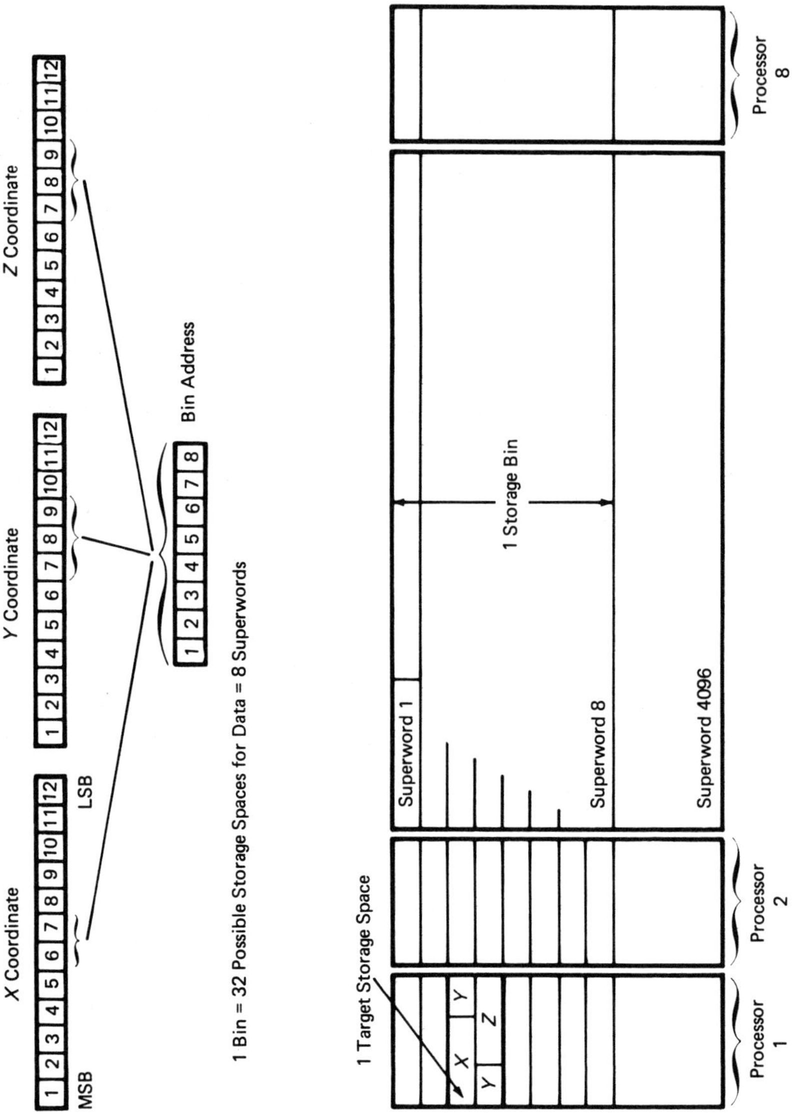

Fig. 9. Memory Organization for Scintillation Test

not compare at any stage is discarded. Figure 10 shows a three-module implementation with the assignment of superwords of memory in each module. For the case shown 12*K* of memory was fully occupied. The processors were utilized 500 milliseconds/second of the time. Less than 8 milliseconds/second of that time was involved in data transfers.

In any randomization scheme such as the one described, the possibility of having an overflow in one of the memory bins must be considered. When the targets have been randomly distributed among the storage bins, the probability of any bin containing a given number of targets follows a Poisson distribution, and from this the probability of overflow may be obtained. For the filter just described, about 3,000 targets were to have been distributed among 256 bins, giving an average of 12 targets per bin. For the probability of overflow to be less than 0.00001, the number of storage spaces per bin must be 32. On the average, 3/8 of the cells in memory will be filled. This means that processing steps (comparisons) have been saved at some cost in memory utilization, raising the question of how these two items are to be traded off.

Figure 11 compares the results of the three mechanizations. Note that the randomization method just described greatly reduces the number of computations, without a substantial increase in the amount of memory required.

In addition to the problem of overflow, another problem arises when associating targets from successive scans. This is the problem of an ambiguous location in storage, arising from uncertainties in the position of the target. There is always a certain amount of error associated with measuring the position, due to noise in the radar system, interference from jamming, uncertainties in predicting positions, etc. For this reason, the least significant bits of the coordinates are disregarded when making comparisons, or when computing an address in memory from a given target position. As a result, the storage bin corresponds to a finite volume in space, (or a set of disconnected volumes), rather than a point as small as the best resolution of the radar. The ambiguity problem arises when the target is near the boundaries of this volume. When a second scan is made, the target may very well appear in a neighboring volume, and hence be referred to a different storage bin where it will not compare with the data stored from the first scan.

In the case of the scintillation test, the number of lost targets is not a serious consequence, since the prime concern is to reduce the number of targets to be examined later by more refined methods. However, for

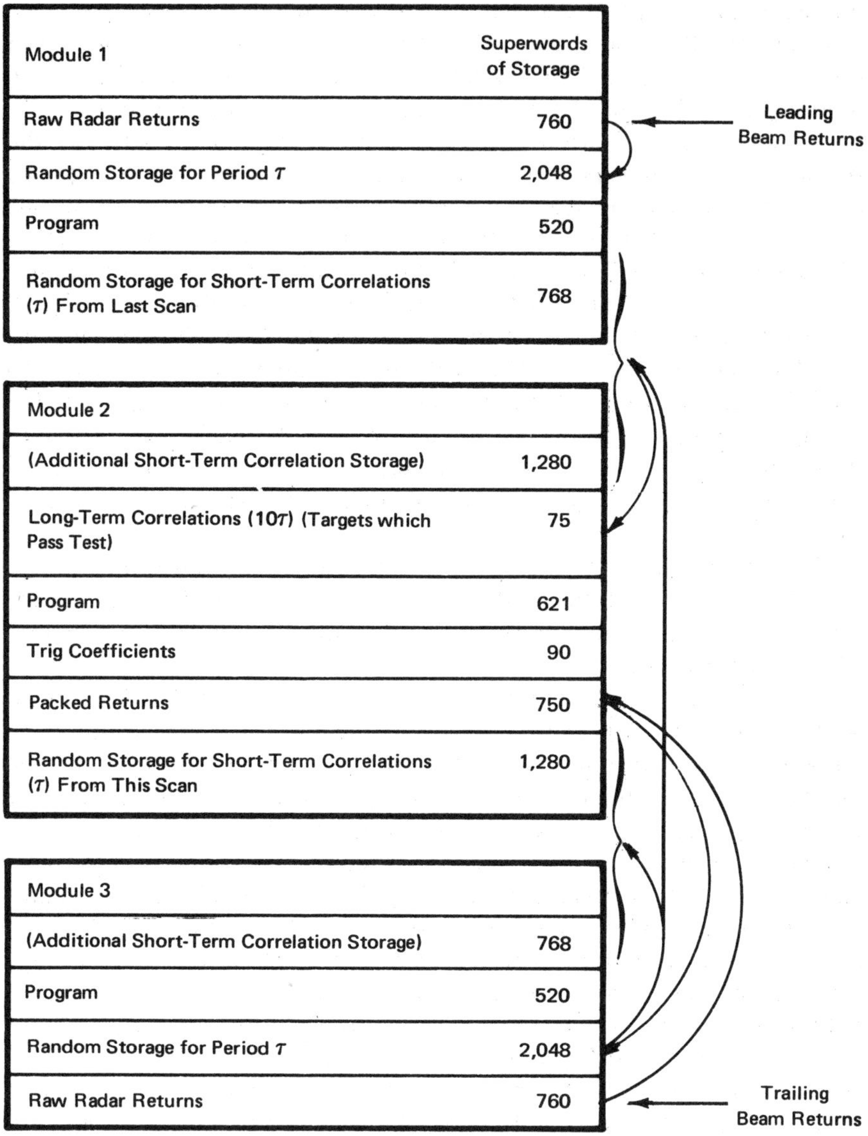

After each period T, modules 1 and 3 interchange roles

Fig. 10. Parallel Processing of Scintillation Test

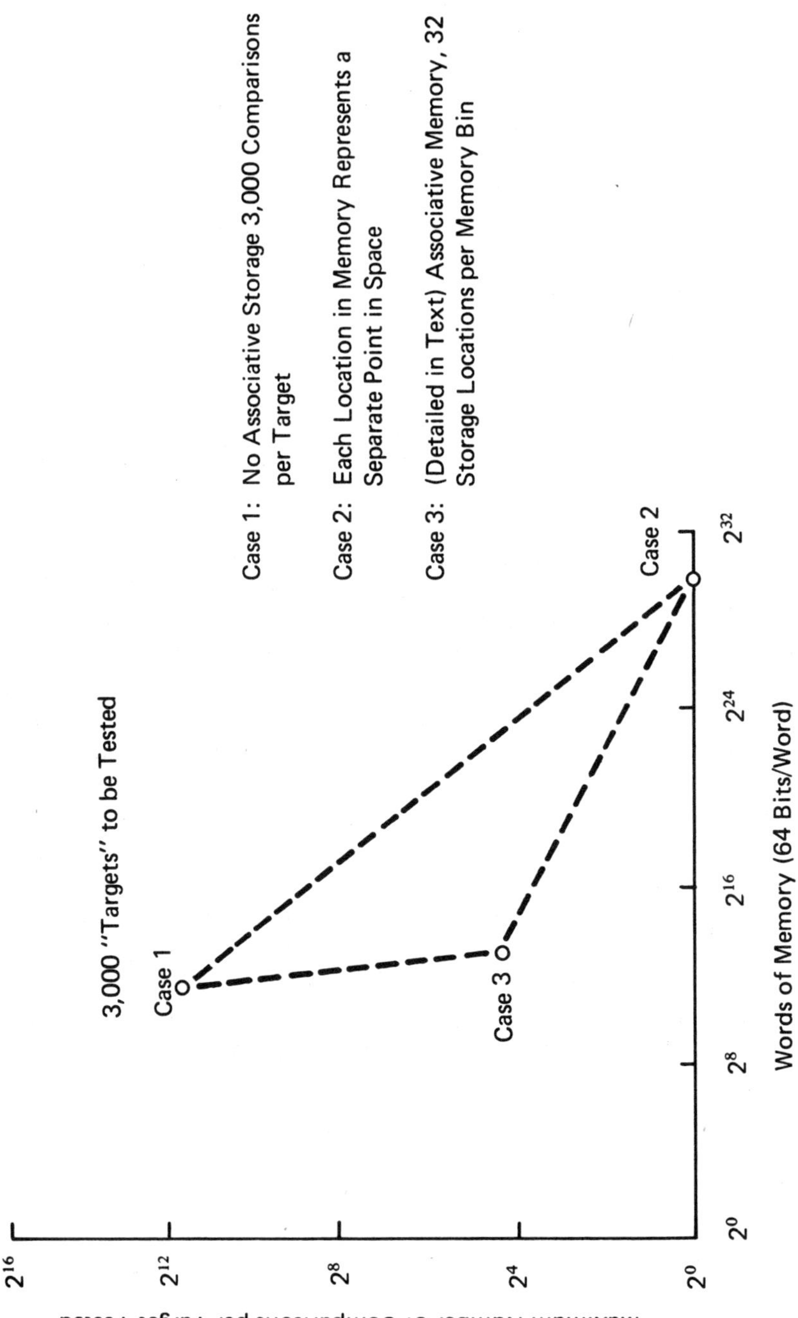

Fig. 11. Comparison of Mechanizations for Scintillation Test

filters which occur later in the chain where the targets are subjected to more demanding criteria, each of the objects being tracked has a higher probability of being an actual threat, and it is important not to lose any through processing deficiencies.

For the case of storage bin overflow, target loss can be very nearly eliminated by providing overflow bins. These bins hold the overflow from any of the associatively addressed bins. Since the number of overflows is small compared to the total number of returns being handled, the extra processing load is relatively small. For the case of addressing ambiguities, a provision can be made to place a given return in more than one storage bin if a possibility of an ambiguity arises.

The problem of addressing ambiguities may be greatly alleviated by computing bin addresses from radar coordinates (range and angle) rather than rectangular (x, y, z) coordinates. (Target data will still be *stored* in rectangular coordinates in many cases, since use of these coordinates simplifies other aspects such as tracking algorithms.)

The reason that addressing by radar coordinates alleviates the ambiguity problem has to do with the relative accuracy with which range and angle are computed, and the fact that a target falling outside the beam will be missed in any event, and addressing ambiguities are immaterial. The target position in space may be measured more accurately in the range dimension than in the angle dimension. The bin in memory should therefore correspond to a volume in space that is rather pancake-shaped: thin in the range dimension, and wide in the angle dimension.

By thus making the volume represented by a bin correspond to the likely direction of measurement error, the frequency of ambiguous jumps from bin to bin will be minimized. When an ambiguity does arise, it will apply only in the range dimension, so that a target usually will have to be stored only in two bins to avoid ambiguity losses.

If an error in angle measurement or prediction is made, it usually results in the target being missed because the target was not illuminated, rather than from some ambiguity of storage.

CHAPTER 5
AN ASSOCIATIVE MEMORY PARALLEL DELTIC REALIZATION FOR ACTIVE SONAR SIGNAL PROCESSING

Richard M. Bird

Librascope
Los Angeles, Calif.

Abstract

An Associative Memory Parallel Deltic processor is described which is capable of correlating multibeam active sonar signals with doppler replicas. For comparison purposes, a conventional delay line Deltic is hypothesized with a configuration and data rates similar to presently operating systems for active sonar data processing. The hypothesized conventional Deltic processes 12 input beams with 15 doppler channels per beam making up a combined total of 180 data channels. The AM Parallel Deltic processor is configured to replace the hypothetical conventional delay line Deltic. It is shown that the AM Parallel Deltic has an equivalent cost but yet has the flexibility for performing other related processing functions without additional complexity or cost. Further, it is suggested that sonar detection performance can be enhanced by increasing the number of doppler channels dividing the expected signal into classes. As the number of parallel channels increase, the advantages the AM Parallel Deltic has over conventional realizations significantly increase. However, of great significance is the flexibility of the AM Parallel Deltic whose processing operations are variable under program control. For these reasons, the AM Parallel Deltic processor could very well find a prominent place in multibeam active sonar processing systems of the future.

INTRODUCTION

Introductory Remarks

The concept of an associative memory has been in existence for some thirteen years now, being first conceived by Slade and McMahon in 1956.[1] Since then, a myriad of applications have been proposed for associative memories.[2,3] However, only a few operational associative memories have been built to date. The principal reason is cost, since electronic hardware is required for each word in the memory.

We are now approaching a new technological era where with MSI and LSI technology, many of the associative memory realizations and applications can be reexamined in a new light. An associative memory has a LSI compatible configuration since it requires identical electronics at each word level in the memory. Further, it is possible to realize the memory itself with active registers. Thus, an entire associative memory can potentially be realized using the LSI structure.

The major advantage of an associative memory is its ability to search the entire memory in parallel. Further, with some minor additions in the word electronics, it is possible to perform parallel arithmetic operations as well. Such a configuration has been named an "associative parallel processor" or APP.[4] Several applications for such an APP configuration have been investigated and reported,[5] including (1) pattern recognition, (2) network flow and resource allocation, (3) data stabilization of multipreformed beam sonar systems, and (4) in-flight processing of ELINT reconnaissance data. This paper presents yet another application, namely an AM Parallel Deltic processor for active sonar signal processing.

Background

Deltic, the acronym for Delay Line Time Compressor, was first described in 1954 and was first implemented by Anderson in 1955.[6,7] Since then, the Deltic correlator has become an essential tool for acoustic research and signal processing. The Deltic samples the polarity of the signal under study at regular intervals and stores these samples in a circulating memory. The speed at which the circulating memory can cycle through a signal sample compared to the sampling period, represents a time compression. Thus, because of time compression, it is possible to generate a correlation function between a *moving time series* and a *stationary time series* in real-time. Either an autocorrelation function or a cross correlation function may subsequently be plotted depending upon whether the *moving time series* and the *stationary time series* are derived from the same signal or from different signals.

Early Deltic realizations employed magnetostrictive delay lines for the circulating memory which is typified by circulation speeds of one microsecond per bit. More recent Deltic realizations have employed glass delay lines and integrated circuit shift registers as the circulating memory which are typified by circulation speeds in these applications, of 100 nanoseconds per bit. In all of these cases, external circuitry is required to form the correlation requirement.

An associative memory (AM) can simulate the action of a circulating memory and can thus be used as a Deltic. A plated wire AM has an equivalent circulation speed as fast as 50 nanoseconds per bit.[8] This type of realization begins to look attractive when it is necessary to perform the correlation process simultaneously on many *stationary time series.* The additional logic required in the above conventional realizations can be made a part of the typical word electronics of an AM. Thus, no additional logic need be required over what typically exists in the word electronics of an AM.

It will be shown that the AM Parallel Deltic has the flexibility for performing other related processing, without additional complexity or cost. Conventional Deltics would require additional equipment and thus additional costs to perform these same other functions. It is roughly estimated that the cost of the AM Parallel Deltic would be approximately the same as the equivalent Conventional Deltic which would not have these other processing functions.

Further, it is conjectured that a Deltic employing many references, corresponding to different classes of expected signals, could provide significant improvement in detection performance over presently employed conventional Deltic systems. The basis for this conjecture is work sponsored by ONR.[9] It will be shown that the AM Parallel Deltic, because of its flexibility and relatively small incremental increase in weight, size, and cost as the number of parallel channels becomes increasingly large, is a feasible, practical realization for such a polyclassed Deltic system. In contrast, such systems realized with presently used hardware techniques, would have a weight, size, and cost prohibitively large and would not have the flexibility nor the processing capability of the AM Parallel Deltic.

Purpose and Contents of Report

Thus, of the memory elements available today, it is generally agreed that plated wire has the best characteristics for realizing an associated memory.[10] The purpose of this paper is to investigate the feasibility of employing a plated wire AM configuration as a parallel Deltic processor for active sonar signal processing. In order to effect a comparison between an AM Parallel Deltic and a conventional Deltic, a hypothetical conventional Deltic is configured. The configuration is similar to existing operational active sonar Deltic processors. A description of the hypothetical conventional Deltic processor is contained in the section, "Hypothetical Model for a Conventional Deltic."

A description of the equivalent AM Parallel Deltic is contained in the section, "AM Deltic Realization." A description of the AM hardware realization is first made in subsection, "Hardware Description;" processing algorithms are then described in the following subsection. A discussion of the physical limits on extending the size of the described AM hardware follows in subsection, "Limits of Extending Size of Described AM." Finally, a cost estimate is made. This report ends with a summary and conclusions.

HYPOTHETICAL MODEL FOR A CONVENTIONAL DELTIC

Functional Operation of Deltic Processor

The functional operation of the Deltic processor is illustrated in Fig. 1. As shown in (*a*) an active signal sonar pulse, $R(t)$, is transmitted, whose pulse width, P, is assumed to be 1/2 seconds in the hypothetical model. Its spectrum characteristic may have several forms such as, single frequency, frequency modulated, and continuous and random over a frequency interval. The transmitted pulse traveling at the speed of the transmitted frequency in water, which is approximately 5,000 feet per second, strikes the target, which acts as a reflecting object, and the signal pulse is thus returned toward the source. The receiver is continually receiving a signal $S(t)$ consisting of noise and signal pulse noise, as shown in Fig. 1(b). The receiver listens for the signal over an interval T, which is typically an interval of many seconds. The signal has been idealized in (b) for demonstration purposes. There are typically false signal returns due to such sources as sea background reverberations, and false targets. Also for long range targets, the raw signal, as it would appear on an oscillogram, would be indistinguishable from the noise. The principal problem is to determine g_t, the signal transit time to and from the target, as shown in (b).

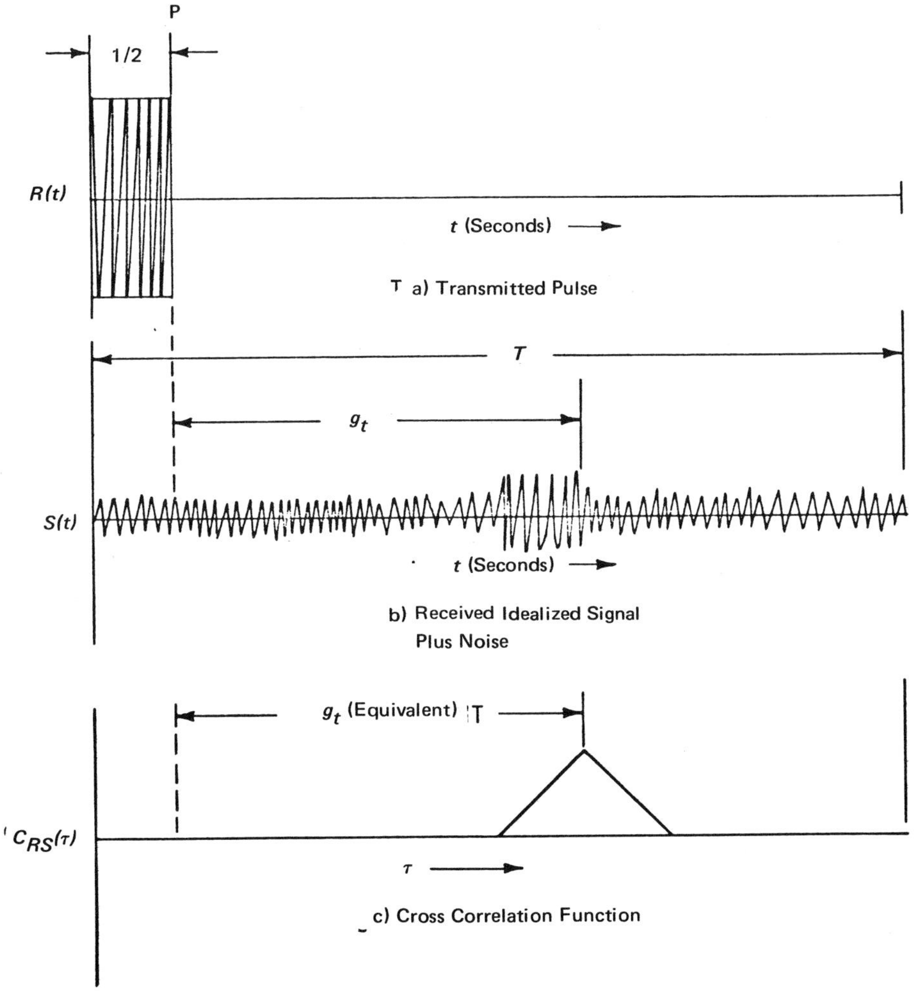

Fig. 1. Graphs to Indicate the Function of an Active Sonar Correlator

The determination of g_t for low signal-to-noise ratio signals can be assisted by the technique of correlation. The continuous cross correlation function, $C_{RS}(\tau)$, over the interval T, between the transmitted signal, $R(t)$, and the received signal, $S(t)$, can be defined as

$$C_{RS}(\tau) = \frac{1}{T} \int_0^T R(t)\, s\,(t - \tau)\, dt \ , \tag{1}$$

where $R(t)$ for $t > T$ is defined as having zero value. A typical cross correlation function that would result is shown in Fig. 1(c). The peak of the cross correlation function $C_{RS}(\tau)$ as shown in (c), would define the transit time, g_t .

A discrete Deltic cross correlation function, $\bar{C}^*_{RS}(n\Delta\tau)$, over the interval, T, may be defined where, $\bar{R}(t)$ and $\bar{S}(t)$ are binary valued, Deltic quantized (infinitely clipped) versions of $R(t)$ and $S(t)$, such that:

$$\bar{C}^*_{RS}(n\Delta\tau) = \sum_{m=0}^{M} \bar{R}^*(m\Delta t)\bar{S}^*(m\Delta t - n\Delta\tau) \ , \tag{2}$$

where

$$m = 0, 1, \ldots, M \ , \qquad n = 0, 1, \ldots, N, \text{ and}$$

$$\Delta t = \Delta\tau = \frac{T}{M} = \frac{T}{N} \ ,$$

and the * indicates a sampled function. But, since $R(t)$ has zero amplitude beyond the interval P, $\bar{C}_{RS}$ need not be integrated over the interval T but over the interval P

$$\bar{C}^*_{RS}(n\Delta\tau) = \sum_{q=0}^{Q} \bar{R}^*(q\Delta t)\bar{S}^*(n\Delta t + q\Delta t - n\Delta\tau) \tag{3}$$

where

$$q = 0, 1, \ldots, Q \ ,$$

$$Q = P/\Delta t \ , \text{ and}$$

n, N, $\Delta\tau$, and Δt are as defined in eq. (2).

Equation (3) above, is implemented by the Deltic. A replica of $\bar{R}^*(t)$ is stored in one circulating memory. The last $Q\Delta t$ seconds of $\bar{S}^*(t)$ is stored in another circulating memory. Then, if the circulating memory is at least Q times faster than the sampling rate, it is possible for the discrete cross correlation function $\bar{C}^*_{RS}(n\Delta\tau)$, to be generated in real-time.

There are several additional factors causing complications for the correlation process; these include:

1) relative velocity between source and target
2) aspect and motion of target
3) aspect and motion of source
4) return signals from more than one target and false targets
5) background noise and variations thereof
6) background reverberation and variations thereof
7) acoustic path (bottom bounce or direct)
8) signal strength due to path attenuation and range.

Because of (1), (2), and (3) above, there will be doppler shifts in the signal transmitted. Therefore, some means of accounting for these doppler shifts is required in the correlator. Because of (4), as well as the need to obtain azimuth information for the target, the receiving field of view is divided into beam segments. In the hypothetical model, the field of view is assumed to be divided into twelve beams. Each beam is further assumed to be divided into fourteen doppler shift increments in addition to the fundamental transmitted signal. Therefore, there are a total of 12 X 15 or 180 different output correlation functions generated.

Configuration of Hypothetical Model

Figure 2 is a block diagram of the Deltic portion of the assumed hypothetical active sonar signal processor. Clipped signals from 12 sonar beams in the frequency range of 85 to 215 hertz are presented to the input. The 12 beams are multiplexed into groups of 4 beams, and are interlaced onto 3 signal delay lines. Each of the 3 signal delay lines are 1,415-bits long, and the signals propagate at 1-bit microsecond. The 12 beams are sampled at 700 hertz rate per beam. As shown in Fig. 2, there is a fourth delay line where the reference replica is stored.

The interlaced signals for 4 beams on each signal delay line are logically combined with the reference signal in the correlation network

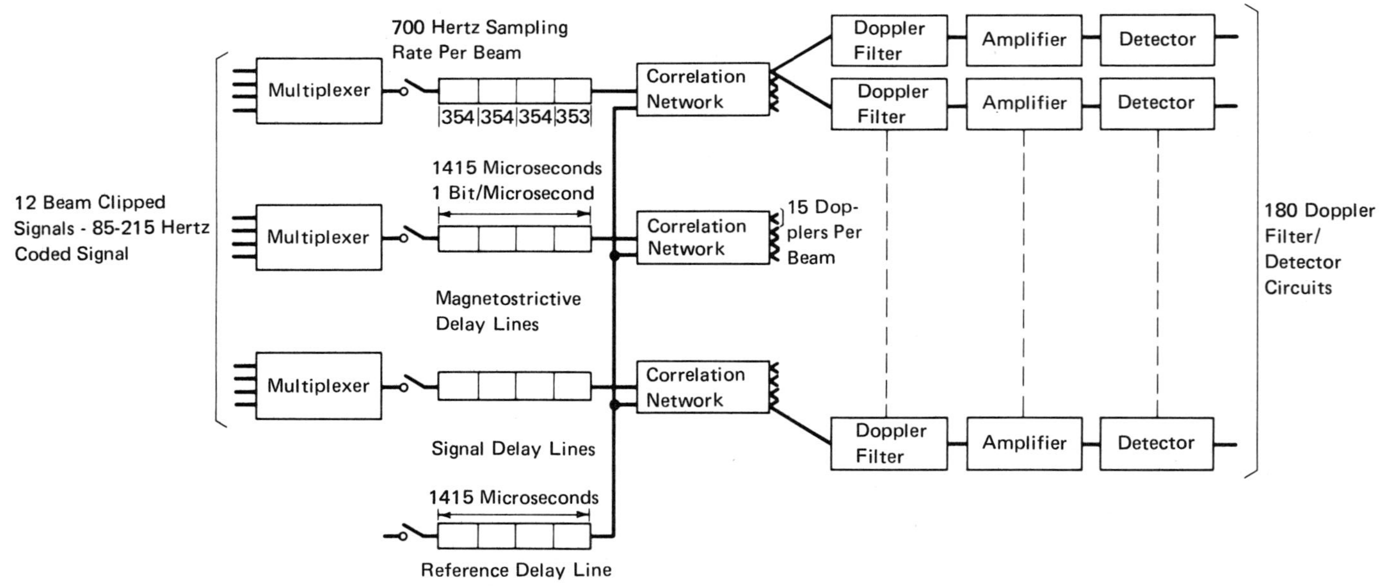

Fig. 2. Assumed Hypothetical Delay Line Active Sonar Deltic Processor

as shown, generating the cross correlation function of the 12 beams with the reference. There are 15 doppler frequencies of interest for each beam making a total of 180 doppler frequencies. As shown, a doppler filter, amplifier, and detector is required for each of the 180 doppler frequencies. The output is a detected analog, 700 hertz pulsing signal.

Although not shown in Attachment *A*, additional circuitry would be required for further processing to accomplish integration of the correlation values, and to prepare the resulting signals for display purposes. The AM realization can function in an equivalent manner up to the output point shown in Fig. 1. However, in addition, the AM realization can perform other functions such as integration of the correlation values, and other possible correlation processing as discussed in the section "Processing Algorithms." It is shown that the additional functions can be performed without increased complexity of the AM hardware.

The following enumerates the hardware in the hypothetical conventional Deltic:

1) three-multiplexers
2) four–1,415 bit, 1 microsecond per bit, magnetostructive delay lines memory units
3) three–correlation networks providing the correlation for 4 beams each
4) 180–doppler filters
5) 180–doppler filter amplifiers
6) 180–doppler detectors

For comparison purposes, the cost of the hypothetical delay line Deltic is estimated below. The largest fraction of the cost is contained in the 180 very narrow band-pass doppler filters, and their associated amplifiers and detectors. These are estimated at between $100 and $200 each. The cost of the delay lines, multiplexers, and correlation networks is roughly estimated at $10,000. Thus, the total cost is estimated at between $30,000 and $40,000.

AM DELTIC REALIZATION

Hardware Description

Fig. 3 represents the associative memory equivalent of Fig. 2. As shown, there are two associative memories; one is the signal memory

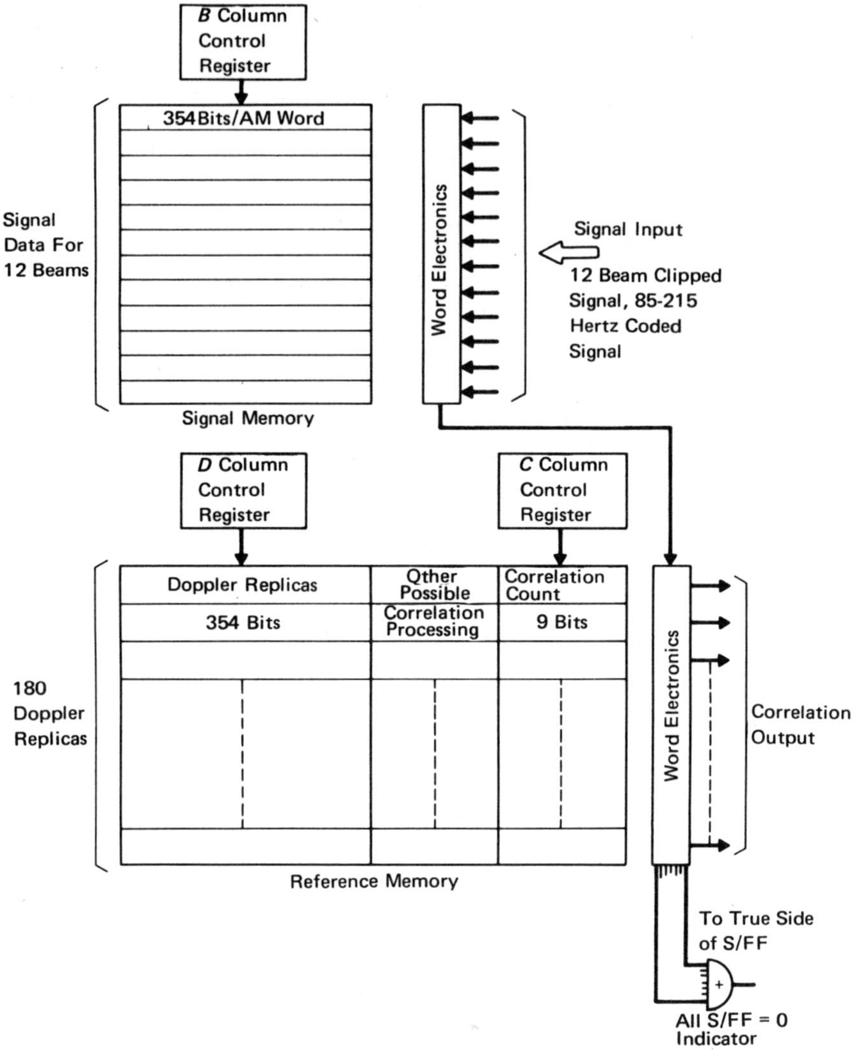

Fig. 3. All Digital Associative Memory Parallel Deltic

and the other is the reference memory. The signal memory has 12 words, one for each beam, and each word has 354 bits. The reference memory has 180 words, one for each of the 180 dopplers. Each reference memory word has 354 bits, for doppler replicas same as for the signal memory, plus 9 bits, to accumulate the correlation sum (integration) for each sample period, plus additional bits for performing other correlation processing. Approximately, 4 microseconds are available for processing each bit in the correlation procedure, when each beam is sampled at the 700 hertz rate.

In performing the correlation procedure, the bits in the signal memory and the doppler replica memory are not shifted as in a delay line but rather, column control registers for each memory are incremented. In this manner, each memory is pseudoshifted. As shown in Fig. 3, the *B* Column Control Register controls the signal memory, whereas the *D* Column Control Register and the *C* Column Control Register control the doppler replica and correlation count parts, respectively, of the reference memory. The column control registers serve to select the particular column on which an operation is to be performed. For instance, "*B*" and "*D*" Column Control Registers can take on values between 1 and 354, and "*C*" Register between 1 and 9. These registers can be incremented on command and will reset themselves back to one after their limit is reached.

The word electronics (WE) for the beam memory and doppler replica memory are illustrated in Figs. 4 and 5 respectively. As can be seen, the WE in the beam memory is simpler than in the doppler replica memory. Each word in the beam memory WE contains the bipolar sense amplifier and a bipolar write driver; no gates are required. Each word in the doppler replica memory contains the same sense amplifier and write driver, but additional gates and flip flops are required. As indicated in Fig. 5, 2 flip-flops (FF) are required, the S/FF and the I/FF. The gates as shown are required for the correlation/integration processing. A few additional gates might be required depending upon the types of the other possible correlation processing one might wish.

Another item shown in Fig. 3 is the giant "or" gate. This giant "or" gate has inputs from the true sides of all S/FF, and provides a logical indication of when all S/FF are logically false or zero.

One other factor in the configuration of Fig. 3 that should have some mention is concerned with the storage of the doppler replicas in the reference memory. It would be necessary to store only 15 doppler replicas rather than 180, if the doppler replicas were alike for all beams. However, 180 AM words would still be required in the reference memory to perform the correlation count and other possible correlation processing functions. Also, some additional logic would be required in the word electronics to sum the outputs from the 15 doppler replicas into the 180 words. This would save 165 X 354 or 58,000 bits of plated wire memory but as mentioned would require more word electronics. Doing this, though, would reduce the flexibility of the AM Deltic. In the configuration shown in Fig. 3, it is possible to store different doppler replicas for each of the twelve beams if desired. This could be very desirable to distinguish between different beams accounting for such factors as, for instance, target speed, target aspect,

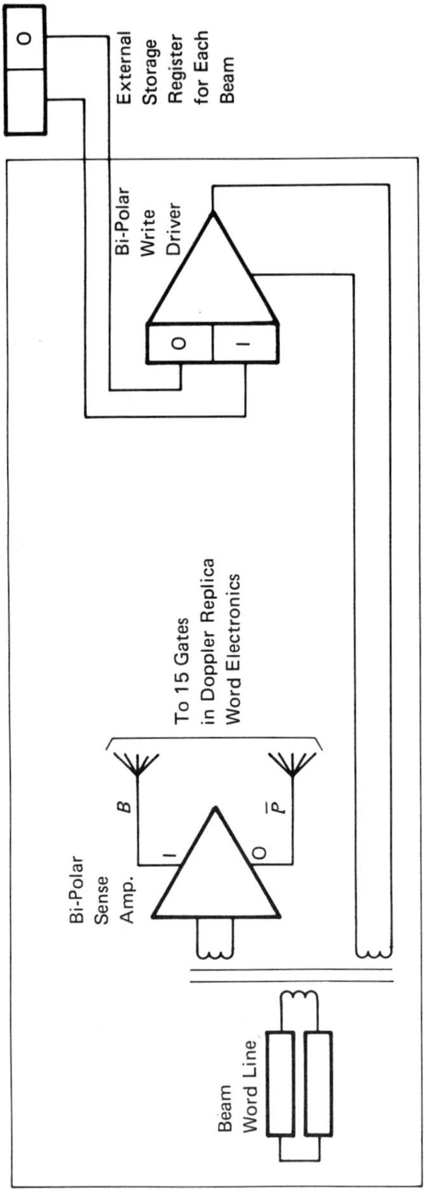

Fig. 4. Beam Memory Word Electronics

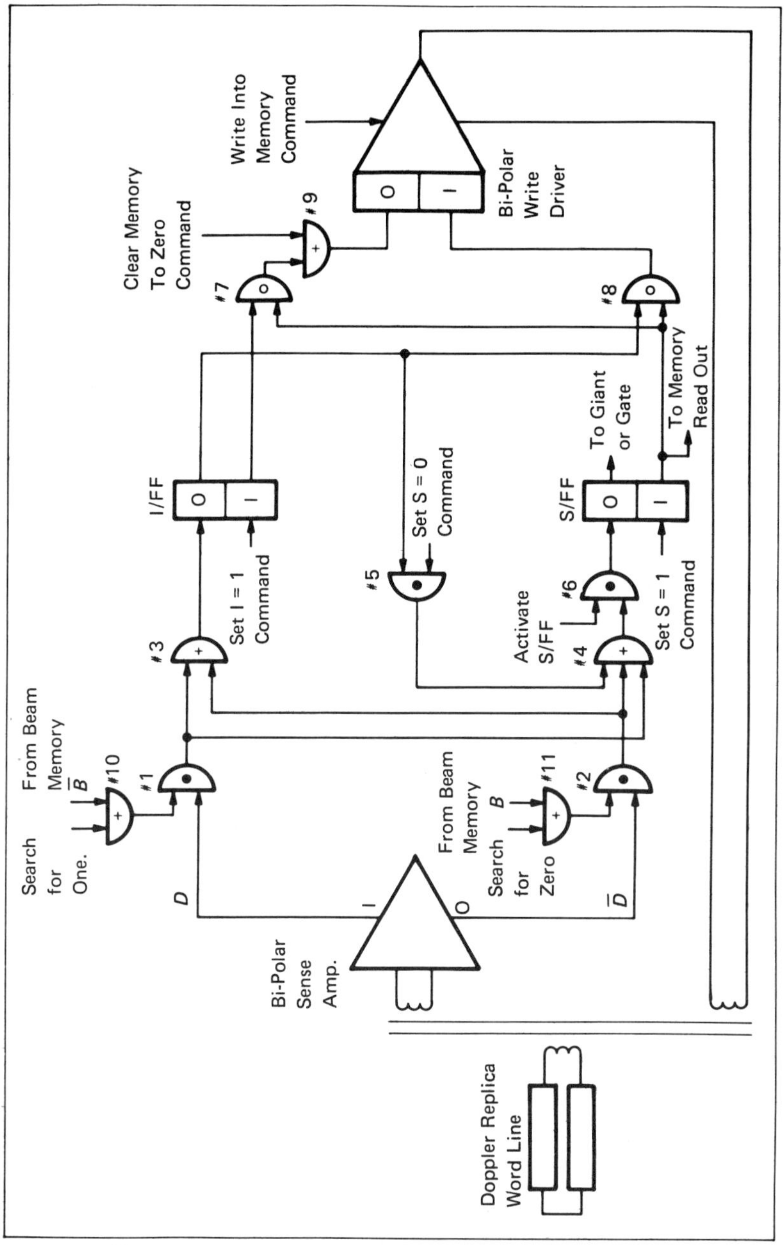

Fig. 5. Doppler Replica Memory Word Electronics

target depth, acoustic path and range. The doppler replicas can be externally loaded and are therefore variable under program control. In contrast, the 15 doppler replica system could store only one class of replicas. Worse yet, in the hypothetical model described, not only do the 15 doppler replicas represent only one class, but the class is fixed by hardware and, therefore, invariable. For these reasons all 180 doppler replicas are configured in Fig. 3.

The processing operations of this hardware should become obvious in the next subsection where processing algorithms are discussed.

Processing Algorithms

Principal Processing Algorithm. The principal processing algorithm performs correlation multiplication for each bit and accumulates the digital sum of the resulting products for all bits in the sample period per eq. 1. This establishes the value of C^*_{RS} $(n\Delta\tau)$ for a particular period. Basically, the operations are as follows:

The word logic in the signal memory is appropriately fanned out into the word logic of the reference memory as inferred by Fig. 4, so that correlation takes place in the word electronics of the reference memory. When the logical correlation for a bit pair $(D \cdot B + \bar{D} \cdot \bar{B})$ is logically true, the correlation counter section in the reference memory is incremented through AM word electronic logic. This operation is, of course, accomplished for all doppler replicas in parallel. In this manner, digital integration is performed. At the completion of the rotation of the 354 bits, i.e., a sample period, the correlation counter contains the correlation value $C^*_{RS}(n\Delta\tau)$ for a particular time sample and thus for a particular value of n. All correlation values for the 180 doppler channels for the particular sample can be made available for outputting in parallel, if desired. Other types of processing are also possible as is discussed in following subsections.

The details for the principal processing algorithm are shown in the flow diagram of Fig. 6. Instructions 1, 2, and 3 set the "*D*," "*B*," and "*C*" column indexing registers to an initial value. "*B*" register may be set to any initial value so the letter "*X*" is indicated. Instructions 4, 5, and 6 clear the contents in the correlation count fields of the AM to zero. Instruction 7 reads the new values for the 12 beams into the signal memory. Instruction 8 increments the *B* register. The AM is preset at this point to begin a correlation processing cycle. Instruction 9 sets the *C* register to one. Instruction 10 sets the S/FF and I/FF to one as

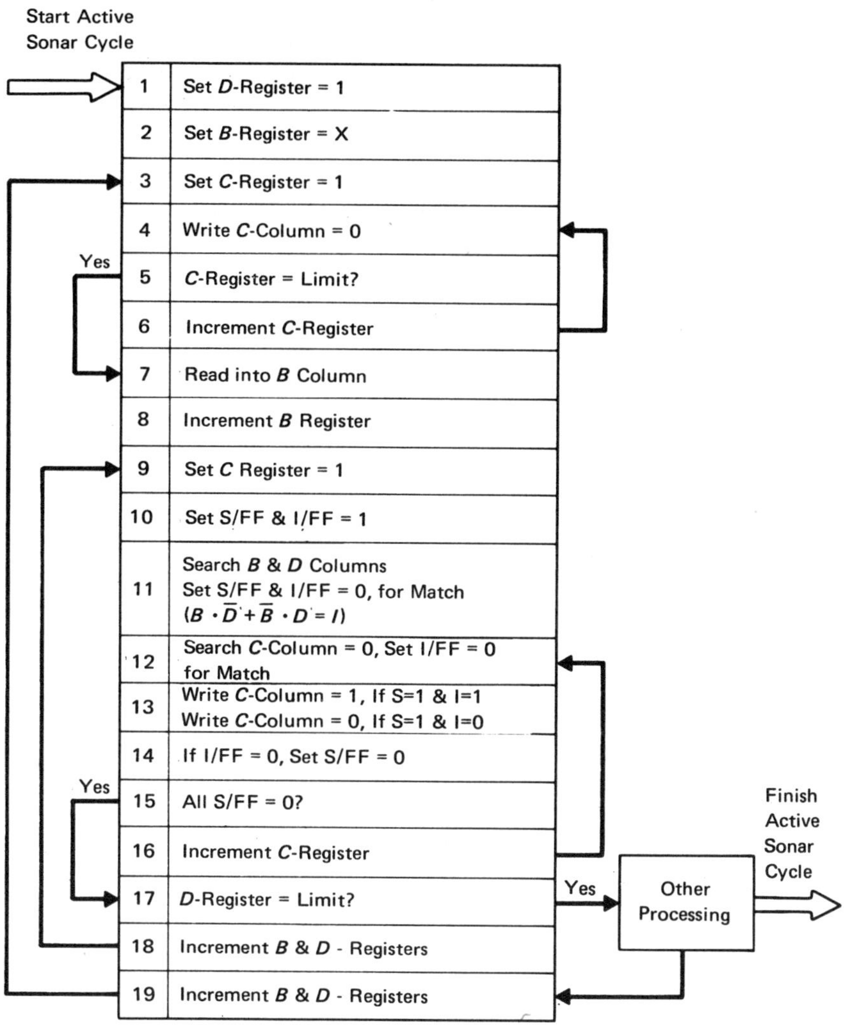

Fig. 6. Correlation Processing Algorithm

illustrated in Fig. 5. Instruction 11 performs the correlation product for a single bit between the beam data in the signal memory and the appropriate doppler replica data in the reference memory. If the logical product is zero, i.e., $(B \cdot \bar{D} + \bar{B} \cdot D = 1)$, the S/FF and I/FF are set to zero by gates numbered 1-4, 6, 10, and 11 in Fig. 5. Instructions 12, 13, 14, 15, and 16 perform the function of adding one into the "C" fields of the reference memory in those instances where the logical product formed in instruction 11 is true. Instruction 12, 13, and 14 perform the carry function one bit at a time. The "*C*" register is incremented in instruction 16, and the carry procedure is iterated by instructions 12, 13, and 14. Instruction 15 determines the termination of the carry ripple procedure. This occurs when all S/FF are zero as determined by the giant "or" gate in Fig. 3. When this occurs, the algorithm jumps to instruction 17. If the *D* register has not reached its limit, the *B* and *D* registers are incremented by instruction 18. The algorithm then jumps to instruction 9, where the correlation procedure continues for the next bit pair in the signal and reference memories.

When the *D* register limit is reached in instruction 17, the correlation summation for a particular sample is concluded, i.e., $\bar{C}^*_{RS}(n\Delta\tau)$ has been determined for a particular value of n. At this point other processing is possible. As an example, the digital value for the resulting correlation for that sample period, i.e., $\bar{C}^*_{RS}(n\Delta\tau)$ for a particular n, might be read out of the memory. Other examples are given in the subsection, "Other Processing Capabilities." At this point a test is made to determine whether the finish of the active sonar cycle has been reached. If the test indicates the processing is not yet concluded, the algorithm proceeds to instruction 19 where the *B* and *D* registers are incremented. The algorithm then jumps to instruction 3. The contents of the *C* fields are again cleared to zero by instructions 3, 4, 5, and 6, and a new value is read into the *B* column by instruction 7. The correlation continues as before with the availability of the data bit from the new beam sample.

Timing Considerations. The factors having an effect on timing will be considered in this section. The first factor stems from the choice of plated wire as the memory element. The fundamental switching speed of the plated wire is in the order of a few nanoseconds. Therefore, the switching speed, as a matter of practicality, is not limited by the plated wire but is limited by the rise and fall time of the interrogation current pulse. Since the drive and plated wire lines in a plated wire memory are transmission lines, the ultimate limitation is posed by the rise and fall time degradation of these lines as viewed as a transmission line. For small memories of one or two hundred words, interrogation times of 50 nanoseconds and write times of 100 nanoseconds are feasible. For larger memories in the order of a few thousand words, interrogation times of 100 nanoseconds and write times of 200 nanoseconds are feasible. For the purpose of this timing analysis, an interrogation time of 75 nanoseconds and a write time of 125 nanoseconds are assumed.

The second factor is the processing algorithm. The principal processing time is required by instructions 11, 12, and 13 in Fig. 6. With 354 bits per beam word, 4 microseconds are available for processing. Instruction 11 would require 75 nanoseconds for execution, an almost negligible amount of time. The major processing time is required to accumulate the sum in the "*C*" fields of the AM. With nine bits required to represent the number 354, instructions 12, 13, 14, 15, and 16 will have to be iterated 9 times in the worst case. The word logic of Fig. 5 has been organized to minimize the amount of processing time to perform the above functions. The "*C*" field is searched once, in instruction 12, and multiwritten into once, in instruction 13 for each iteration. The time required for these functions is 200 nanoseconds based on a 75 nanosecond search and 125 nanosecond multiwrite time. Allowing another 100 nanoseconds for performing instructions 14, 15, and 16, based on 50 megahertz logic, then 300 nanoseconds would be required to perform these functions. In the worst case, then, 2.7 microseconds would be required for 9 bits. The total processing for each beam sample could be completed well within the 4 microseconds available.

Other Processing Capabilities. There are three points in the correlation processing procedure where other processing, as illustrated in Fig. 6, might be performed, namely:

1) other processing at the end of each beam sampling period
2) other processing at the conclusion of the active sonar processing cycle
3) other processing resulting from different initial loading of data.

Examples of case (1) are as follows:

a) All correlation values for the 180 doppler channels for a particular beam sample period can be made available for outputting in parallel.

b) The largest correlation value in the 180 channels could be determined by a parallel maximum value algorithm, and only that value outputted as opposed to all values as in (*a*) above.

c) The same function as in (*b*) could be performed for all dopplers in each beam (15 channels).

d) The peak correlation value of all samples for each doppler frequency can be calculated and stored with its associated time of arrival (TOA) providing sufficient data, for instance, for automatic ranging purposes. This can be accomplished by first storing in the "other possible correlation processing" field in Fig. 4, the first correlation count and the TOA at that instant. On successive sampling cycles, a greater than test is performed between the newly accumulated correlation count value and the previously stored value. If the resulting comparison indicates a greater than condition, then the old value and its TOA is replaced by the new value and its TOA. At the completion of the correlation cycle, the maximum correlation value for each doppler and its associated TOA will be remaining in the memory and are available for outputting. Variations of this procedure are of course possible to reject noise spikes.

The following is an example of case (2) above. Assuming that the procedure in (*d*) above is followed, a maximum value search on all of the maximum correlation values in memory can be performed thus determining the beam, doppler frequency, and TOA for the maximum value of the 180 maximum correlation values.

The following is an example of case (3) above. Work sponsored by ONR[11] indicates that detection performance may be enhanced in active sonars by dividing the expected signal into classes and then processing simultaneously data for each class. While determination of the optimum number of classes awaits further research, the factors that form the basis for the concept of multiple classes can be outlined in qualitative terms. Examples of factors for multiple classes include:

1) target speed (available in presently used doppler replica concept)
2) target aspect
3) target and source depth
4) acoustic path (bottom bounce or direct)
5) range.

Systems that employ many references corresponding to different classes might provide a significant improvement in performance over single class, fixed systems in operation today. The capability of the AM Parallel Deltic which allows variable reference signal loading and variable processing algorithms, both under program control, as well as the simultaneous correlation of the incoming signal with all references, represents an attractive, feasible realization for a multiple class Deltic. Because of the inherent variability of the AM Deltic through program control, such a system could employ adaptive algorithms. Further, the basic data structure of the AM Deltic could also be changed by program control, allowing results obtained from new field data or new intelligence information on enemy ships and weapons to be easily incorporated into the AM Parallel Deltic processing procedure.

Other parallel processing algorithms are of course possible. These are mentioned to serve as examples of what is possible.

Limits of Extending Size of Described AM

The number of words in the signal memory and reference memory of Fig. 3 can be extended to thousands of words. For instance, with a DIMUS parallel beam former, one might wish to process signals from a few hundred sonar beams. This would require a few thousand words for doppler replica representation in the reference memory. This is possible with the AM realization.

The extension of the memory in the other direction, i.e., in the number of bits in a word, is timing constrained. Where the sampling rate is 700 hertz, the allowed processing time for each bit in the signal memory is given by

$$T = \frac{1}{700\,B} \tag{4}$$

where B is the number of bits in a signal memory word.

As discussed previously, the time consuming part of the algorithm is concerned with rippling the carry in the correlation count fields of the reference memory. With the number of bits greater than 511 but less than 1,024, 10 iterations of instructions 12 through 16 in Fig. 6 would be required in the worst case. This would limit the number of bits per word to about 500.

However, the average number of iterations or carry ripples over a sample period will be considerably less than this value. The estimation of the average number of carry ripples over an entire sample period is dependent upon many factors. However, it can be shown that when the number of channels is in the order of 200, the number of carry ripples for a 10-bit register will always be less than 1/2 of the worst case value. The expected number should be significantly below this 1/2 figure. It is concluded, therefore, that the AM parallel Deltic here described could be configured to process many hundreds of doppler channels with as many as 1,000 active bits per beam.

Cost Estimate

Based on the complexity of the word electronics, the system cost in production quantities is estimated at between $100 and $400 per AM word, depending upon the size of the memory and control requirements. Therefore, it is roughly estimated that the realization in Fig. 3 would cost in the order of $30,000, in production quantities. It is further roughly estimated that, with engineering development costs included, unit number one would likely cost around $250,000.

CONCLUSIONS

The following conclusions can be drawn from a comparison of the two configurations.

Processing Capabilities

The AM Deltic can perform all the functions of the conventional Deltic, as well as other desirable processing functions. Examples of other processing functions include (1) the digital integration of the correlation products over the sample period, (2) the selection of that doppler frequency over a sample period with the maximum value, (3) the determination of the maximum integrated correlation value and its time of arrival for the complete correlation cycle for each doppler frequency channel, and (4) the determination of the doppler frequency channel with the maximum value from the maximum integrated correlation values for each doppler frequency channel. All processing is performed in parallel. The resulting data for outputting is available in digital form as opposed to analog form for the conventional Deltic.

Cost

The cost of the AM Parallel Deltic should be at least equivalent and, if anything, less than that of the conventional Deltic. The cost of the AM Deltic for 180 doppler channels is estimated at $30,000.

Flexibility

The AM Deltic has greater flexibility in several respects. First, it has greater processing capabilities as indicated in (1) above, any of which can be modified under program control. Therefore, different processing algorithms can be specified depending upon the operational situation making possible adaptive beam processing schemes. Second, the doppler frequencies are determined by loading the replica memory, and are accordingly variable, as opposed to being fixed as is the case for an analog doppler filter. Because of this flexible doppler replica representation feature, several important performance advantages over present operational systems are possible, such as (1) by bunching the doppler frequencies around a predetermined velocity, greater velocity resolution is possible for the same number of doppler channels, (2) each beam may have different doppler replicas optimized to detect and abstract information on a particular target within that beam, (3) each beam and even each target within each beam can have replica references divided into classes representing factors that influence the statistics of the signal which potentially increase the probability of detection and identification, and, (4) different modes of operation are possible for loading the replica reference channels such as, for instance, to increase the probability of detection, or to increase the probability of a target identification, or to obtain more accurate information such as velocity resolution, and where different modes could be simultaneously in effect on different beams. Third, the sampling rate is not tied to the circulating rate as is the case for a delay line Deltic. Therefore, the sampling rate could be a variable and specified by program control.

Size

The plated wire AM Deltic should be smaller and weigh less than the conventional Deltic.

Environmental Tolerance

The conventional Deltic has analog filters and delay lines which can be significantly affected by temperature and vibration variations. The digital nature and nondestructive read out characteristic of the plated wire AM Deltic give it an inherent environmental tolerance advantage over the conventional Deltic.

The AM Parallel Deltic configuration described can be expanded to accommodate more sonar beams, and doppler reference channels, as well as more bits for representing a signal and doppler replica word. Configurations to handle many hundreds of doppler replica channels are possible for as many as 1,000 bits per beam signal word for a 700 hertz sampling rate.

There are two principal new and novel features in the AM Deltic realization of Figs. 3, 4, and 5. The first feature is the concept of storing data in two AM's as shown in Fig. 3, and then combining information from subsequent searches into one common word electronics as indicated in Figs. 4 and 5. The second is the word logic organization of Fig. 5. This organization provides for the counting function in the "*C*" field. More conventional AM organizations with tag fields would require four searches, and four multiwrite times per bit in the "*C*" field.[12] This organization, then, provides for a four to one speedup.

Based on the material presented in this report, it is concluded that the AM Parallel Deltic compares as good or better in all respects with the conventional delay line Deltic model for the same number of beams and doppler channels, where only the functions of the conventional Deltic are performed. Further, the comparison particularly of size, weight, and cost, would become very significantly in favor of the AM Parallel Deltic, if either the number of beams were increased or the number of doppler references per beam were increased.

However, probably of greater significance is the flexibility of operation capability, and the additional processing operations possible with the AM Parallel Deltic. Because of these features it should be possible to obtain better detection, identification, and classification performance than provided by the inflexible Deltic systems in operation today. Future Deltic systems will likely have more beams, and a requirement for more doppler references per beam. Because of all of these reasons, AM Parallel Deltics could very well find a prominent place in multibeam active sonar processing systems of the future.

References

1. A. E. Slade and H. O. McMahon. "A Cryotron Catalogue Memory System," *Proceeding of the Eastern Joint Computer Conference,* December 10-12, 1956.
2. R. H. Fuller, "Content-Addressable Memory System," *Defense Documentation Center,* Document AD 417 644, June, 1963.

3. A. G. Hanlon, "Content-Addressable and Associative Memory System—A Survey," *IEEE Transactions on Electronic Computers,* XV (August, 1966), 509-21.
4. R. M. Bird, R. H. Fuller, and R. M. Worthy, "Study of Associative Processing Techniques," *Defense Documentation Center,* Document, AD-621516, August, 1965.
5. R. M. Bird, "Applications of Associative Parallel Processors," *ONR/RADC Seminar on Associative Processing* (Washington, D.C., May 3, 1967).
6. *Quarterly Report for January 1 to March 31, 1954*, (University of California Marine Physical Laboratory [March, 1954]).
7. V. C. Anderson, *The DELTIC Correlator* (Acoustic Research Laboratory, Harvard University [January 5, 1956]).
8. R. H. Fuller, J. C. Tu, and R. M. Bird, "A Woven Plated-Wire Associative Memory," *1965 NAECON* (Dayton, Ohio, May 10-12, 1965).
9. C. R. Baker and R. E. Cunningham, *A Comparison of Detectors for Active Sonar Using Experimental Data* (Confidential), (Defense Documentation Center, Document AD 389 908 [March 1968]).
10. Fuller, Tu, and Bird, *op. cit.*
11. Baker, *op. cit.*
12. Fuller, *op. cit.*

CHAPTER 6

MATRIX COMPUTATIONS ON AN ASSOCIATIVE PROCESSOR

Jesse H. Katz

Computer Processes, Inc.
Los Angeles, California

INTRODUCTION

An *associative processor* is a computer subsystem made up of two main parts: a storage unit consisting of an *associative memory* and a logic unit which provides for *parallel computation*. In this paper we investigate the applicability of an associative processor to some familiar matrix problems. Specific aims are (1) to identify the properties of an associative processor that are relevant to the problems under consideration, and (2) to estimate the extent to which an associative processor can speed up solutions.

The following section of this paper traces briefly the evolution of the associative processor; "Specification of an Associative Processor," gives the specification of an associative processor (AP) whose special purpose is the solution of a system of linear equations; the next section, "Solution of a System of Linear Equations," presents in detail an AP solution to a system of linear equations and is followed by an evaluation of the performance of AP on this problem; "Other Matrix Problems" discusses the solution of some other matrix problems on an associative processor; and the last section summarizes the main findings.

The material contained in this paper was developed while the author was with the Bunker-Ramo Corporation, Canoga Park, California.

The author is grateful to W. C. McGee, now with International Business Machines Corporation, for his review of this work.

EVOLUTION OF THE ASSOCIATIVE PROCESSOR

In the earliest computers the programmer could retrieve a word of data from memory only by specifying the *location* of that word. Before long, it was realized that it would be worth a hardware price to be able to retrieve a word of data on the basis of its *contents* rather than its location.

Perhaps the first significant step toward providing circuitry for content-addressing was the inclusion, some 15 years ago, of the *table look-up* feature in the IBM 650. The feature allowed the programmer to specify an *argument* and a beginning *search address;* the computer then retrieved the address of the first word encountered whose contents were greater than or equal to the given argument. The 650 performed this function by examining words in the table in *sequence;* therefore, the function's execution time was variable and proportional to the number of words examined.

While built-in table look-up demands extra hardware, in certain situations it yields significant savings in execution time. Consequently, the facility has been incorporated in a number of computers.

The development in 1955 of the cryotron circuit (a circuit with both storage capabilities and logic capabilities) made possible a major advance. With this circuit, data in memory could not only be content-addressed; but all words in memory could be examined *in parallel.* Much laboratory work has been carried out since then on parallel-search, content-addressable memories, with a variety of hardware components other than cryotrons having been considered. In recent years, these memories have become steadily more attractive—both on economic and utilitarian grounds—with the result that general interest in them has risen sharply.

A parallel-search memory which enables the user to retrieve a word of data according to the contents of an arbitrary *subset* of the word is known as an associative memory. It is evident that this device can be applied with considerable effectiveness to a wide class of information retrieval processes.

To be able merely to *retrieve* data on the basis of its contents is not particularly helpful, however, in many mathematical processes. Much more essential is the facility to *operate numerically*—and in parallel—on words in memory satisfying a specified condition. We call a device having this added facility either an *associative processor* or an *associative computer,* depending on whether the device operates under the control of another computer or strictly under its own control.

SPECIFICATION OF AN ASSOCIATIVE PROCESSOR

We now give the specification of an associative processor (AP) whose special purpose is the solution of a system of linear equations. The specification is intended as a functional representation rather than a hardware representation.

Components of AP

A diagram of AP is given in Fig. 1. The components of this associative processor are as follows:

1) The memory proper M is an m-by-3 matrix, representing m words and 3 fields. For reasons discussed later, (p. 139), the fields have the symbolic labels i, j, and a_{ij}.
2) Associated with M is a set of 7 vectors. Vectors v_1 and v_2 are 2 dimensional, the components being in one-one correspondence with the i and j fields of memory. Vectors v_3, v_4, v_5, and v_6 are 1 dimensional, being in correspondence with the a_{ij} field of memory. Vector t is m-dimensional, the components being in one-one correspondence with the m words of memory.
3) Three "black boxes" T, C, and R, representing the functions *tag*, *compute*, and *retrieve*, govern the general action within AP. These functions result in changes to t, M and v_6 respectively. Generally,

$$t = T(\mathrm{M}, v_1, v_2) ,$$

$$\mathrm{M} = C(\mathrm{M}, v_3, v_4, v_5),$$

and

$$v_6 = R(\mathrm{M}, t) .$$

These functions are specified in detail in the next section.

Driving AP is a controlling computer. We assume data can be transferred between the computer and AP, and among the components of AP, according to the lines drawn in Fig. 1.

FUNCTIONS OF AP

Function T

The tag function T sets the components of tag vector t to 1 or 0, depending on whether or not the corresponding words in memory

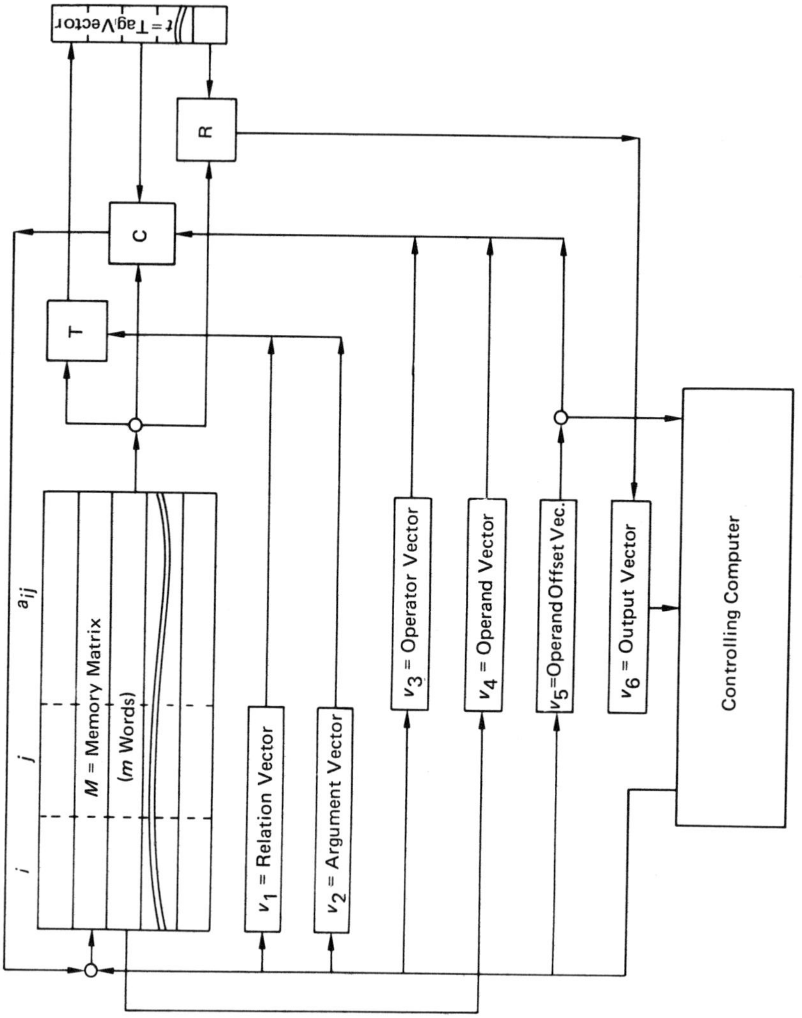

Fig. 1. Functional Diagram of AP

satisfy the conditions specified by v_1 and v_2. T examines all m words of memory in parallel.

The components of relation vector v_1 can take on the values $=, >, <$, representing the usual binary relations *equality, greater than, less than.* The components of argument vector v_2 consist of the arguments with which the i and j fields of each memory word are compared.

Thus, if we let*[1]

m_{ij} =	j^{th} field of ith word of memory,	$i = 1, \ldots, m, j = 1, 2, 3,$
t_i =	i^{th} component of t,	$i = 1, \ldots, m,$
v_{ij} =	j^{th} component of ith vector,	$i = 1, 2, j = 1, 2,$

and $(m_{ij}\, v_{1j}\, v_{2j})$, $j = 1, 2$, signify a Boolean function which is evaluated as *one* if the indicated relation holds and *zero* otherwise, then

$$t_i = (m_{i1}\, v_{11}\, v_{21}) \cdot (m_{i2}\, v_{12}\, v_{22}), \quad i = 1, \ldots, m. \tag{1}$$

Equation (1) represents the specified action of function T.

Function C

Let m_i denote the ith word in memory, and let a "tagged" word refer to a word m_i for which $t_i = 1$. The compute function C operates numerically—and in parallel—on the a_{ij} field of all tagged words in memory in accordance with v_3, v_4, and v_5. In a final step C resets all tag bits.

Operator vector v_3 can take on the values $\times$, $\div$, $\ominus$, $\otimes$, representing multiplication, division, offset subtraction, and offset negative multiplication. The operators are defined as follows:

1) multiplication ($\times$): replace the a_{ij} field of each tagged word in memory by the product of this field and operand vector v_4.
2) division ($\div$): replace the a_{ij} field of each tagged word in memory by the quotient obtained by dividing this field by operand vector v_4.
3) offset subtraction ($\ominus$): subtract from the a_{ij} field of each tagged word in memory the a_{ij} field of the memory cell "l" locations away (lower), where l is the value in operand offset vector v_5.

---

*The reader is cautioned that i and j here are used differently than in the section "Components of AP"

4) offset negative multiplication (ⓧ): subtract from the a_{ij} field of each tagged word in memory the product of operand vector v_4 and the a_{ij} field of the memory cell "l" locations away (lower), where l is the value in operand offset vector v_5.

Thus

$$m_{i3} = \begin{cases} m_{i3} v_4 , & v_3 = \times , \\ m_{i3}/v_4 , & v_3 = \div \\ & \\ m_{i3} - m_{i-v_5 3} , & v_3 = \ominus , \\ m_{i3} - m_{i-v_5 3} v_4 , & v_3 = \otimes , \end{cases} \qquad \text{for } i \in (1, \ldots, m) \ni t_i = 1, \qquad (2)$$

followed by

$$t_i = 0, \qquad i = 1, \ldots, m . \qquad (3)$$

Equations (2) and (3) represent the specific action of function C.

Function R

The retrieve function R transfers in sequence the a_{ij} field of each tagged word in memory to the controlling computer via output vector v_6. In a final step R resets all tag bits.

Thus, the specific action of function R is represented by

$$v_6 = m_{i3} , \qquad \text{for } i \in (1, \ldots, m) \ni t_i = 1 \qquad (4)$$

in sequence, followed by

$$t_i = 0, \qquad i = 1, \ldots, m . \qquad (5)$$

SOLUTION OF A SYSTEM OF LINEAR EQUATIONS

Consider the system of linear equations

$$
\begin{aligned}
a_{11}x_1 + a_{12}x_2 + \ldots + a_{1n}x_n &= a_{1\,n+1}\\
a_{21}\,x_1 + a_{22}x_2 + \ldots + a_{2n}x_n &= a_{2\,n+1}\\
&\ldots\\
a_{n1}x_1 + a_{n2}x_2 + \ldots + a_{nn}x_n &= a_{n\,n+1}\,.
\end{aligned}
$$

Numerical analysts have developed two classes of methods for solving such systems: direct and iterative. The method that is most appropriate in a given instance depends both on the numerical properties of the equations and on the properties of the computer being used. The classic direct method—the Gauss method—consists of successive elimination of the unknowns. We shall investigate in detail the applicability of AP to this method.

The Gauss Method

The Gauss method consists of two stages: the forward stage and the backward stage. The forward stage is the major stage and serves to produce a triangular matrix; the backward stage then completes the solution. We explain the method by example; this gives us a particular problem to observe later on in "An AP Program for the Gauss Method."

Consider the system

$$
\begin{aligned}
2x_1 + 3x_2 - 4x_3 &= 3\\
-3x_1 + 5x_2 - 4x_3 &= 3\\
3x_1 + 2x_2 - 3x_3 &= 4\,.
\end{aligned}
$$

The given matrix is thus

$$
A = \begin{pmatrix} a_{11} & a_{12} & a_{13} & a_{14}\\ a_{21} & a_{22} & a_{23} & a_{24}\\ a_{31} & a_{32} & a_{33} & a_{34} \end{pmatrix} = \begin{pmatrix} 2 & 3 & -4 & 3\\ -3 & 5 & -4 & 3\\ 3 & 2 & -3 & 4 \end{pmatrix}.
$$

We begin by constructing the first row of the triangular matrix,

$$
b_1 = (b_{11} \quad b'_{12} \quad b_{13} \quad b_{14}),
$$

where

$$b_{1j} = \frac{a_{1j}}{a_{11}}, \qquad j = 1, \ldots, 4 .$$

Thus

$$b_1 = \left(1 \ \frac{3}{2} \ -2 \ \frac{3}{2}\right) .$$

Next, construct an auxiliary matrix

$$A^{(1)} = \begin{pmatrix} a_{11}^{(1)} & a_{12}^{(1)} & a_{13}^{(1)} \\ a_{21}^{(1)} & a_{22}^{(1)} & a_{23}^{(1)} \end{pmatrix},$$

where

$$a_{ij}^{(1)} = a_{i+1\ j+1} - a_{i+1\ 1}b_{1\ j+1}, \quad i = 1, 2, \quad j = 1,2,3 .$$

Thus

$$A^{(1)} = \begin{pmatrix} \frac{19}{2} & -10 & \frac{15}{2} \\ -\frac{5}{2} & 3 & -\frac{1}{2} \end{pmatrix} .$$

Now repeat the above process, using $A^{(1)}$ in place of A. We get

$$\left(1 \ -\frac{20}{19} \ \frac{15}{19}\right)$$

for the second row of the triangular matrix and a new auxiliary matrix of

$$A^{(2)} = \left(\frac{7}{19}, \frac{28}{19}\right) .$$

As a last step we get

$$(1, 4)$$

for the third row of the triangular matrix. Thus, the full triangular matrix is

$$\begin{pmatrix} 1 & \frac{3}{2} & -2 & \frac{3}{2} \\ 0 & 1 & -\frac{20}{19} & \frac{15}{19} \\ 0 & 0 & 1 & 4 \end{pmatrix},$$

representing the system of equations

$$\begin{aligned} x_1 + \frac{3}{2}x_2 - 2x_3 &= \frac{3}{2} \\ x_2 - \frac{20}{19}x_3 &= \frac{15}{19} \\ x_3 &= 4. \end{aligned}$$

The backward stage now consists merely of solving for x_2 and x_1, working from the third equation to the first. We get $x_2 = 5$ and $x_1 = 2$.

An AP Program for the Gauss Method

The main trick in solving a matrix problem on an associative processor consists simply of storing in memory the subscripts associated with each matrix element. This enables the subscripts to enter into the search condition, and provides the key whereby an entire row (column) of a matrix can be operated upon in parallel. It is for this reason that memory in AP is partitioned into three fields: i, j and a_{ij}.

In this section, the Gauss method is programmed for the AP defined in the section "Specification of an Associative Processor." To simplify matters we assume arithmetic to be in floating point, so that scaling problems can be omitted. Also we omit handling the occurrence of division by zero. Following are program conventions and initial conditions:

1) Let $[w]$ denote the contents of field 3, i.e., the a_{ij} field, of word w.

2) Let i, k, and r be integer variables.
3) Let the triple (i, j, a_{ij}) be stored in AP word $(n + 1)(i - 1) + j$.
4) Let $i = 1$ initially.

Table 1 gives the program code. The effect of this code on the example of the section "The Gauss Method" is shown in Table 2.

EVALUATION OF AP

In the matrix problems being considered, the execution time is roughly proportional to the number of multiplications and divisions required for a solution. This follows from the fact that multiplications and divisions are dominant operations in these problems—and execution times for these operations are lengthier than execution times for more basic operations, e.g., load, store, add.*[2]

The above consideration motivates the introduction of a statistic called a *step.* This statistic will be used as the basis for comparing the performance of a conventional computer with that of AP. A step is defined as:

1) a single multiplication or division on a conventional computer, or
2) one or more multiplications or divisions executed in parallel on AP.

If a given problem takes n_1 steps on a conventional computer and n_2 steps on AP, then the *speed-up factor* for the problem is estimated to be $n_1/c\ n_2$, where c is the ratio of step-times (AP to conventional).

Despite the roughness in this approach, the results can be expected to represent a reasonable first approximation.

Conventional Computer Performance

The number of steps needed for the Gauss method on a conventional computer is $\frac{n}{3}(n^2 + 3n - 1)$. This results from the following:

1) The forward stage requires the development of a sequence of $(n - 1)$ auxiliary matrices. The construction of the ith auxiliary

*The ratio of the execution time of the floating point multiplication (division) operation to that of a typical basic operation is generally lower nowadays, however, than it was on earlier computers.

Table 1
AP Code for Gauss Method

Operation	Remarks
1. $(=, >) \to v_1$	Set relation vector
2. $(i, i) \to v_2$	Set argument vector
3. T	For specified i, tag a_{ij}, $j > i$
4. $(\div) \to v_3$	Set operator vector
5. $[i(n+2) - n - 1] \to v_4$	$[i(n+2) - n - 1] = (a_{ii})$
6. C	Compute row i of triangular matrix
7. If $i = n$, go to 21	If $i = n$, forward stage complete
8. $i + 1 \to r$	Initialize r variable
9. $(\otimes) \to v_3$	Set operator vector
10. $(n+1) \to v_5$	Initialize operand offset vector
11. $(r, i) \to v_2$	Set argument vector
12. T	For specified r tag a_{rj}, $j >$ specified i
13. $[(n+1)(r-1) + i] \to v_4$	$[(n+1)(r-1) + i] = (a_{ri})$
14. C	Compute row $(r-i)$ of $A^{(i)}$
15. If $r = n$, go to 19	If $r = n$, computation of $A^{(i)}$ complete
16. $r + 1 \to r$	Step r variable
17. $v_5 + (n+1) \to v_5$	Modify operand offset vector
18. Go to 11	Compute next row of $A^{(i)}$
19. $i + 1 \to i$	Step i variable
20. Go to 2	Perform next cycle of forward stage
21. $n \to k$	Begin backward stage: initialize k variable
22. $(<, =) \to v_1$	Set relation vector
23. $(k, k) \to v_2$	Set argument vector
24. T	For specified k, tag a_{ik}, $i < k$
25. $(x) \to v_3$	Set operator vector
26. $[k(n+1)] \to v_4$	$[k(n+1)] = x_k$
27. C	Use x_k
28. $(k, n+1) \to v_2$	Set argument vector
29. T	For specified k, tag $a_{i\ n+1}$, $i < k$
30. $(\ominus) \to v_3$	Set operator vector
31. $(n + 1 - k) \to v_5$	Set operand offset vector
32. C	Compute x_{k-1}
33. If $k = 2$, go to 36	If $k = 2$, backward stage complete
34. $k - 1 \to k$	Step k variable
35. Go to 23	Compute next x_k
36. $(n+1, n+1) \to v_2$	Set argument vector
37. T	Tag $a_{i\ n+1}$, $i = 1, \ldots, n$
38. R	Retrieve $x_1, \ldots, x_n$ in turn.

Table 2
Example of Gauss Method on AP

Cell	Initial Conditions			Changes to a_{ij} in Forward Stage						Changes to a_{ij} in Backward Stage			
	i	j	a_{ij}	row 1 of T	row 1 of $A^{(1)}$	row 2 of $A^{(1)}$	row 2 of T	row 1 of $A^{(2)}$	row 3 of T	Use x_3	Get x_2	Use x_2	Get x_1
1	1	1	2										
2	1	2	3	$\frac{3}{2}$								$\frac{15}{2}$	
3	1	3	-4	-2						-8			
4	1	4	3	$\frac{3}{2}$							$\frac{19}{2}$		②
5	2	1	-3										
6	2	2	5		$\frac{19}{2}$								
7	2	3	-4		-10		$-\frac{20}{19}$			$-\frac{80}{19}$			
8	2	4	3		$\frac{15}{2}$		$\frac{15}{19}$				⑤		
9	3	1	3										
10	3	2	2			$-\frac{5}{2}$							
11	3	3	-3			3		$\frac{7}{19}$					
12	3	4	4			$-\frac{1}{2}$		$\frac{28}{19}$	④				

Notes: 1) T = triangular matrix
2) Encircled numbers constitute solution

matrix $A^{(i)}$, $i = 1, \ldots, n-1$, calls for $(n - i + 1)$ divisions and $(n - i + 1)(n - i)$ multiplications.*[3] Also, a final division is required to obtain the final row of the triangular matrix. Thus, the number of multiplications and divisions in the forward stage is

$$1 + \sum_{i=1}^{n-1} [(n-i+1) + (n-i+1)(n-i)] = \frac{n(n+1)(2n+1)}{6} .$$

2) The number of multiplications in the backward stage is

$$\sum_{i=1}^{n-1} i = \frac{n(n-1)}{2} .$$

AP Performance

The number of steps required for the Gauss method on AP is $\frac{n^2 + 3n - 2}{2}$. This is obtained as follows:

1) The $(n-i+1)$ divisions required in the development of $A^{(i)}$ can all be done in parallel. The $(n-i+1)(n-i)$ multiplications required in developing $A^{(i)}$ can be done in $(n-i)$ steps, each step consisting of $(n-i+1)$ multiplications in parallel. Also, as before, a final division is required to obtain the final row of the triangular matrix. Thus, the number of steps in the forward stage is

$$1 + (n-1) + \sum_{i=1}^{n-1} (n-i) = \frac{n(n+1)}{2} .$$

2) The $\frac{n(n-1)}{2}$ multiplications required in the backward stage can be done in $(n-1)$ steps.

Comparative Performance

From the above we see that AP offers a speed-up factor of

*By computing reciprocals, divisions can be traded for multiplications. For our purposes we do not distinguish between these operations.

$$\frac{\frac{n}{3}(n^2 + 3n - 1)}{\frac{c}{2}(n^2 + 3n - 2)},$$

which approaches $2n/3c$ for large n. As an example, in solving a system of 50 linear equations, use of an AP with $c = 3$, say, would result in approximately an 11-fold speed-up. Since the step-time of a particular hardware implementation of AP is (more or less) constant, the speed-up factor for solving a linear system increases in (more or less) direct proportion to the size of the system.

OTHER MATRIX PROBLEMS

Solution of a Linear System by Pivotal Condensation

In solving a linear system, it is often desirable to employ a variant of the Gauss method, viz., the method of pivotal condensation. This method mitigates the cumulative round-off problem associated with matrices that are not "well behaved." The refinement introduced by this method is that the order in which unknowns are eliminated during the forward stage is data-dependent.

The method of pivotal condensation requires the same number of multiplications and divisions as the basic Gauss method. The extra computation is associated with the development of the ith auxiliary matrix $A^{(i)}$, $i = 1, \ldots, n - 1$, in the forward stage; specifically this computation consists of finding the "pivot element," i.e., the maximum element, in an $(n - i + 1)$-order matrix.

It is pertinent to note that algorithms have been developed for retrieving the AP word having the maximum entry in a specified field [1, 2]. Thus, a slight extension of the capability of the AP defined in the section, "Specification of an Associative Processor" would result in an associative processor with special facility for solving a linear system by the method of pivotal condensation.

The problem of finding the maximum entry in a w-word table on a conventional computer requires $(w - 1)$ comparisons (plus overhead). Therefore, in solving a system of n linear equations, the number of comparisons required for pivot element searches is

$$\sum_{i=1}^{n-1} \left[(n - i + 1)^2 - 1\right] \simeq \frac{n^3}{3} \text{ for large } n .$$

These comparisons, together with associated overhead, can be replaced by $(n - 1)$ searches for maximum on an associative processor.

Evaluation of a Determinant

The evaluation of an nth-order determinant by a procedure corresponding to the Gauss method [3] requires $\frac{(n-1)(n^2+n+3)}{3}$ multiplications and divisions, which can be done in $\frac{(n+4)(n-1)}{2}$ steps on an AP. Hence, for this problem, the associative processor also yields a speed-up factor of $2n/3c$ for large n.

Inversion of a Matrix

Solving a system of linear equations is closely connected with the problem of inverting a matrix. In fact, if $A = (a_{ij})$ is a given matrix and $A^{-1} = (q_{ij})$ is its inverse, then

$$\sum_{k=1}^{n} a_{ik} q_{kj} = \begin{cases} 1, & i = j, \\ 0, & i \neq j, \end{cases} \qquad i, j = 1, \ldots, n ,$$

so that the inversion of an n-by-n matrix is equivalent to solving n systems of n linear equations.

A more efficient method for inverting a matrix involves the use of *pivot operations* [4]. If $A = (a_{ij})$ is a given matrix, then a pivot operation on the element a_{rs} results in the construction of a matrix $B = (b_{ij})$ as follows:

$$b_{ij} = \begin{cases} a_{ij} - a_{rs}^{-1} a_{is} a_{rj} , & i \neq r, \quad j \neq s, \\ a_{rs}^{-1} a_{is} , & i \neq r, \quad j = s, \\ -a_{rs}^{-1} a_{rj} , & i = r, \quad j \neq s, \\ a_{rs}^{-1} , & i = r, \quad j = s . \end{cases}$$

An nth-order matrix can be inverted by n such operations.

Each pivot operation requires n^2 steps on a conventional computer as opposed to $(n + 2)$ steps on an AP. Thus, for matrix inversion by this method, the associative processor offers a speed-up factor of $n^3/cn(n + 2)$, which approaches n/c for large n.

Determination of Dominant Eigenvalue

A problem arising in mathematical physics is the following: Given an nth-order real, symmetric matrix A, finding the eigenvalue λ of largest absolute value.

A straightforward iterative technique for obtaining the required eigenvalue is as follows [5]: Let $x^{(0)}$ be an arbitrary nonzero vector.*[4]

Let

$$x^{(i+1)} = Ax^{(i)}, \qquad i = 0, 1, \ldots,$$

and let

$$\sigma_i = \frac{\sum_{j=1}^{n} x_j^{(i)} x_j^{(i+1)}}{\sum_{j=1}^{n} x_j^{(i)} x_j^{(i)}}, \quad i = 0, 1, \ldots,$$

where $x_j^{(i)}$ is the jth component of $x^{(i)}$. Then

$$\lim_{i \to \infty} \sigma_i = \lambda \quad .$$

On a conventional computer, each stage of the iteration process requires n^2 multiplications to obtain a new vector; $2n$ multiplications to obtain two scalar products; and 1 division. On an AP the $(n^2 + 2n)$ multiplications can be done in $(n + 2)$ steps. Therefore, the overall speed-up factor resulting from use of an AP is

---

*The scalar product of $x^{(0)}$ and the orthonormal eigenvector associated with λ must be nonzero.

$$\frac{n^2 + 2n + 1}{c(n+3)},$$

which approaches n/c for large n.

SUMMARY AND CONCLUSIONS

Computer designers have devised a variety of hardware techniques for carrying out computation in parallel. These techniques have been incorporated in computer systems from the *micro*level to the *macro*-level. The associative processor represents one such technique. It could play a significant role in computer systems of the future.

An attractive feature of the associative processor is that it can effect parallel processing without imposing special burdens on the programmer. This conclusion is based on the fact that the program code of Table 1 was no more difficult to write than a conventional computer code.

We have derived a rough approximation to the speed-up offered by an associative processor on some matrix problems. The results are summarized in Table 3. It is seen that an appropriately defined AP can speed up these problems to a remarkable extent. Note that these gains in speed are obtained merely by taking advantage of the parallelism inherent in existing methods, rather than by making any changes in the methods themselves.

While we have examined here only a few problems, it is evident that an AP can be applied effectively to a wide class of matrix problems. The following points are noted specifically:

1) Pivot operations of the type described in the section "Inversion of a Matrix" are at the heart of the Simplex method of linear programming; therefore an associative processor could be applied with considerable success to that important problem.
2) The speed-up factor of $2n/3c$ will be attained in a class of matrix problems. The underlying reason is the following: In various procedures on an n-order matrix, one develops successively an $(n-1)$-order matrix, an $(n-2)$-order matrix, etc., down to a 1-order matrix. In developing an i-order matrix, $i = 1, \ldots, n-1$, the AP is required to perform (approximately) i^2 multiplications, which it can do in (approximately) i steps. Therefore, the speed-up factor for such a procedure approximates

Table 3
Speed-up Offered by AP

Problem	Method	Number of Steps on Conventional Computer	Number of Steps on AP	Speed-up Factor for Large n
1. Solution of a system of n linear equations	Gauss method	$\frac{n(n^2 + 3n - 1)}{3}$	$\frac{n^2 + 3n - 2}{2}$	$\frac{2n}{3c}$
2. Evaluation of an nth-order determinant	Variant of Gauss method	$\frac{(n - 1)(n^2 + n + 3)}{3}$	$\frac{(n + 4)(n - 1)}{2}$	$\frac{2n}{3c}$
3. Inversion of an nth-order matrix	Pivot operations	n^3	$n(n + 2)$	$\frac{n}{c}$
4. Determination of dominant eigenvalue of real, symmetric n^{th}-order matrix	Iterative method	$(n + 1)^2$ per iteration	$n + 3$ per iteration	$\frac{n}{c}$

Notes: A *step* on a conventional computer is defined as a single multiplication or division.
A *step* on an AP is defined as one or more multiplications or divisions executed in parallel.
The constant $c = \frac{\text{step-time on an AP}}{\text{step-time on a conventional computer}}$.

$$\frac{\sum_{i=1}^{n-1} i^2}{c \sum_{i=1}^{n-1} i} \simeq \frac{2n}{3c} \text{ for large } n \, .$$

3) The speed-up factor of n/c will also be attained in a class of matrix problems. This is due to the fact that the main procedure in such problems consists of the development of a sequence of n-order matrices.

In conclusion, the associative processor should serve as a stimulus for much needed research on two broad fronts: (1) the exploration for parallelism in existing mathematical methods, and (2) the development of brand new mathematical methods for use on hardware capable of exploiting parallelism. In all likelihood, the associative processor will yield major improvements in speed for an extensive class of mathematical processes.

References

1. G. Estrin, and R. Fuller, "Algorithms for Content-Addressable Memories," *Proc. Pacific Computer Conference IEEE,* 1963, pp. 118-30.
2. A. D. Falkoff, "Algorithms for Parallel-Search Memories," *J. ACM, 9:* (1962), 488-511.
3. V. N. Faddeeva, *Computational Methods of Linear Algebra,* (New York: Dover Publications, Inc., 1959), pp. 72-5.
4. D. T. Finkbeiner, *Introduction to Matrices and Linear Transformations,* (San Francisco: W. H. Freeman and Co., 1960), pp. 100-5.

Additional References

H. S. Wilf, *Mathematics for the Physical Sciences,* (New York: John Wiley and Sons, Inc., 1962), p. 34.

DEVICE TECHNOLOGY UTILIZATION IN PARALLEL PROCESSORS

CHAPTER 7

A MISSION ORIENTED ASSOCIATIVE PROCESSOR USING PLATED WIRE

W. C. Meilander

Surveillance Data Systems Engineering
Goodyear Aerospace Corp., Akron, Ohio

J. Garrett and M. Bialer

Materials Technology, Electronics Branch
Wright-Patterson Air Force Base

INTRODUCTION

Associative processors (AP) have been shown to be superior to conventional processors in their application to radar and command control data processing as a result of recent studies at Goodyear Aerospace Corporation. This ability is due to the fact that an associative processor can efficiently manipulate large data bases at the speed required by real-time systems. A functional associative processor employing plated wire as the storage medium is in the final stages of test and evaluation. Salient features of the plated wire AP design, fabrication and demonstration program* are discussed in this paper.

*Work performed under Goodyear Aerospace Corporation in-house funding and Air Force Material laboratory WPAFB, Contract F33615-68-C-1555.

PROGRAM OBJECTIVES

One objective of this program was to determine the requirements for a plated wire, associative array module as determined from an application study of a real-time Air Force radar data processing requirement. A method for using an associative processor to distinguish between ground and airborne targets was devised. Algorithms satisfying the specific application were derived, and preliminary programming was carried out to quantify the number and type of operations needed. A description of the radar application is given in the following section.

A second objective was to construct a plated wire associative array module that would satisfy the requirements determined from the radar application study. A control subsystem was to be constructed that would be capable of exercising the plated wire AP array module according to the functional requirements and of demonstrating associative operations.

AP APPLICATION OPERATIONS

The following is a summary of requirements established to satisfy program objectives.

AP Application Operations

The following operations are required in the application of the AP to the radar problems:

1) Search Operations
   a) maximum value
   b) minimum value
   c) between limiting comparands
   d) search flag
   e) exact match with comparand
   f) mismatch with comparand
   g) greater than comparand
   h) less than comparand
2) Logic Operations
   a) set response toggles
   b) reset response toggles
   c) complement response toggles
   d) shift response toggles
   e) write flag from response
   f) write common to selected words

3) Arithmetic Operations
   a) add common argument
   b) subtract common argument
   c) add memory fields
   d) subtract memory fields
   e) multiply memory fields
   f) multiply by common argument
   g) divide memory fields
   h) divide by common argument

Functional Requirements

The associative array matrix and electronics functional requirements include the following:

1) The number of words in the associative array module shall be 256 (128 words to be electronically instrumented).
2) The number of bits in each word in the associative array module shall be 256 (128 bits to be electronically instrumented).
3) The read-out and interrogate modes will be nondestructive with an interrogate time of 100 η sec. per bit.
4) The multiwrite mode will write the contents of the response stores into the same bit position of any or all words simultaneously at a rate of 300 η sec. per bit.
5) The write mode will write a sequence of input data words bit-parallel in a 16-bit byte into memory words as a function of the sequential selection of all words in the memory from a lowest indexed point to the highest in no more than 300 η sec. per 16-bit byte.

Demonstration Requirements

The following requirements will be met for system demonstrations:

1) The radar data processor outputs will be simulated on punched paper tape.
2) The AP array module performance requirements shall be demonstrated by means of an exerciser.
3) The exerciser will operate under control of a paper tape input.
4) The exerciser output will be displayed on an oscilloscope.

APPLICATION DESCRIPTION

An interesting application of an associative processor, although limited in scope, is provided by an airborne interceptor radar system. Without the associative processor, the *airborne moving target indicator* (AMTI) display is difficult to interpret because of range-ambiguous targets, noise targets, and moving ground targets contained in the output. With an associative processor, a clean output display is provided.

By eliminating range-ambiguous and noise targets and providing the capability to select airborne targets only, moving ground targets only, or both, the associative processor provides an easily interpreted display. These results are achieved by the associative processor while using less than 20 per cent of available real-time. Because of the nature of the associative processor, the processing time is independent of the number of targets.

The radar system considered in this application is made up of a multiple *pulse-repetition frequency* (PRF) range-ambiguous radar, a *digital doppler signal processor* (DDSP), and a display system. The *associative processor* (AP) is inserted between the DDSP and the display system (Fig. 1). The radar employs a sector scan and operates with 3 staggered PRF's, which yield an average unambiguous range of approximately 10 naut. mi. The maximum ambiguous range is 40 naut. mi.

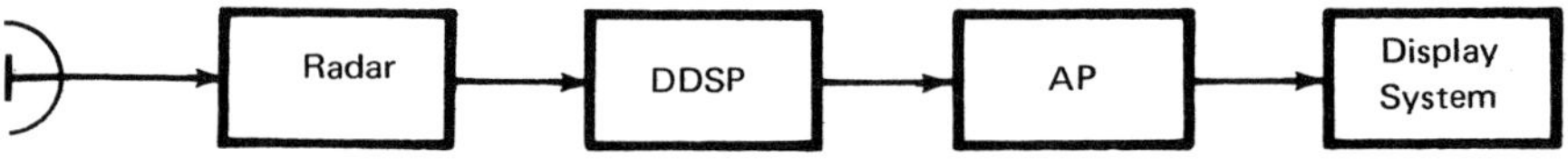

Fig. 1. Airborne Radar System

In the AMTI mode of the radar, a B-scan display is presented that has a dimension of 0 to 40 naut. mi. and an azimuth dimension of ±45 deg. Within this display, all targets are displayed that fall inside a range window. The position of the range window within the 40 naut. mi. range may be moved by a manual control.

The DDSP outputs target hits to the display system during the part of each range sweep that corresponds to the range window interval. Within the range window, the unambiguous targets are displayed along with the ambiguous (or ghost) targets and false noise targets. The difficulties in display interpretation are compounded when moving ground targets appear with the airborne targets.

The functions of the AP are:

1) To resolve target range ambiguity; this prevents the displaying of ambiguous and false noise targets, and allows only unambiguous targets to be displayed.
2) To discriminate between ground and airborne moving targets to give the operator the choice of displaying airborne targets only, moving ground targets only, or both.

The AP is made up of two major subsystems, a control subsystem and an associative subsystem. The control subsystem contains a high-speed conventional memory in which are stored the system program, microprograms, and microinstructions necessary to exercise the associative subsystem. The processor size for this application is 128 words by 128-bits per word.

The AP operations are performed in the high-speed associative subsystem. The associative subsystem receives microinstructions and data from the control subsystem and outputs the resulting responses to the control subsystem.

The essential target parameters stored in each word are: range (R), azimuth (θ), sine θ, cosine θ, and PRF identity. In addition to these parameters, each associative memory word contains tag fields for housekeeping purposes and scratch fields for arithmetic operations. The AP input/output function is performed by the control subsystem. The data transfer into the AP is initiated by the activation of an interrupt signal.

The data processing required for target range resolution and ground moving target discrimination falls into one of the following: (1) within-beam processing, or (2) scan-to-scan processing. Within-beam processing determines target existence and true range of the target. Scan-to-scan processing provides a predicted range and azimuth for each declared target stored in the AP. The AP must be capable of processing the DDSP output within a given time period. For the radar considered, this time is approximately 17 μsec. A timing analysis shows this processing can be performed in the AP in 15 μsec. The updating of the sin θ and cos θ fields is performed in the period between each display range sweep. The processing time required for this updating is small compared with the expected available time. The time required by the AP to perform scan-to-scan processing is 1052 μsec.

PROGRAM STATUS

Associative Processor Array

When used in an associative processor (AP), plated wire is subjected to severe operational constraints not encountered when the plated wire is used in destructive read-out or nondestructive read-out (DRO and NDRO) applications. These constraints are caused by the nature of the write operation in an AP, coupled with the limited time available for performing the write operation and by the type of disturbs encountered in the AP operation. In a DRO or NDRO memory, the write operation is performed a word at a time. In an AP memory, writing can be performed either on a word or field of a word or on a single bit in many words. This last write requirement prevents the use of higher currents for writing than reading. Further, the short write time allocated prevents the use of multiple passes to accomplish the write operation.

The effort applied has been directed toward determining the conditions under which the wire will operate satisfactorily in AP applications. An evaluation of plated wire was made with regard to its physical properties and switching properties.

A plated wire evaluated for use in AP applications was tested with a pulse pattern which would be the worst-case sequence expected in system operation. The test patterns consisted of a long history cycle, a single write operation, adjacent bit and read disturb cycles, and a read operation.

An evaluation of strap geometry was made to determine the optimum bit strap geometry consistent with the system requirements and circuit philosophy. Several types of keepers have been evaluated in regard to their effectiveness on system performance.

Experimental results show that plated wire can be employed in large associative processor arrays. The tests indicated that wide operating margins and relative insensitivity to disturbs is attainable. The optimum results were obtained with a two-turn bit strap, bipolar word currents and special keepers.

Array Module Electronics

The word driver is a bipolar driver that must deliver 40 ma. It must be a high impedance during interrogate time so that it does not affect the wire, which is used as a sense line. Drivers for two words are packaged in a 3/4 x 1 inch flat pack containing 6 IC chips. This driver can be made monolithic with present technology.

The bit driver must drive 400 ma. into an 80 Ω load. Four-bit drivers are being packaged in a 3/4 x 1 inch pack containing 6 IC chips. The driver load is a two-watt resistor, deposited on a *BeO* substrate. Heat is removed from the substrate by a cold plate.

The minimum system signal level was set at 3 *mv*, based on data from an on-line wire tester and from the various test arrays.

The sense amplifier presently being used is a special monolithic device built for GAC by Motorola. It was designed for use with a coupling capacitor. In this application, a-c coupling is not allowable because circuit recovery would be too long; therefore, a zener diode coupling network is used.

Response store circuits are located in the array module exerciser logic nest. The next generation AP array module will employ MSI response store circuits packed integrally with the wire drivers and sense amplifiers.

A number of tunnel structure assemblies were evaluated for array fabrication. The materials included polyethylene, polyethylene-mylar, Videne,* Videne-Mylar, polyvinyl fluoride, and epoxy-glass. Several epoxy-glass prepregs have been evaluated, and offer the best tunnel material at present.

Control Subsystem

The memory array module exerciser (Fig. 2) is a program-controlled system that controls array module tests and evaluation under conditions dictated by functional requirements. Four modular cabinets house the system logic, power supplies, maintenance and test panel, input unit, and output-display unit. The array module is mounted on retractable slides within one of the cabinets (Fig. 3).

A block diagram of the system is shown in Fig. 4. System logic is contained on 42 multilayer printed circuit cards and is implemented with Motorola emitter-coupled logic gates. A microprogram memory accepts instructions, data, and macroprograms from a punched paper tape reader. Microprograms are executed in 50-*ns* steps. Memory contents can be displayed on a storage cathode ray tube or by means of lamps on the maintenance panel. A visual display of microprogram-operation elapsed time is presented on a high-speed counter.

---

*TM The Goodyear Tire and Rubber Company.

Fig. 2. Memory Array Module Exerciser

PERFORMANCE DEMONSTRATION

System tests and performance demonstrations have confirmed the accomplishment of the functional objectives of a mission oriented associative processor using plated wire.

Initial tests and evaluations centered about the AP array module, the plated wire array and array electronics were subjected to the maximum operational requirements expected in associative processor applications. Worst case data patterns were tried to test the disturb sensitivity of the AP plated wire array geometry. Memory contents are displayed on a storage tube for observation. Errors are internally detected and their locations in memory are automatically recorded. A total error count is recorded at the end of each test. All data and operating instructions are programmed and inserted by means of a paper tape. Worst case testing of the instrumented array uncovered a small number of bad bits, mostly wire oriented. These defects were corrected by replacing the defective plated wires. It is believed that the wires were damaged while being inserted into the tunnels. Bit patterns were inserted in bytes of 1 through 16 bits. The memory was operated within the required 400 η sec. write-read cycle time and 10 MH_z interrogate rate. Multiwrite characteristics were also demonstrated.

An associative processor application to a representative radar display problem was demonstrated. Program tapes containing search and arithmetic operations were assembled according to the range ambiguity and target discrimination algorithms. No actual radar-AP interface exists; therefore, target and navigation data was simulated and formatted for paper tape. Included in the simulation were the results of target calculations normally handled in a sequential machine arithmetic unit. Because of the slow data and instruction handling characteristics of the paper tape reader, a greatly magnified problem-solution time

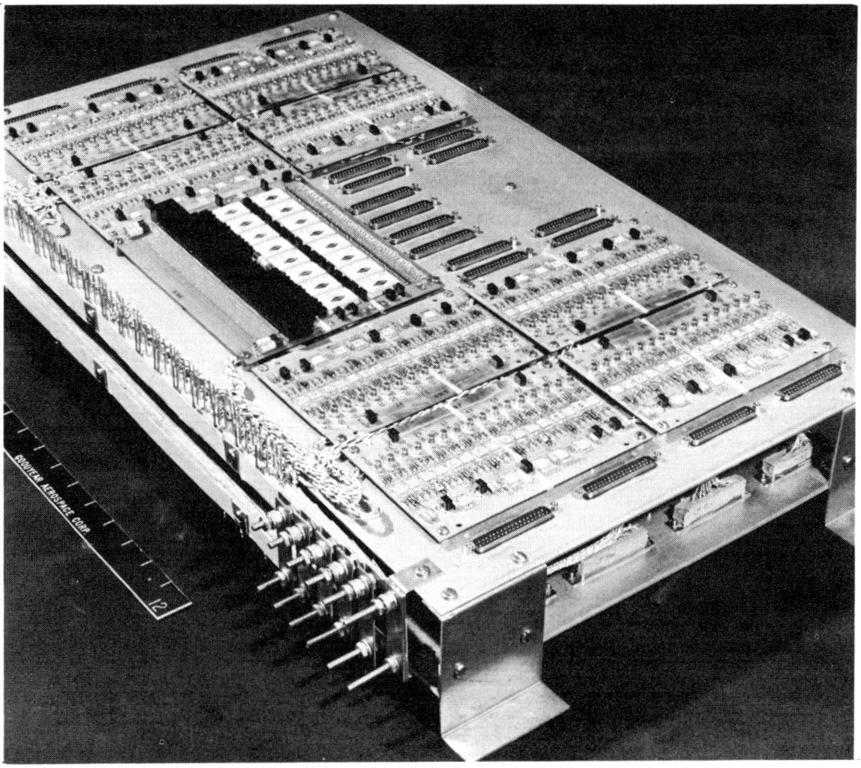

Fig. 3. Plated Wire Memory Array Module

would be observed. To give a true measure of problem-solution time, a 20 MH_z electronic counter capable of counting the 50 η sec. program steps is tied into the program control on an interrupt basis. Only the time necessary to perform parallel search, parallel arithmetic, and write

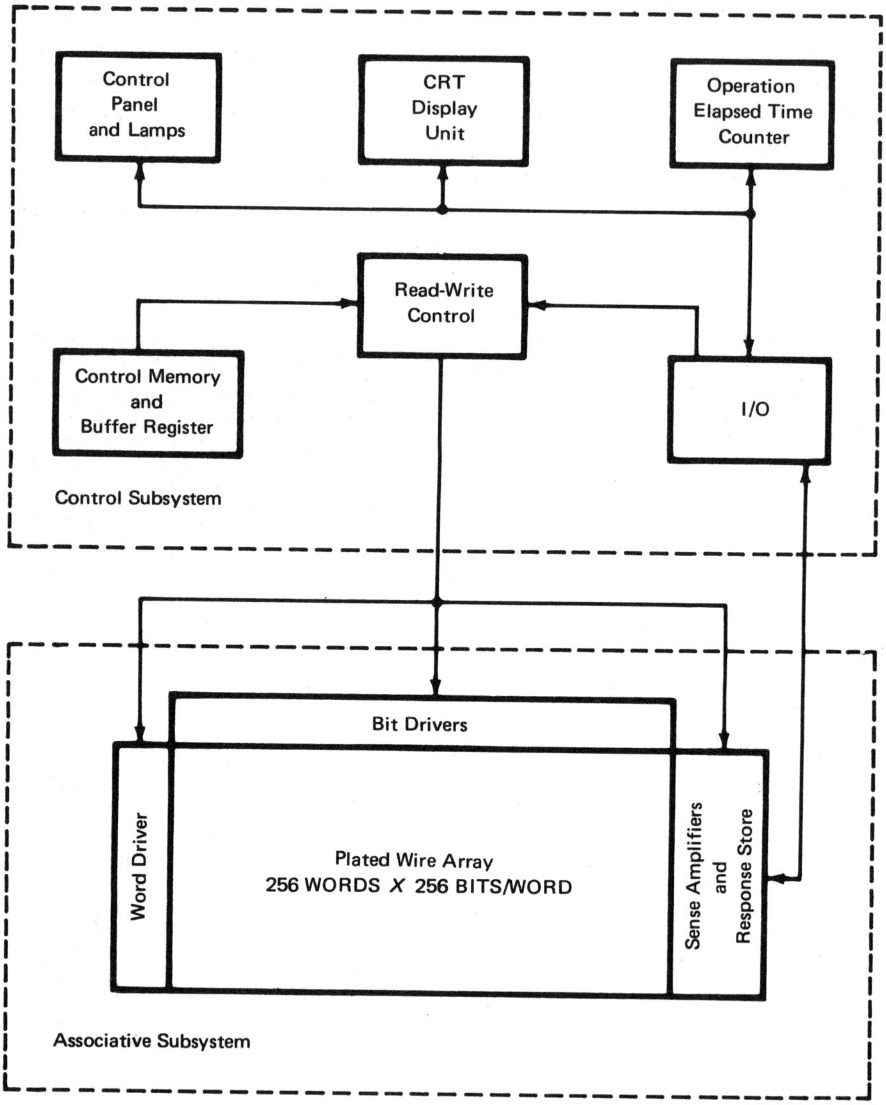

Fig. 4. AP System Block Diagram

operations is counted. Typical search times were measured to be 3.9 microseconds for a 32-bit field search of 128 words, including writing response tags. Adding two 16-bit fields of 128 words required 18.8 microseconds. Multiplying two 16-bit fields of all 128 words required 336.0 microseconds.

A target situation is displayed on a storage screen. Various symbols differentiate target types according to whether they are ground-moving or air-moving targets. A display control permits the selection of target type. The display scope will display real and false targets prior to the employment of the associative processor operations. Switching in the processor results in a dramatic reduction of display confusion. From the demonstrations performed on the mission oriented associative processor, with nickel-iron plated wire used as the storage medium in the associative array, it is apparent that plated wire associative processors can be built and will perform in every way expected.

CHAPTER 8

INTERACTION BETWEEN LSI AND COMPUTER SYSTEM ARCHITECTURE

Rex Rice

Fairchild Camera and Instrument Corporation
Palo Alto, Calif.

Abstract

The advent of integrated circuits and now of Large Scale Integration has already drastically changed the logical design optimization parameters used in digital system design. It becomes increasingly apparent that interconnections (i.e., busing local control decoding, and lavish use of array-like logic) will generate and dominate a new design philosophy. The lavish use of array-like LSI elements will, in turn, cause and enhance functional modularity and, to some degree, a standardization of array-type control components at a very high level.

The availability of LSI functional units, LSI memory, and LSI bus multiplexors will, in turn, make parallelism economically attractive. Parallelism is herein broadly defined to mean two or more identical functional units in a single system. This allows inclusion of a range of equipment from a system such as ILLIAC IV to a PDP-8 with two arithmetic units.*

INTRODUCTION

With discrete components and with early integrated circuits, most designers tailored systems to minimize the "circuit count" in a system. Even today textbooks still espouse this theme. The advent of batch fabrication of integrated circuits into interconnected arrays (i.e., LSI)

*The work on the ILLIAC IV PEM is supported by the Advanced Research Projects Agency, Rome Air Development Center, under contract USAF 30(602)4144. Fairchild is a supplier of LSI memory systems to the Burroughs Corporation, the prime contractor on the ILLIAC IV computing system.

has forced us into the realization that *interconnections between the arrays* is the parameter which should be optimized even at the expense of not using or "wasting" circuits within the array.[1]

The author believes that bus-oriented system designs for data paths, logic, and controls will emerge in the new generation.[2] Local decoding of operation codes will be used and will result in more than a minimum of circuits but will vastly reduce the mechanical interconnection problem for a large net cost savings. More "functional slices" will appear in designs; thus providing to some degree a new class of "standard" LSI components. The main storage, the basic CPU logic, certain regular code translation logic, and some control logic components will emerge as "off the shelf" LSI items. The use of read-write (or read-only) LSI memory for storing control-code sequences will regularize much of control logic, thus taking advantage of the availability of LSI.

LSI provides an inherent component speed-up by using lower power within the array, by having much shorter signal transmission paths, and by utilizing more complex circuit cells at little extra fabrication expense. Taken alone, this will give the normal system speed-up expected of faster circuits. *A more revolutionary change will result from the use of LSI memories where the memory cycle can approach the clock cycle of the system.* Our thinking on the use of external or "internally stored" registers can be altered as we realize the advantages of 100 to 200 nanosecond memory cycles.

With magnetic core, film, or plated wire memories, low costs are achieved by having very large magnetic arrays driven by a small amount of electronics. Our present generations of systems are consciously or unconsciously built around this consideration (i.e., one large main memory). Very large systems such as the CDC 7600 clearly show this type of architecture.[3] LSI memory prices will soon approach the price per bit of magnetic memories in the 600 nanosecond or faster range. *Most important, however, is the fact that the cost/bit is almost linear from a small memory to a large memory.* This second aspect of the use of LSI memories will drastically alter our concepts of system architecture. This implies we can break up our single large memory (when using magnetics a large array is necessary to reduce cost) in any way we choose and at no cost penalty. We can multiplex more memories for speed, or distribute memory to more locally autonomous, processing functional units, etc. In a later generation, the use of more exotic memories such as multipart, associative, etc., will create system designs presently only hazily visualized.

The point here is that batch fabricated component arrays will force more functional splitting of designs and consequently can present us with an economically attractive way to achieve parallelism in the use of identical components and in paralleling processing operations. This will be true from small systems to large systems.

The author is not a purist and still believes a mixture of LSI, integrated circuits, etc., should be used as required. Figure 1 presents the idea that we can optimize designs using LSI (or any other components) for cost, for speed, or for some intermediate design objective. If we investigate some applications of LSI at the cost end of the scale and others at the speed end of the scale, it follows that other intermediate objectives can be met.

It is common practice to equate the advantages of LSI with a circuit-to-circuit cost comparison where the LSI array circuit is replacing an equivalent integrated circuit. As shown in Fig. 2, this is a very misleading comparison. The total functional price-to-performance ratio of the complete computing system, not the cost per bit or the cost per circuit of the LSI circuit replacing the normal integrated circuit, is the thing upon which we should focus attention. As illustrated, the LSI array will replace the interconnections which normally occur on printed circuit boards, the connectors which connect the integrated circuit boards, large numbers of wires which normally cost about $.06 each installed, and finally the integrated circuits and components which are the only items normally counted. The cost comparison is therefore not just a circuit-for-circuit replacement comparison but a complete cost-replacement comparison for all the hardware including nuts and bolts and circuits replaced by LSI. Further, consider the implication of a system operating with a 100 nanosecond memory cycle when comparing performance of the LSI computer.

The author only dimly perceives the exciting changes to come. To illustrate some concepts, let us briefly consider several examples. First, let us consider some LSI functional components which may be used to construct systems.

Conceptually, we also allow the inclusion of certain integrated circuits and other discrete components on printed circuit boards to make an operable functional unit. In principle, however, as much as possible of the functions will be in LSI componentry. After considering the several elements so developed, let us then hypothesize several systems using these elements.

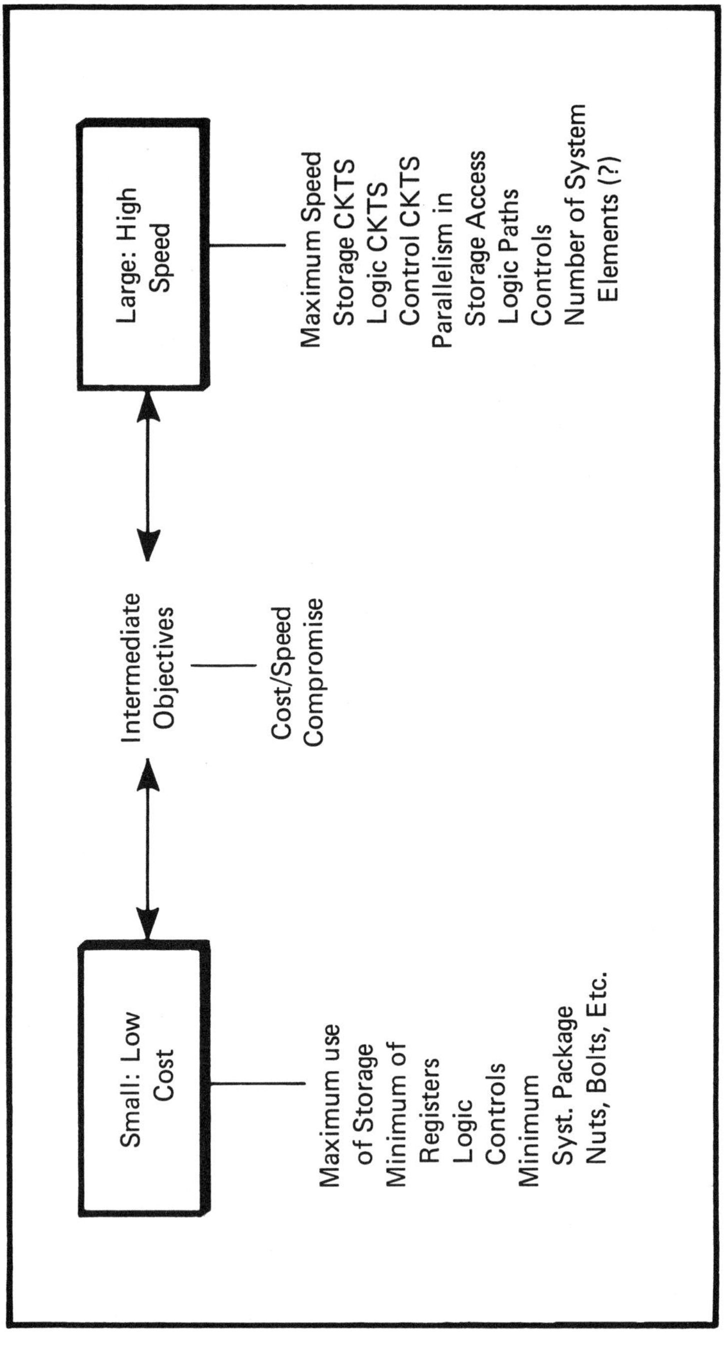

Fig. 1. LSI Designs Optimized for Cost and/or Speed

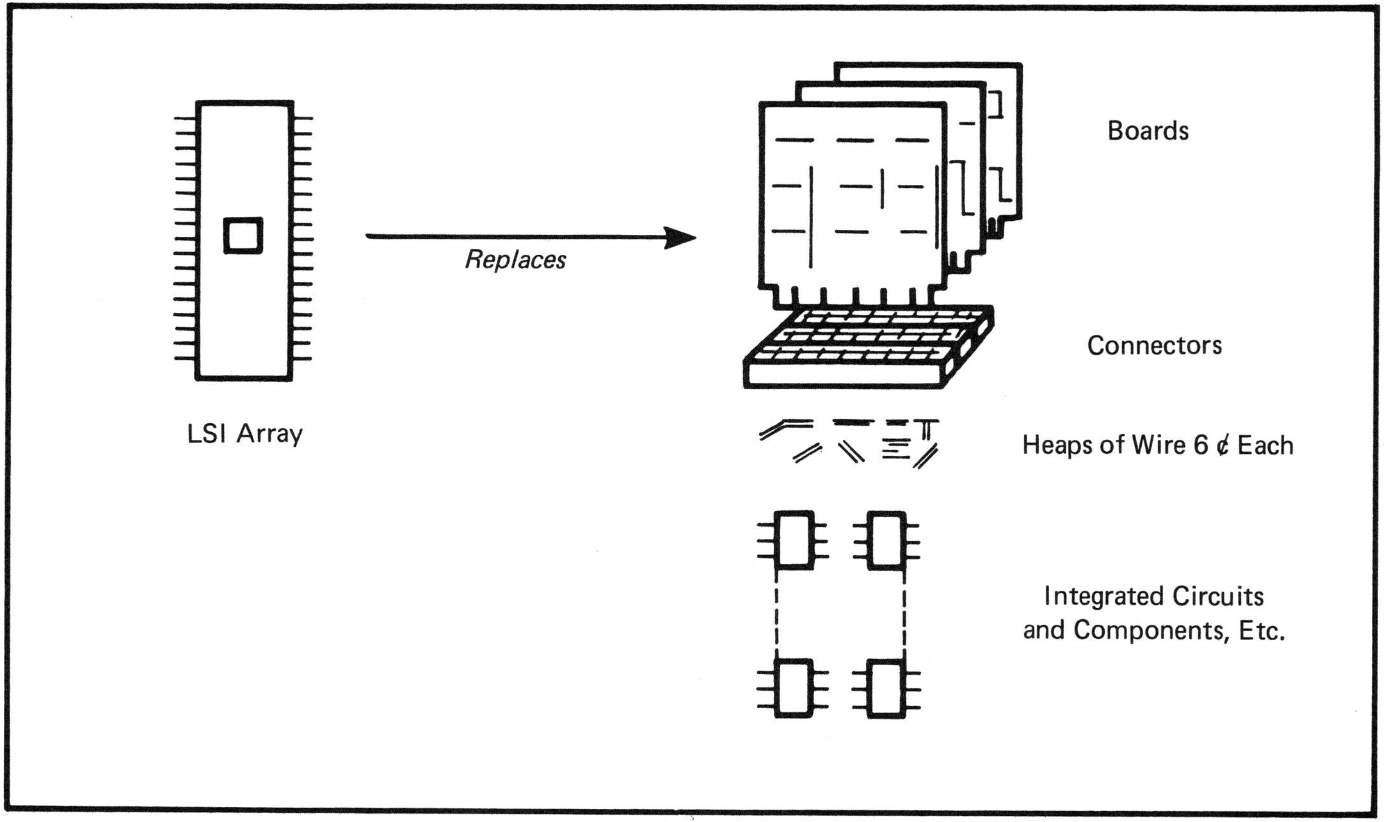

Fig. 2. Total Functional Price/Performance not Cost/Bit or Cost/Circuit

LSI MAIN MEMORY AND BUS SWITCHES

The first element of interest, an LSI memory, is illustrated in Fig. 3. The first difference to be remembered about LSI memories is that they can operate at what is normally considered to be system logic clock speeds. A second item is that they have linear incremental costs. That is to say, an LSI memory may be modular in the number of words or in the number of bits per word, and above some minimum size it doesn't particularly matter if you double, triple, or quadruple the size of the memory. The new increment will cost the same amount of money as the previous increment.

As defined herein, an LSI memory is comprised of the memory array, address register(s), address decoders, and output encoders including necessary drive. With LSI arrays, the address decoding takes a little time but once addressed, a nondestructive read occurs and the stored word register so addressed remains available. The addressed word can therefore be used for either (or both) reading and writing as though it were a register in the machine system. This is not true of magnetic memories where a nondestructive read occurs and the data must be stored immediately in some other register in order to be retained for both regeneration and for other use. LSI memory designs can be made so a bit mask is inherent within the memory to allow selective reading or writing of bits into and out of a memory word. Such a bit mask loadable directly from a stored word is proposed as an integral part of the LSI memory. Memories similar to the one described herein are being constructed by Fairchild Semiconductor under contract to the Burroughs Corporation for use in the ILLIAC IV computer. These memories replace the film memories originally planned.

A needed LSI component not now available is a special bus switch. This is illustrated in Fig. 3 as a device which allows the memory output/input to be switched onto one or more bidirectional data buses. A family of such LSI devices organized as bit slices for a number of channels should be made available. Modularity in the number of bits per bus and in the number of buses serviced should be provided. Although this function is not directly a memory function, it is so closely related to use of a memory it is included as a requirement.

Finally, a complete LSI memory function should provide for at least one dual-rank address register which is connected to a bidirectional bus. A second process-element-type register which has a counting ability built in is also needed. This allows incrementing addresses for both immediate addressing use and for storing the incremental address thus

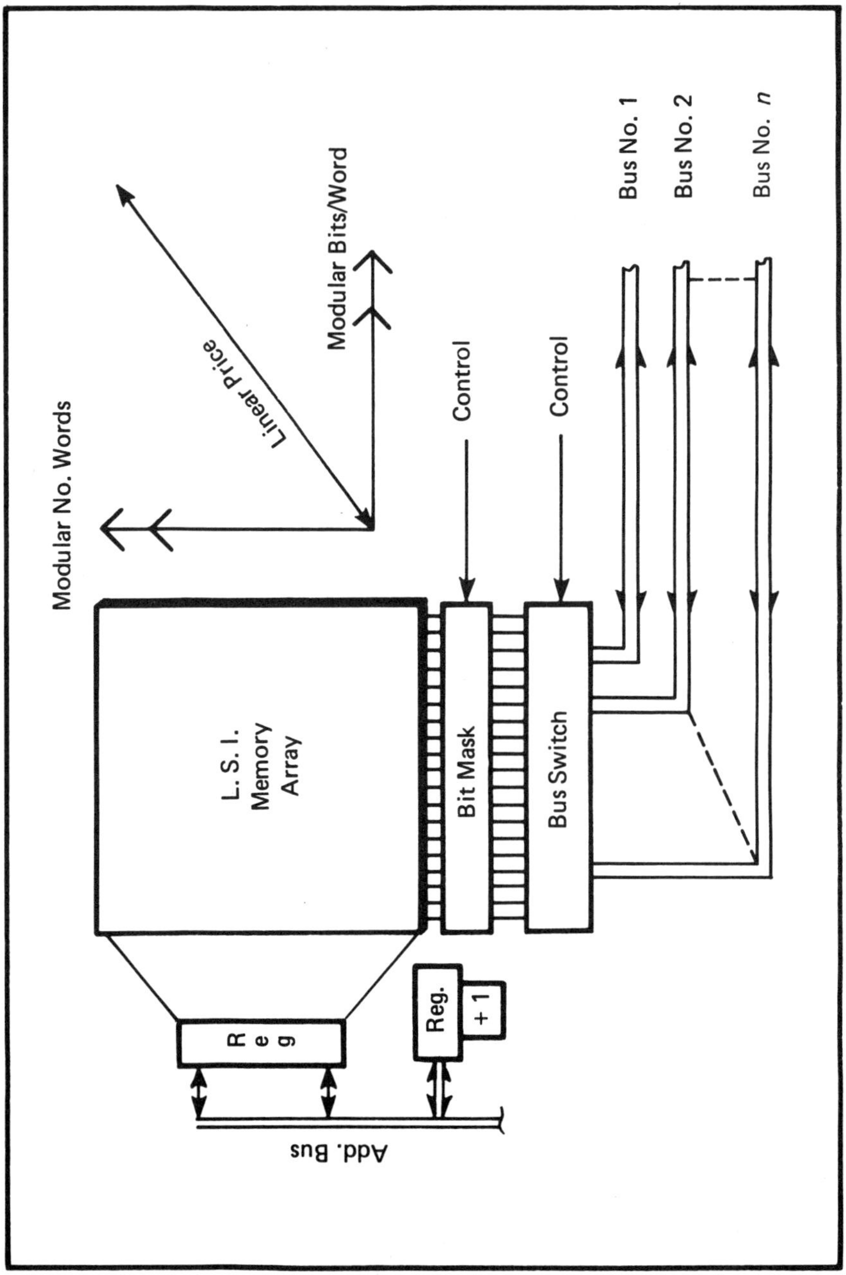

Fig. 3. An LSI Memory Operating at Clock Speeds and Having Linear Incremental Costs

obtained. The use for this "address processing function" will be presented in examples later.

LSI PROCESSING FUNCTIONS

LSI processing functions which have compatible clock cycles with fast LSI memory will provide a new dimension for architectural design. For example, Fig. 4 illustrates an LSI process element with the following specifications:

1) General:
   - *a*) LSI function; 100 < speed < 200 nanoseconds
   - *b*) 2 (or 4) bits per dual in-line package
   - *c*) Bipolar technology
2) Basic Functions (Cost Oriented):
   - *a*) And - (Bus) to (Accumulator)
   - *b*) Or - (Bus) to (Accumulator)
   - *c*) Exclusive Or - (Bus) to (Accumulator)
   - d) Not - Invert (Accumulator)
   - e) Shift - (Accumulator) left
   - *f*) Shift - (Accumulator) right
   - *g*) Add - (Bus) to (Accumulator)
   - *h*) Subtract - (Bus) from (Accumulator)
   - *i*) Load - (Bus) into (Accumulator)
   - *j*) Store - (Accumulator) to (Bus)
   - *k*) Clear - Reset (Accumulator) to zero
3) (Optional LSI functions at extra cost):
   - *a*) Fast Multiply
   - *b*) Fast Divide
4) The basic operation controls:
   - *a*) Operation Code Bus Connection
   - *b*) Local operation decoding and timing where an operation takes more than one clock cycle
5) The basic data connections:
   - *a*) Memory Bus connections
   - *b*) Shift Left and Shift Right connections.

This process unit (assume it is two-bits wide and in a 16 lead dual in-line package) is modular and can be made to match any word length. Note that the shift connections are brought out of the package. A full

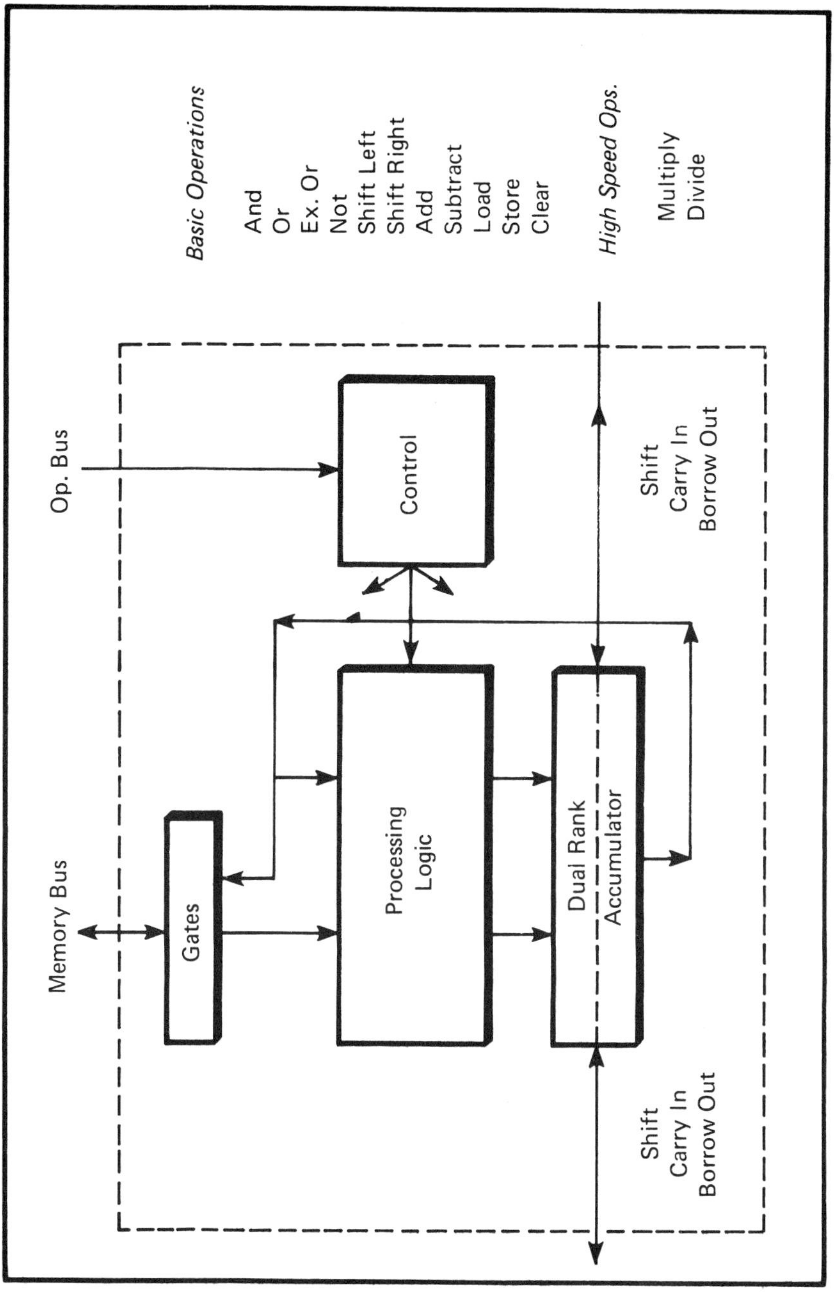

Fig. 4. LSI Process Functions Operate at Clock Speeds and Are Modular

process function for a 16-bit system would, for example, include 8 of these LSI functions, a small printed circuit board, and a few integrated circuits to complete the carry or testing operations needed for addition, subtraction, multiplication, division, and for sign control.

In a fast process unit an additional fast carry generator can be incorporated together with enough logic and registers to provide for fast multiplication and division. Internal clocking for multiple control cycles and an interlock line can be provided. Obviously this LSI function is more complex and will be more expensive. Probably less bits of process function will be contained in a single package.

Present technology has some examples of LSI functions heading in the direction indicated above. For example, the Fairchild 3800 8-bit Parallel Accumulator MOS Integrated Circuit performs many of these operations.* The author believes that although MOS is interesting for low-speed operations in the microsecond range, the true impact of LSI power will not be apparent until fast LSI (i.e., 100 to 200 nanosecond) components are available. When mixtures of core memories and LSI functions are used, MOS components can be attractive; however, when true high-speed LSI memories (probably bipolar) are available, matching LSI processing speeds are necessary. In this paper fast memories and processing components are assumed.

AN EXAMPLE PROGRAM IN AN LSI MEMORY

Assume that the LSI memory described earlier has a program and some data loaded into it. Assume further that the necessary LSI processing functions and controls are available in the system. It is interesting to study the sequencing of the program in the LSI memory. Figure 5 *a, b, c,* and *d* illustrates a simple but essentially complete instruction stream. Typical memory cycles will sequence as follows:

1) First: Obtain address of next instruction (Fig. 5*a*). Access, use, and increment stored program counter. (In modern time-sharing systems, it is desirable to have a number of stored program counters). Note that the stored address in the program counter is retained for instruction addressing while the incremental address is replaced for use on the next cycle.

---

*3800, 8-Bit Parallel Accumulator, MOS Integrated Circuit, Fairchild Semiconductor, 313 Fairchild Drive, Mountain View, California.

2) Next: Access instruction, then access data addressed by instruction, and execute operation.
   a) Type 1: Data Access Operation (Fig. 5*b*).
      i) Access instruction at location specified by program counter (before incrementing).
      ii) Send operation code to System Operation Bus.
      iii) Send data address to memory.
      iv) Access data addressed by instruction and move to System Data Bus.

      Note: Type 1 is used for "A LD" and "*B* +" in the example.
   b) Type 2: Data Store Operation (Fig. 5*c*). This operation is the same as Type 1 above except that data is written into memory from the System Data Bus.

      Note: Type 2 is used for "C ST" in the example.
   c) Type 3: Branch Operation (Fig. 5*d*).
      i) Access instruction at location specified by program counter (before incrementing).
      ii) Send operation code to System Operation Bus.
      iii) Send branch address to a memory addressing register and save for later use.
      iv) Access stored program counter and write branch address into it.
      v) Return control to next access instruction cycle.

      Note: Type 3 is used for "*X BR*" in the example.

Even with the stored program counter (which is optional in a given design), this represents only 12 memory address decoding times. Assuming a 100 nanosecond memory system, this routine takes 1.2 microseconds. Note that as soon as the memory output starts on the "*B* +" cycle, addition can start and need not be complete until the address of "*C* ST" has been accessed. In a 100 nanosecond system, this allows about 300 nanoseconds for addition. In a simple PDP-8 configured but speed oriented LSI system where the program counter is not stored, the memory access cycles drop to 8, including branching. The 100 nanosecond LSI memory systems allow a very respectable theoretical maximum instruction rate of 5,000,000 per second in a serial (non-multiplexed) system.

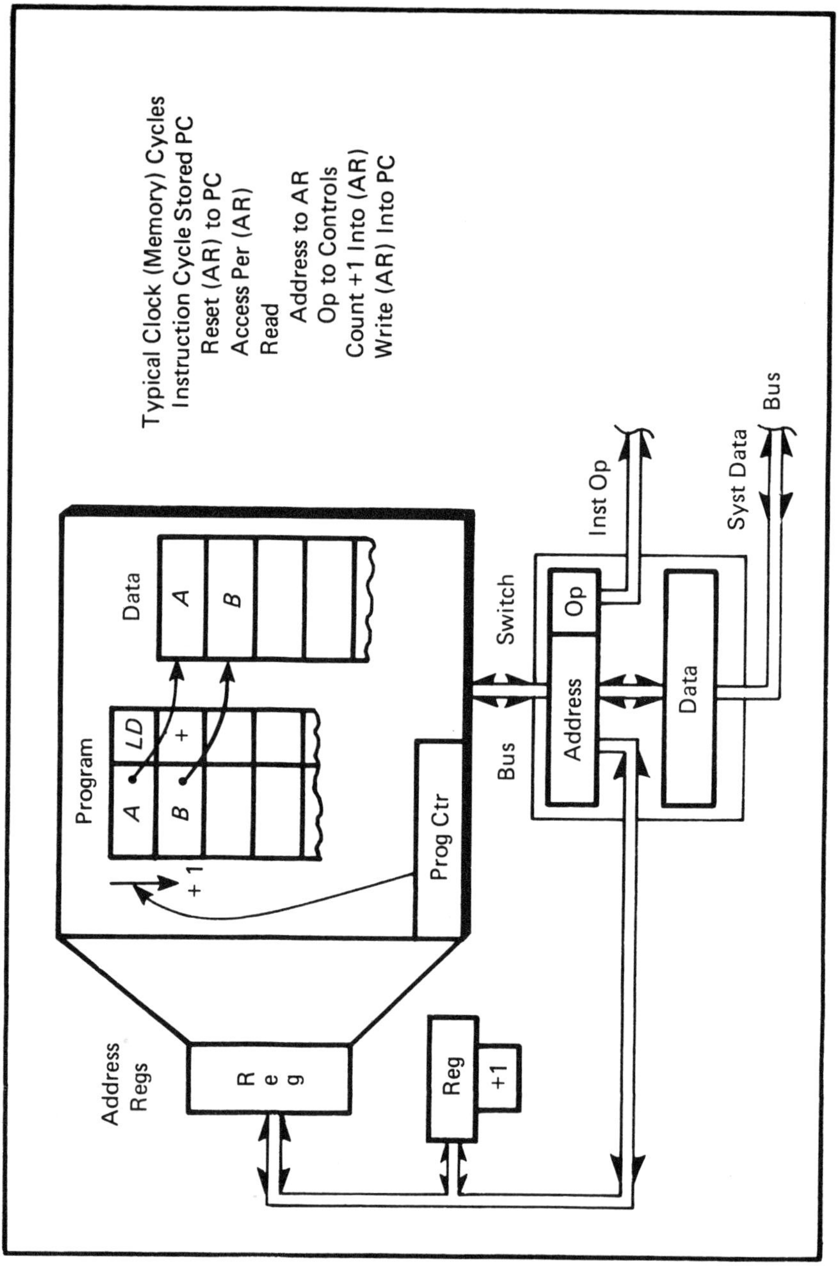

Fig. 5a. LSI Memories with < 200 Nanosecond Cycles Permit New Logical Organizations with Little or No Performance Loss

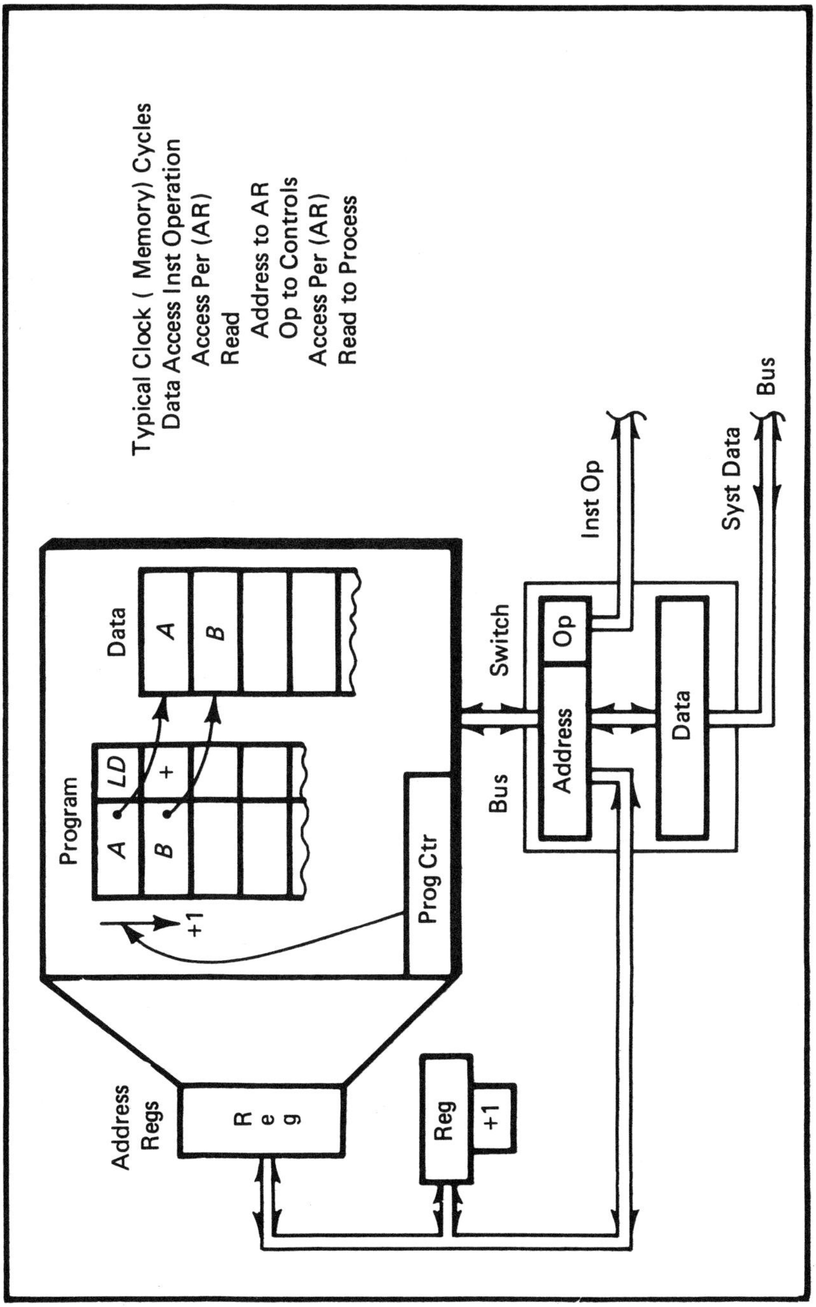

Fig. 5b. LSI Memories with $<$200 Nanosecond Cycles Permit New Logical Organizations with Little or No Performance Loss

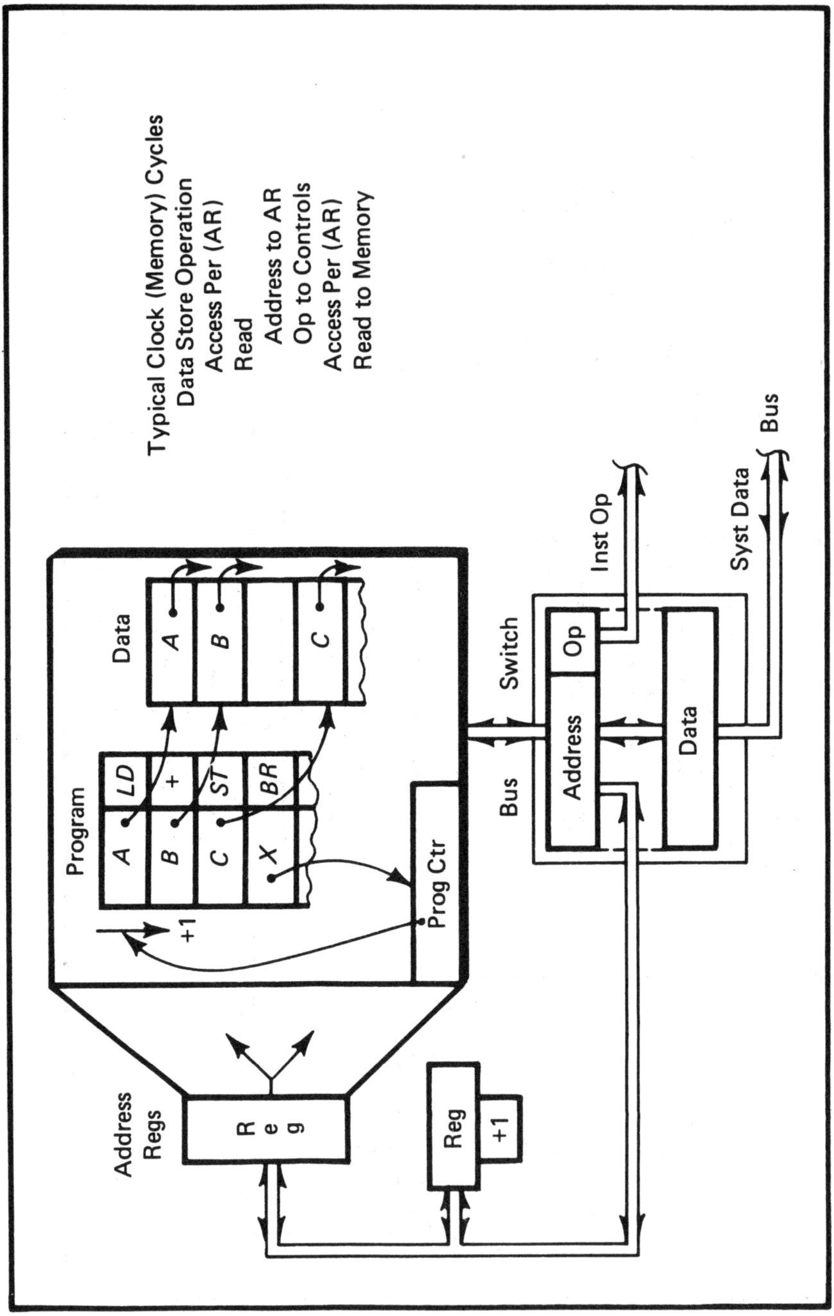

Fig. 5c. LSI Memories with <200 Nanosecond Cycles Permit New Logical Organizations with Little or No Performance Loss

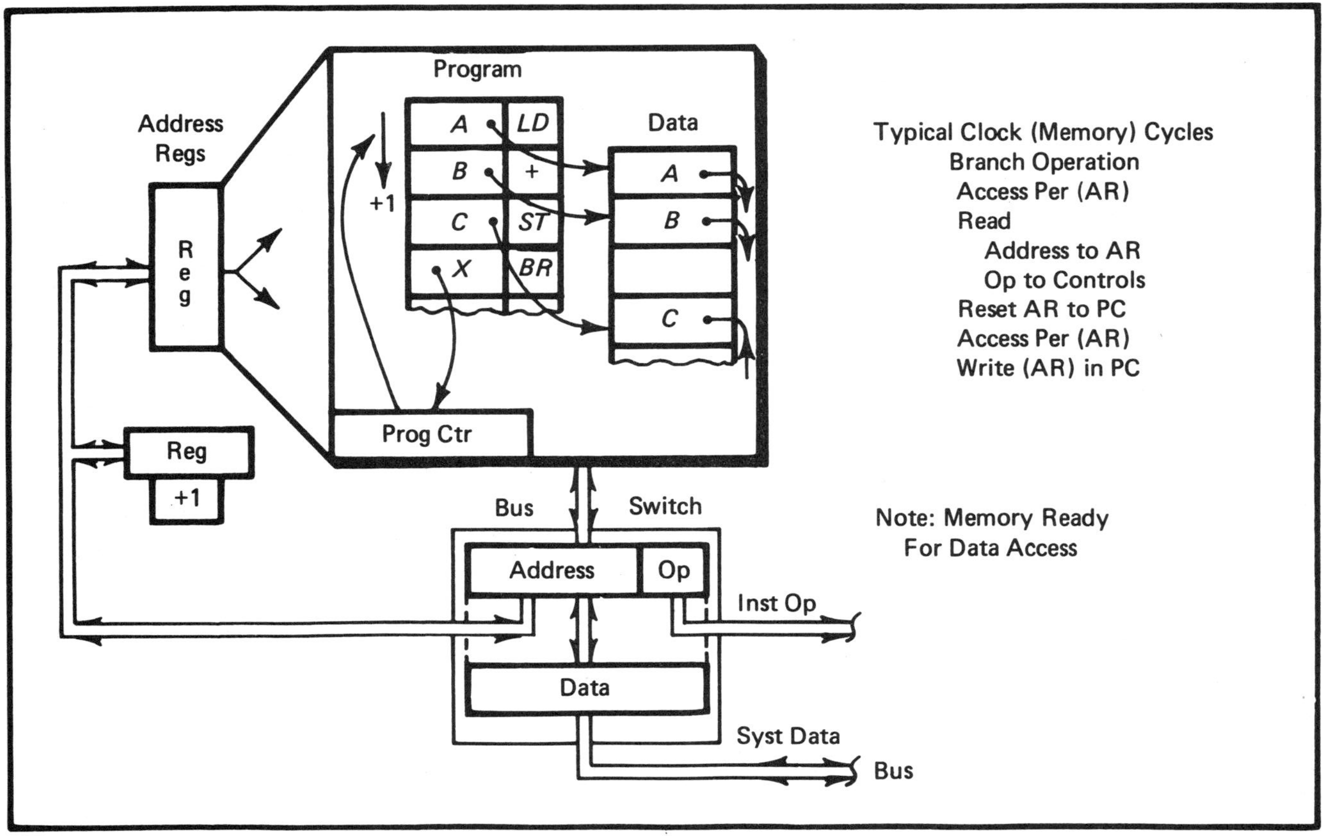

Fig. 5*d*. LSI Memories with ≤200 Nanosecond Cycles Permit New Logical Organizations with Little or No Performance Loss

In the routine illustrated the following sequence of operations occur:

Cycle		*Operation*
Instruction Address		Access PC and increment
Instruction Fetch		Access "Address of A" and "LD"
Data Fetch		A Load in Accumulator
Instruction Address		Access PC and increment
Instruction Fetch		Access "Address of B" and "+"
Data Fetch		Add B to (Accumulator)
Instruction Address	+	Access PC and increment
Instruction Fetch		Access "Address of C" and "ST"
Data Fetch		C to Memory
Instruction Address		Access PC and increment
Instruction Fetch		Access "Branch Address" and "BR"
Branch Store (PC Address)		Store Branch Address in PC
Instruction Address		- - - - -

.
.
.
Etc.

In this paper, indexing is not presented. However, a process unit designed for indexing may be used on the address bus to enhance indexing speeds. An alternative is to use the regular process unit for this operation. It will be necessary to occasionally store and retrieve the contents of the accumulator if the single process unit is used.

In the example process unit described, a combination of LSI process components and integrated circuits were assumed to be mounted on a PC board. Communication (bidirectional) to both a data bus and an operation code bus was provided.

Test/Branch instructions and other types which need to sample the sign of overflow, and other accumulator conditions can be simply implemented in the system operation controls or another functional unit. When the operation code is detected on the bus, the process unit can put the necessary bit pattern on the data bus, and sampling can occur in the control function to initiate further action. Many such possibilities exist and are a designers choice. Note that with the process functions enumerated in Fig. 4 plus testing, masking, indexing, and branching, the spectrum of necessary basic computing operations for any system are available. Similar techniques can be used to construct Input/Output functions.

AN LSI READ/WRITE MEMORY FOR CONTROL

Given an LSI memory with addressing, read/write capability, a bit mask, and bus switches, together with LSI processing functions, the next subject of interest is an LSI means to control the system. Note that each LSI function proposed connects to an operation code bus and has local decoding. (This was assumed for the memory but not explicitly illustrated in Fig. 3.) It is proposed herein that an identical read/write LSI memory can be used for microprogramming. Assume that the "basic" operation codes (i.e., +, –, etc.) can be directly used by the LSI functions and are locally decoded and executed as required. Assume other desired codes require a sequence *of basic operations* to complete the desired action, thus generating a *macro-operation.* This is analogous to microprogramming used in current machines. In a small system "add" (or "subtract") and "shift" would be basic and through coded procedures would be used to generate multiply, divide, etc.

Fig. 6 illustrates the same type LSI memory described before as a microsequence control memory. The instruction operation from the main instruction sequence is conducted through the system timing and control to the bus switch of the control memory and then to the address register. It is used to access the starting instruction of a procedure. (In some machines an encoding of the instruction to a different address may be optionally included.) This initial microsequence access is illustrated in the Start arrow in Fig. 6. The word thus accessed is brought out through the bus switch (Note: boxes in bus switch are merely to illustrate information transmitted and are not registers) to the System Operation Bus. As soon as the code is available, the address register of memory is advanced by +1 and the next microcode is brought out for use on the next instruction cycle. By this technique a program of operation codes of any desired number can be produced. To completely generalize the system it is necessary to give the microinstruction the ability to branch. This is accomplished by a stored Microbranch suboperation code which switches the operation bits directly to the address register to be used as an address, thus causing a branch. The information and arrows shown on the left of Fig. 6 illustrate the actions required. Finally, a stored "Microstop" code is used to return control to the programmed instruction stream.

The Read/Write type LSI memories will have tremendous advantages in control systems. The microinstruction procedures

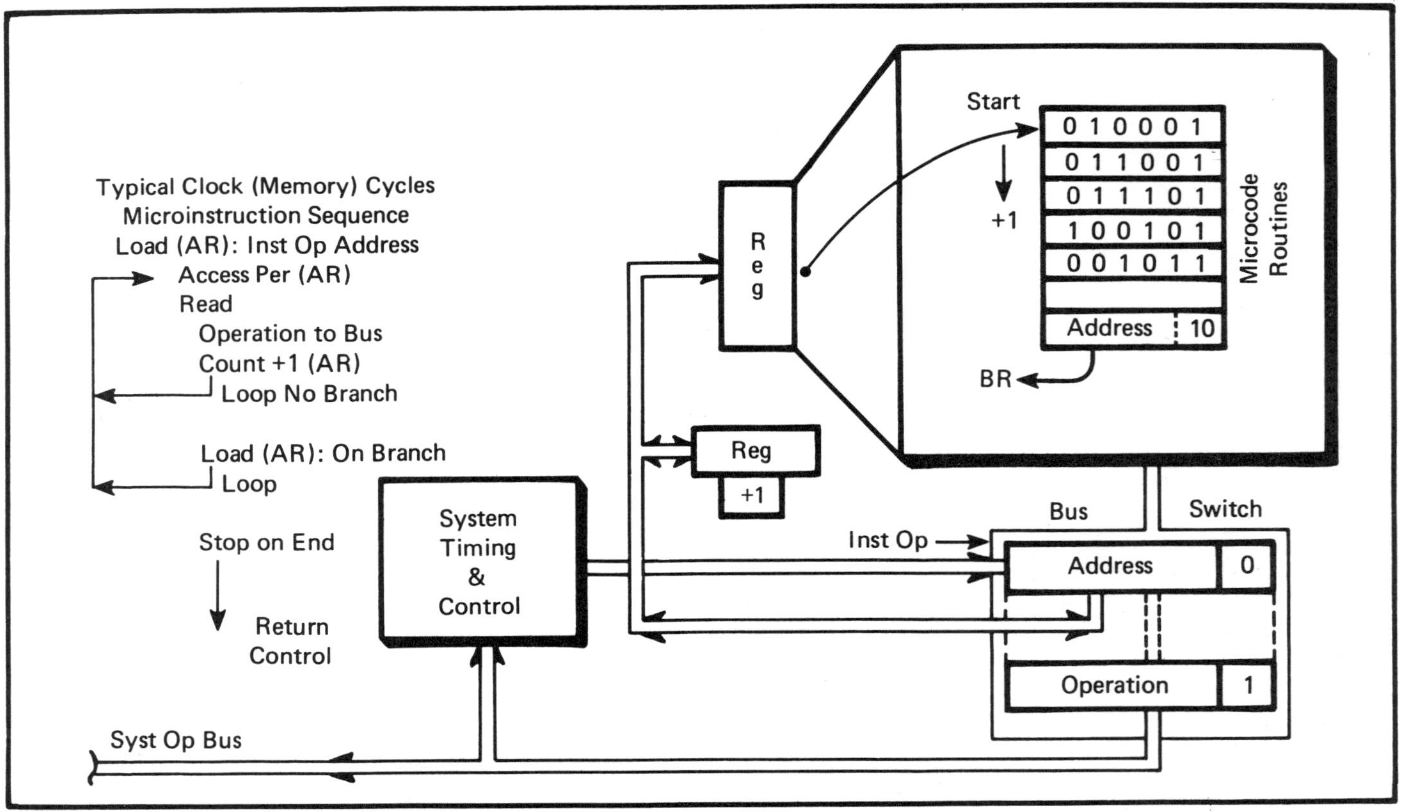

Fig. 6. LSI Memories with <200 Nanosecond Cycles (without Change) Are Excellent for Microprogram System Control

can be paged in and out of the memory "at will," thus giving a tremendous boost in flexibility and capability over the read-only control memories. The perennial problem of dynamic language improvements is thus enhanced through this type of LSI system.

LSI COST ORIENTED SYSTEMS

Let us next construct some example systems using the previous components. These are hypothetical and greatly simplified systems. As shown in Fig. 7, the author proposes that LSI components in the 100–500 nanosecond range can replace small and medium computer systems. The components used together are an LSI memory, a bus switch, a basic process unit, and a basic control (probably constructed out of regular I.C.'s.) This configuration could be compared to a machine class which is roughly comparable to the PDP-8 type machine. One major difference exists. This is a PDP-8 type machine with 200 nsec complete operating cycle, not a microsecond or greater operating cycle. This LSI COS, that is Cost Oriented System, will store both the programs and the data together with the microcontrol in the same memory. Competent designers will be able to compress both the microcode memory and the main memory into the same addressing elements of hardware. Since LSI is linear in cost and modular by word, we can obtain any size necessary to get the data and instruction programs as well as microcodes in the same memory. This is desirable in a small machine since by paging all three in and out we can minimize the total amount of memory required for a given application. In further examples, we will refer to this as an LSI COS, or Cost Oriented System.

LSI SPEED ORIENTED SYSTEMS

Identical LSI component designs in either the same speed range or in a faster speed range can be assembled to make LSI Speed Oriented Systems. In Fig. 8, the author proposes that LSI in the 50–150 nanosecond range with parallel architecture will replace fast and large systems. Here in the Speed Oriented System, we may use multiplexed LSI memories for overlapped access and the fastest LSI processing functions. LSI permits memory modularity in the number of words and in the number of bits/word, thus allowing a wide range of speed/size choices.

In the Speed Oriented System, the program instructions and the data may be stored in one memory and the microinstructions in another so that multiplexing of the two will give faster operation.

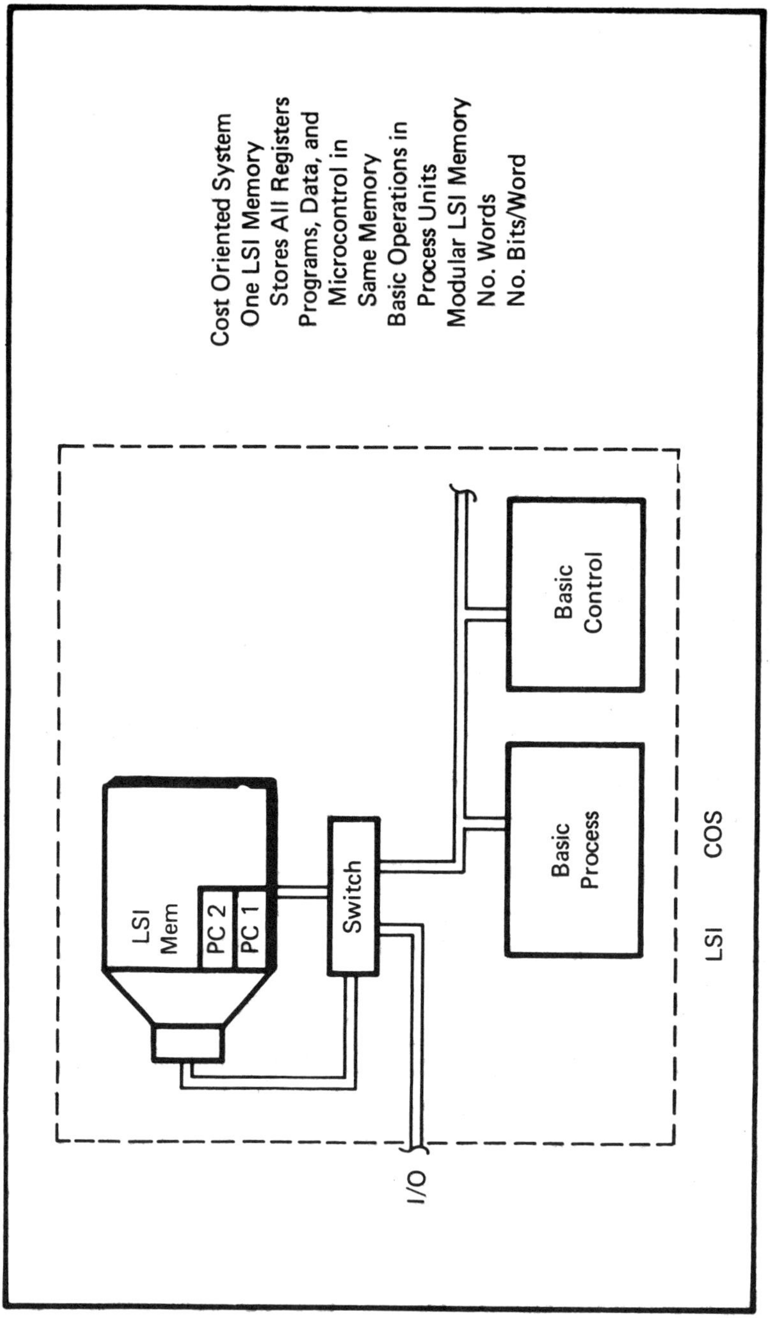

Fig. 7. LSI in 100-500 Nanosecond Range with "Serial" Architecture Will Replace Small and Medium Systems

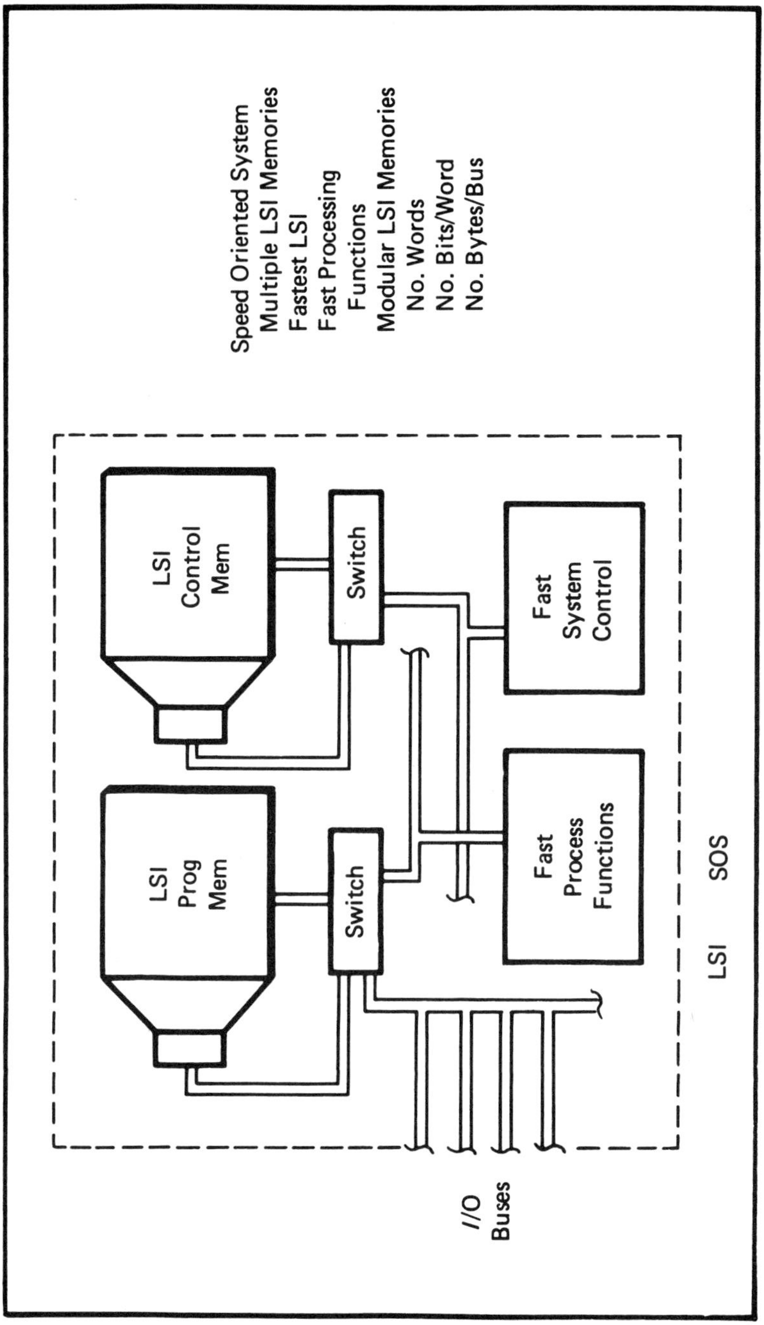

Fig. 8. LSI in 50-150 Nanosecond Range with "Parallel" Architecture Will Replace Fast Systems

Obviously, a design option would be to store interchangeable program data and microinstructions in both memories and to multiplex all operations. These variations are left as designers choices which can be obtained with the use of LSI memory. In Fig. 8, several bus switches are shown connected to fast process functions and fast system control. Although one of each type of processing function is shown, it follows that specialized process functions could be put on the data buses and operation buses and would be responsive to particular operation codes so that paralleled or specialized operations could be obtained as desired. Note also that a stacking of standard bus switches can be used to get multiple input/output buses into the Speed Oriented System. As stated earlier, the author only dimly perceives the variety of desirable logical organizations which can result from the proper combinations of these and other building blocks.

LSI MATRIX COMPUTING SYSTEMS

A different arrangement for using LSI computers is illustrated in Fig. 9. Here, LSI Speed Oriented Systems may be configured into a matrix computer organization (for example, the ILLIAC IV system).[4] Here, an array of LSI Speed Oriented Systems are illustrated as rectangles with the lower attachment to each rectangle representing a bus switch. In an ILLIAC type system, each computer communicates with a north, a west, an east, and a south adjacent computer. These connections are shown as NS, EW bus connections going between the bus switch which is a part of each computer. In addition, in an ILLIAC type system, it is necessary to get data from an external large store into the array so that each computer can be loaded when it is appropriate. This is shown as a bus traveling horizontally past rows of LSI-SOS computers and connected into each appropriate bus switch. Thus, the same kind of LSI components discussed herein can be organized into matrix computers as easily as we have seen them configured into other forms of high-speed or low-speed systems.

LSI CDC TYPE SYSTEMS

Another very interesting class of machine organizations is typified by the 7600 and 6600 type systems produced by Control Data

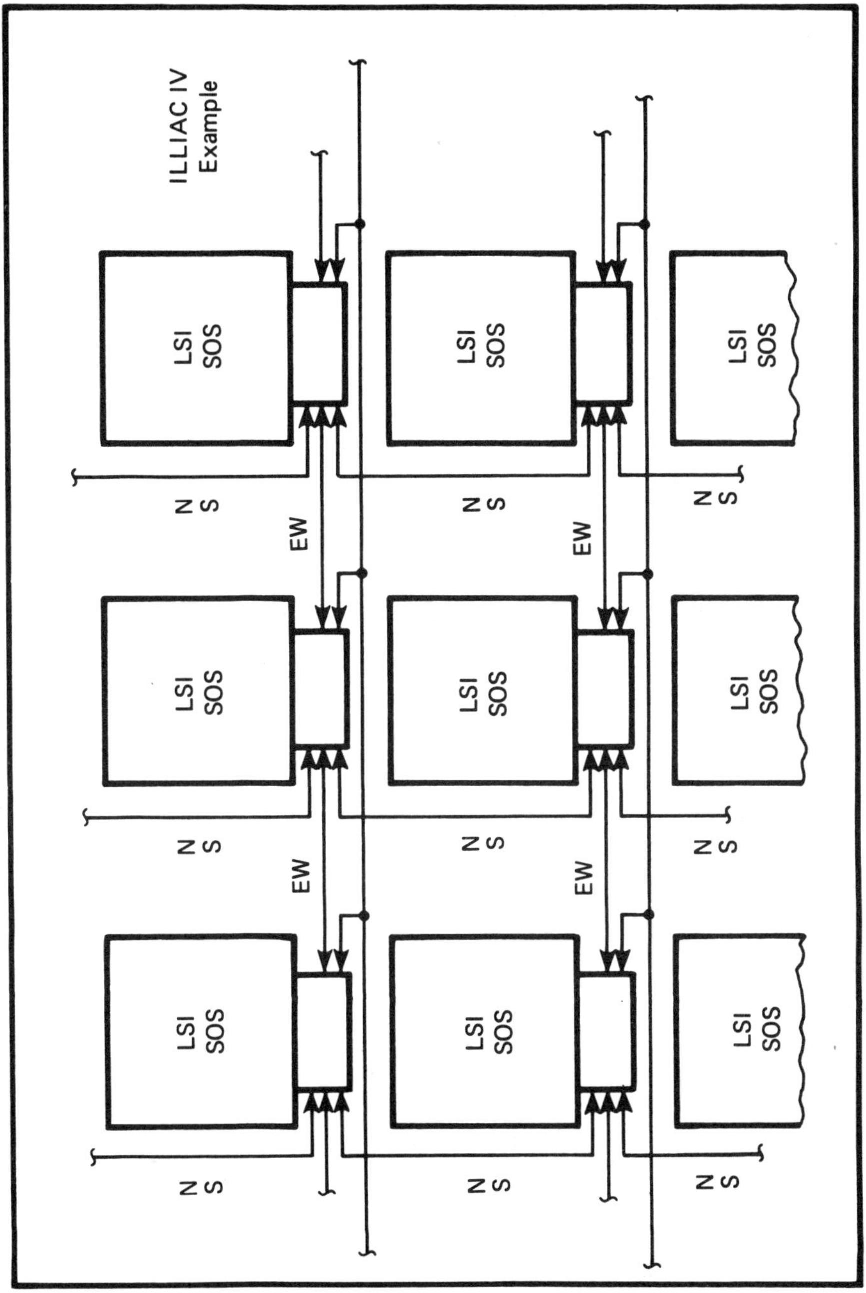

Fig. 9. LSI Speed Oriented Systems May Be Configured Into Matrix Computer Organizations

Corporation (Fig. 10).[3] LSI Speed Oriented Systems may be configured into processing elements for CDC-type large and fast architecture. LSI Cost Oriented Systems can be organized for use as the peripheral process units working into a stack of bus switches. In the CDC type organizations, it is desirable to have a very large inexpensive memory which is not necessarily very fast. This memory is illustrated in this particular figure as a Large Core Array also working into the bus switches. Out of the bus switches may come a multiplicity of data channels which are connected to a series of multiplexed LSI memories to get overlapped operation. The system control could be made up of one (or more) LSI-SOS computers configured with appropriate bus switches to direct operations to other paralleled LSI-SOS computers doing processing tasks. The standard Control Data configurations are normally established as they are because the magnetic memories have a hierarchy of speeds. In this particular system, the main memory and system clock cycle speeds of all of the main portion of the computer could be identical; therefore, the designer should adjust his thinking about architecture to take advantage of the new speeds provided by LSI. In other words, the author believes LSI gives other advantages not presently available when using core memory and regular circuit components.

CONCLUSION

The author believes that a selected small number of appropriately designed LSI functional elements can be constructed into a large variety of computer systems. These can range from small, low-cost systems which approximate 100 nanosecond PDP-8 type systems to a very large high-speed serial or parallel systems. Very large systems typified by CDC type designs or matrix systems like ILLIAC IV may be assembled from LSI components. Since systems at both ends of the size/speed scale can be assembled, it follows that many other very useful, interesting and economically attractive system organizations lie between these extremes. The usefulness of this more or less standard set of LSI components coupled with a small number of integrated circuits is limited only by the imagination and desire of the system architect attempting to put together new machine organizations (See Fig. 1).

The preparation of this paper to discuss the relationship of LSI and parallel processing turned out to be a pleasant surprise and provided some architectural concepts for the author. It was surprising to realize

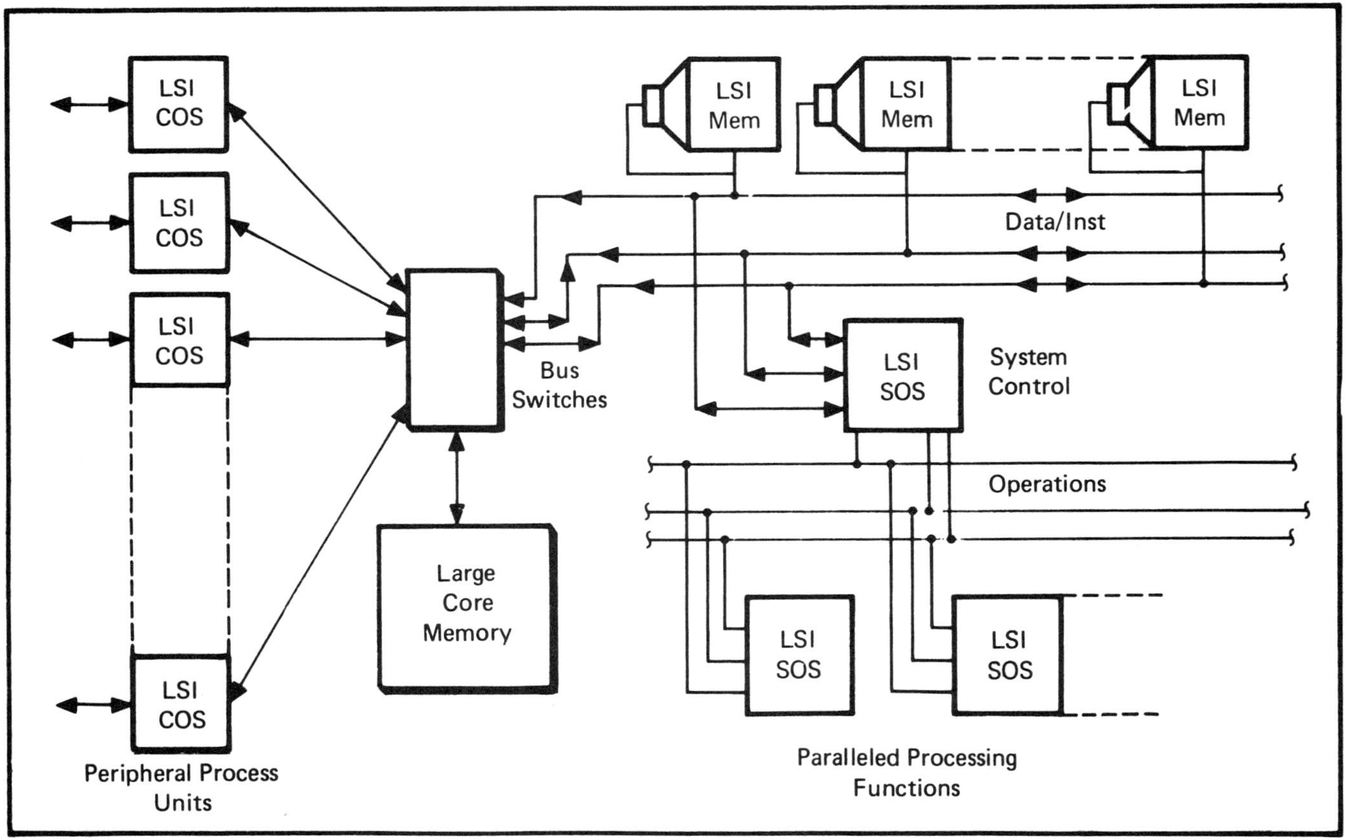

Fig. 10. LSI Speed Oriented Systems May Be Configured into CDC-Type Large Fast Architecture

how few types of LSI components were needed. The concept of functional regularity and bus oriented systems through the use of standard functions is extremely powerful. The author was further surprised at the variety of systems which could be constructed from a very elementary set of "off the shelf" building blocks. The performance of any of these hypothetical systems was not only respectable but surprisingly fast.

References

1. R. Rice, "Impact of Arrays on Digital Systems," *IEEE Journal of Solid State Circuits,* SC-2: No. 4 (December, 1967).
2. S. Levy *et. al.,* "System Utilization of Large-Scale Integration," *IEEE Transactions on Electronic Computers,* EC-16: No. 5 (October, 1967).
3. P. Bonseigneur, "Description of the 7600 Computer System," *IEEE Computer Group News,* 2: No. 9 (May, 1969), 11.
4. G. Barnes *et. al.,* "The ILLIAC IV Computer," *IEEE Transactions on Computers,* C-17: No. 8 (August, 1968).

CHAPTER 9
INTERCONNECTION/ECONOMIC TRADEOFFS WITH LSI

Wendell B. Sander

Fairchild Camera and Instrument Corporation
Palo Alto, Calif.

If interconnections were free, LSI would be economically uninteresting. To see that this is true, consider that the best yield in the silicon sense is with the smallest die. Assuming interconnections are free, then the only cost remaining is the silicon cost and the lowest cost will occur using very small silicon chips. Therefore, economically there would be no point to large-scale integration. There may still, under these assumptions, be interest due to other reasons such as reliability, size, weight, etc.

In considering the realities of interconnections, there is the cost of mounting the silicon chip in its package and then assembling the package into the total system. The system packaging cost is roughly proportional to the number of pins to be connected, and component packaging cost is roughly proportional to the number of pins on the package. Therefore, the cost of interconnections in the system is proportional to pin count. To get a better feel for this, consider a two-sided (i.e., layers of copper) printed circuit board with plated-through holes. The spacing between the dual in-line packages in such a configuration is determined by the number of leads required to interconnect the system. Certainly using more layers will allow placing the packages closer together. However, this, in turn, increases the cost. Therefore, for a chosen system interconnection technology, in this case two-sided boards, the area of board required is proportional to the number of packages. If each package contained twice as many pins and were twice as long, it would take at least twice as much board area to interconnect them, and possibly even more area because the placement problem is more difficult for optimum wire routing. The assertion that component packaging cost is proportional to the number of pins is more difficult to prove and possibly a less accurate assumption; however, in practice it is a reasonably good assumption.

Thus in assessing the cost of using an LSI chip, we can now look at the total system cost involved where the cost items we need to deal with are:

1) the number of LSI package pins and packaging cost per pin
2) the cost of silicon used to create the system where the cost of each silicon chip is a function of the area of the chip in the logic function of interest.

Therefore, cost of an LSI module in the system is the cost of the silicon plus the system cost per pin times the number of pins on the logic function.

To study LSI economic tradeoffs we will consider:

1) the silicon area used per pin for several classes of LSI modules
2) silicon area per pin vs. function per pin
3) function per dollar vs. pins per package

To avoid problems in technology dependence of silicon area, we will define a unit of silicon area which is technology free as follows: *A grain of active silicon area is defined as the minimum active silicon area required to form a general purpose three-input gate with a given circuit and semiconductor technology.* With this definition we can now look at the relation between the function, the silicon area to perform the function, and the number of pins required by that particular class of function. Figure 1 plots silicon area versus pins for several classes of function. First, assume a function such as memory. Then for a given number of pins we can establish how much silicon area is consumed by the maximum number of bits of memory that can be put into that package. Since the amount of function contained in the package is related to the amount of silicon in the package, this then is a first step toward getting a handle on the amount of function per pin.

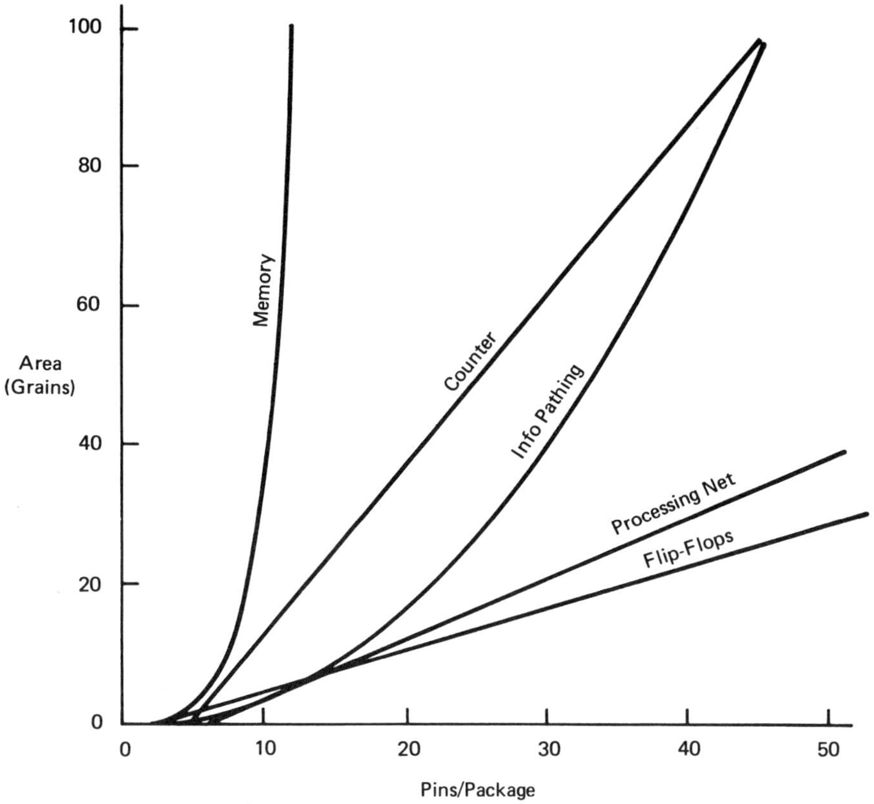

Fig. 1. Pins/Package vs. Area for Several LSI Function Types

Looking in detail at Fig. 1, it can be seen that a memory function with very few pins can contain a great deal of silicon. This is because each time a pin is added on a memory package and is used for addressing, the amount of memory that can be put in the package is doubled, and therefore double the amount of silicon can be used.

At the other extreme, with flip-flops, at least four pins are required for each extra flip-flop. Here a linear relationship exists between pins and area because adding a flip-flop adds a fixed number of pins and a fixed amount of area. Other linear networks in the pin vs. area sense are processing networks such as an adder or a counter where the slope is steeper than for general flip-flops. This is true because for each flip-flop added in a counter only one pin is added. Thus, more useful silicon area per pin can be contained within a counter than within general purpose flip-flops. Information pathing as illustrated in Fig. 1 is a network which accepts data paths in and distributes data paths out with a crossbar-like interior network. With such an element, the area tends to go up as the square of the number of pins because adding one pin would add a row and the next pin add a column.

A meaningful relationship between silicon area to quantity of function can be defined as follows: *A quantity of function contained in a given module or component is defined as the minimum grains of area required to perform the exact function desired of the given module or component in the given application.* Note that in general a component is not idealized for the application so that it contains more grains of area than the minimum necessary for the function. Also, a quantity of function has meaning only in a given application and with a given component. To get a better understanding of function as it relates to pins and area, Fig. 2 illustrates the case of generalized logic. Here a particular logic network is assumed so that a specific optimum solution is possible. Now consider the pins per package versus the area contained in each of those packages for three cases.

First, the optimized logic solution determined by designing specifically the logic necessary to do the job in the most efficient manner gives the optimized logic curve of Fig. 2.

Second, the read-only memory solution which has the characteristic that the package will contain the same amount of function with the same number of pins as the optimized solution except that more silicon area is required. Thus, silicon area and quantity of functions are not equal.

The third solution is to use individual gates and interconnect them as necessary to achieve the logic function. In this case, all of the silicon area is used to give function so that the quantity of function equals the quantity of silicon; however, the function per pin is very low since many more pins are required than in the optimized logic solution.

The cost of an LSI package in the system is determined by two parameters: the pins on the package and the silicon area required to accomplish the function within that package. We have seen that on a total system basis an assumption of a uniform cost of pins per package is a fairly reasonable assumption. The silicon cost per unit area for various sized chips has been fairly well defined by B. T. Murphy.[1] Using these figures and these characteristics for silicon yield and a fixed cost for the number of pins, then knowing

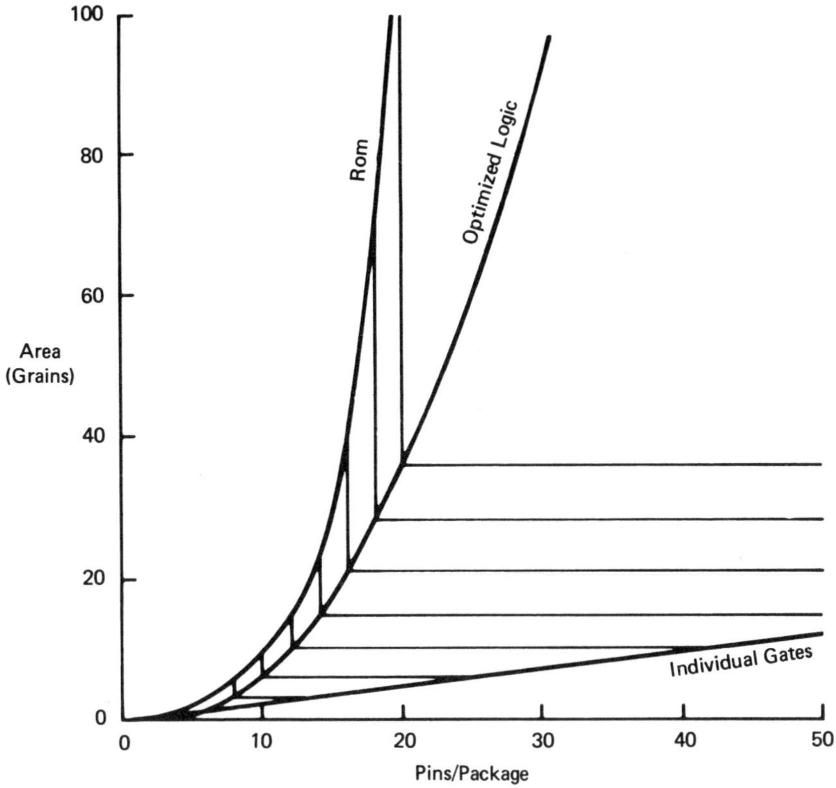

Fig. 2. Pin/Area Tradeoffs for Logic Implementation

for any given package its chip size and number of pins, we can make an approximation of the cost or at least get an idea of the shape of the cost function for these pins. The function per dollar, then, is simply the amount of function the package contains divided by the cost of the package in the system which is related not to the function area but to the true silicon area within it. Taking the case of the logic net as is described in Fig. 2 and looking at the function per dollar, we can arrive at the curves of Fig. 3. This figure is obtained by simply taking the curves of Fig. 2 for read-only memory, optimum logic net, and gates, and converting the points on that curve to function per dollar. The gate curve is nearly flat. What this means is that by adding gates to the package we really gain nothing in terms of function per dollar. This would be expected to be true of all functions that were linear in Fig. 1 because whether you make 4 gates in a package with 16 pins, or 8 gates in a package with 32 pins, the cost per gate is not affected very much. The read-only memory solution and the optimum logic net solution in this case were fairly close at small numbers of pins. However, the optimum logic net showed an advantage at larger numbers of pins of about 2:1. It should be stressed that tradeoffs for logic are very sensitive to the choice of the particular logic being considered. It is not necessarily safe to say that a read-only memory is a good solution for general logic problems. However, it does say that read-only memories can be a good solution for logic problems and it is probably more efficient in function per dollar to deal with a number of small read-only memories in small packages than a few large read-only memories.

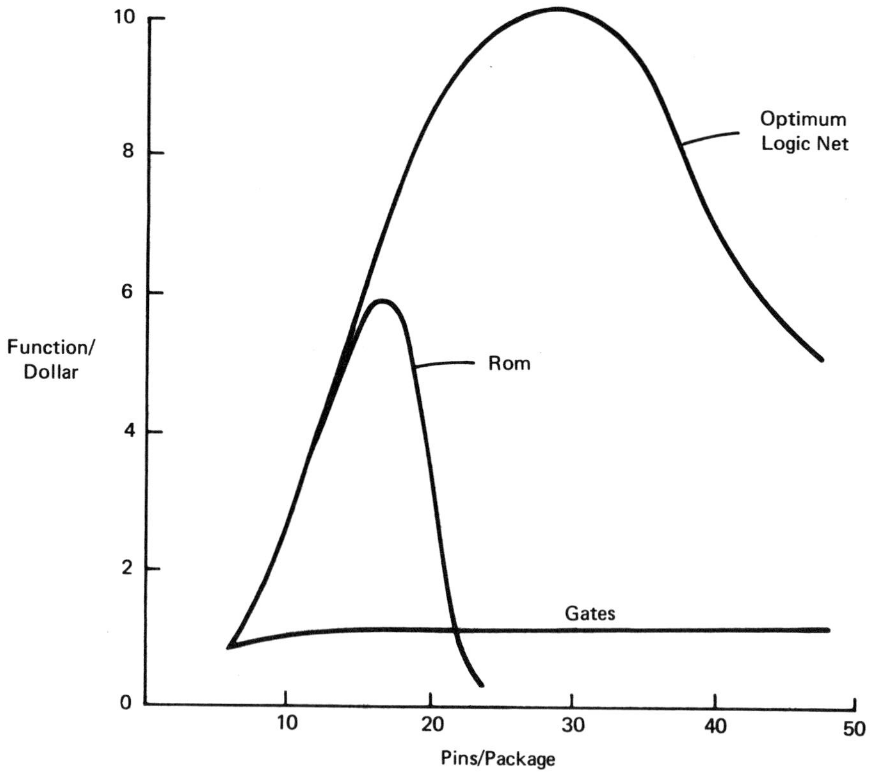

Fig. 3. Function/Dollar vs. Pins/Package for Logic Implementation

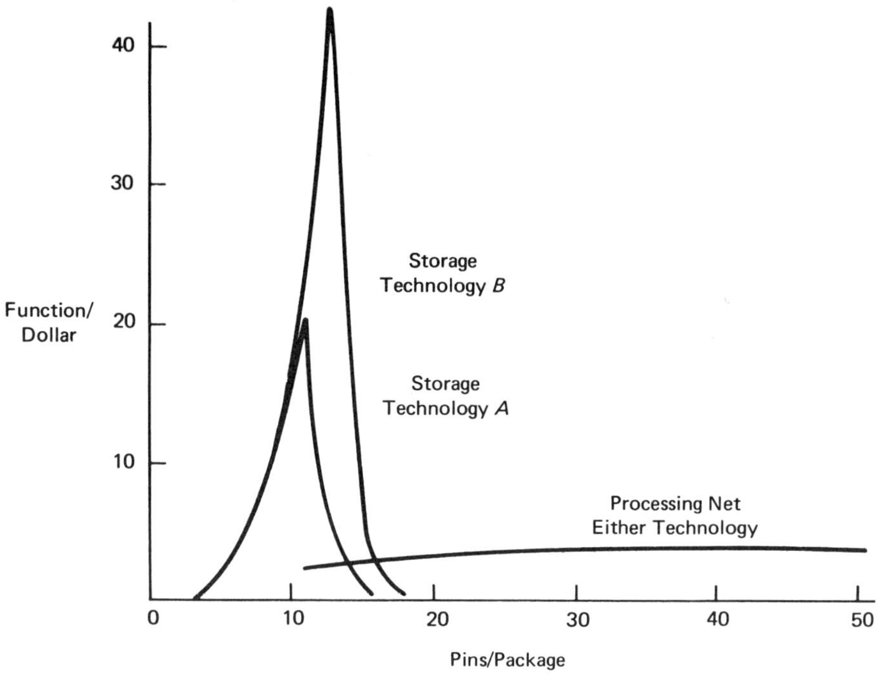

Fig. 4. Function/Dollar vs. Pins/Package for Two Functions with Two Technologies

The curves of Fig. 3 tend to rise and then fall. The early part of the curve is almost entirely pin dominated in cost; i.e., the cost of the silicon is small with respect to the package. In this region, the read-only memory and optimum logic net follow exactly the same curve since they use the same number of pins. As the function is made larger, the silicon area gets so large that the chip gets expensive, and this region is reached earlier with the read-only memory than the optimum logic net.

Figure 4 illustrates function per dollar for two of the curves in Fig. 1. In this case, function per dollar vs. pins per package is plotted for two different technologies for both the storage function and processing net function. In the case of the processing net, either technology leads to the same curve because the package is pin dominated even up to 50 pins. However, in the storage technology we very quickly reach the point where the chip cost dominates the package cost. The assumed difference in technology for this example was about a 4:1 variation in process yield. This is roughly the variation that might be expected in a running semiconductor production line. Note that as you get past the pin-limited part of the curve into the chip-limited part you can reach points where the function per dollar varies by an order of magnitude with process variation. This is the variation the manufacturer sees in cost per dollar operating in that region. Thus, it shows something of the hazard of pushing the technology too far, because it can mean that the production of such a device is an extremely variable parameter and that the cost to the manufacturer to make the device is highly unpredictable. This explains many of the situations where a part is built in the laboratory and a few made in production, but ability to produce in volume is extremely difficult to achieve.

Figure 5 illustrates the effect of taking different functional approaches to the choice of LSI. In this case, we are contrasting the use of separate processing net functions and flip-flop functions in terms of function per

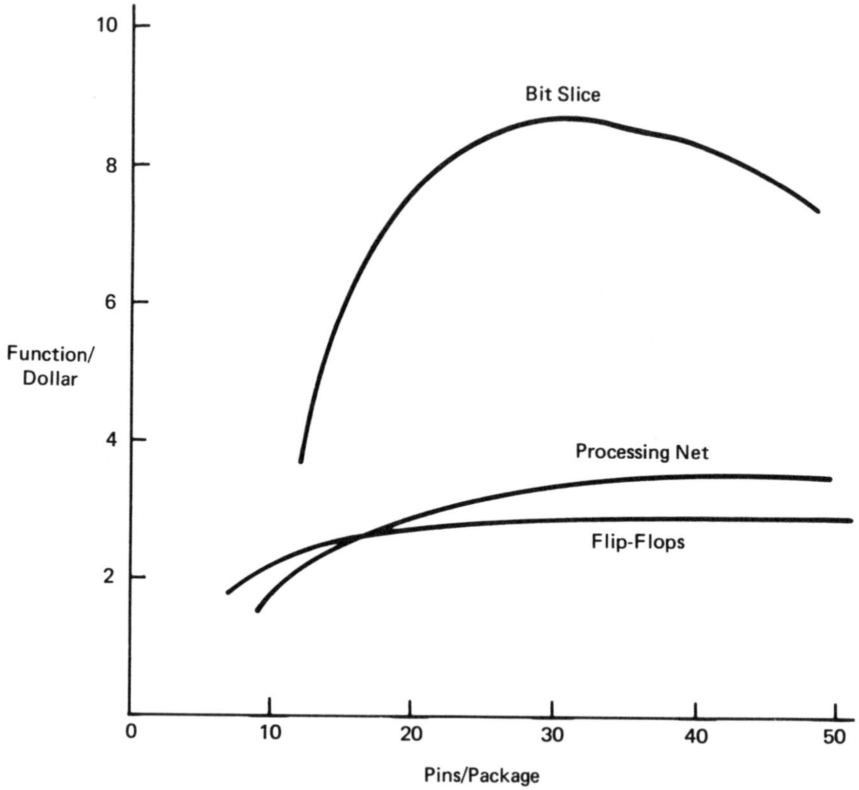

Fig. 5. Function/Dollar vs. Pins/Package for Two Approaches to System Design

dollar vs. pins per package against the alternative of combining these together to form a bit/slice approach to the total functional processing network. The separate processing network and flip-flops are less efficient by a factor of about 3:1 in terms of the amount of function gained per dollar spent. Observe also that function per dollar is fairly constant for a wide range of pins per package. Therefore, there is not a great deal to be gained in this case by increasing the number of pins per package. In fact, in general the gains beyond 24-pin packages are not particularly dramatic. There turns out to be very few cases where the extra pins can be effectively utilized in the function per dollar sense with single-chip technology as we use it today.

In conclusion, the key insight that can be gained from this analysis is that functional design approaches can provide significant gains in function per dollar. Storage is one of the most efficient function per dollar producers of all LSI functions, as was intuitively clear.

Also, LSI does not necessarily imply enormous number of pins per package to provide good function per dollar. The cost of custom design was not considered in this analysis. The philosophy here is that within these constraints of economics, and within the system design constraints, it is necessary for the component manufacturer to devise a family of standardized functions that can be sold at a high enough volume to allow the function per dollar economics shown herein to be meaningful.

Reference

1. B. T. Murphy, "Cost-Size Optimum of Monolithic Integrated Circuits," *Proceedings of the IEEE,* 52: (December, 1964), 1537–45.

SYSTEM AND MACHINE ORGANIZATION

CHAPTER 10
THE ADVANCED AVIONIC DIGITAL COMPUTER

Ronald S. Entner

Naval Air Systems Command
Washington, D. C.

The Advanced Avionic Digital Computer (AADC) development program represents an organized attempt to establish an integrated digital computer system capability for naval airborne application in the 1975–85 time frame. The AADC, as it is termed, is the title of a class of incrementally variable computer systems, all configured from a limited number of standard, byte-functional modules. These modules are designed to take maximum advantage of forthcoming monolithic LSI technology. Each byte-functional module will perform a major computer function, such as an arithmetic or control function, on single bytes of data. Full word processing will be performed by groups of parallel stacked byte processors, or by the recirculation of byte data within single modules. It is believed that this concept will permit the fabrication of various avionics digital computer systems, each having a different basic word length, yet assembled from the same set of standard modular building blocks. By "standard," it is meant that the electrical and mechanical interface, as well as the functional performance of each module, will be defined well enough to insure interchangeability within type. Figure 1 illustrates an arithmetic and control unit configured with byte-functional modules.

The computer organization depicted in Fig. 2 represents the baseline AADC macroorganization as of March 16, 1969. It is believed that this

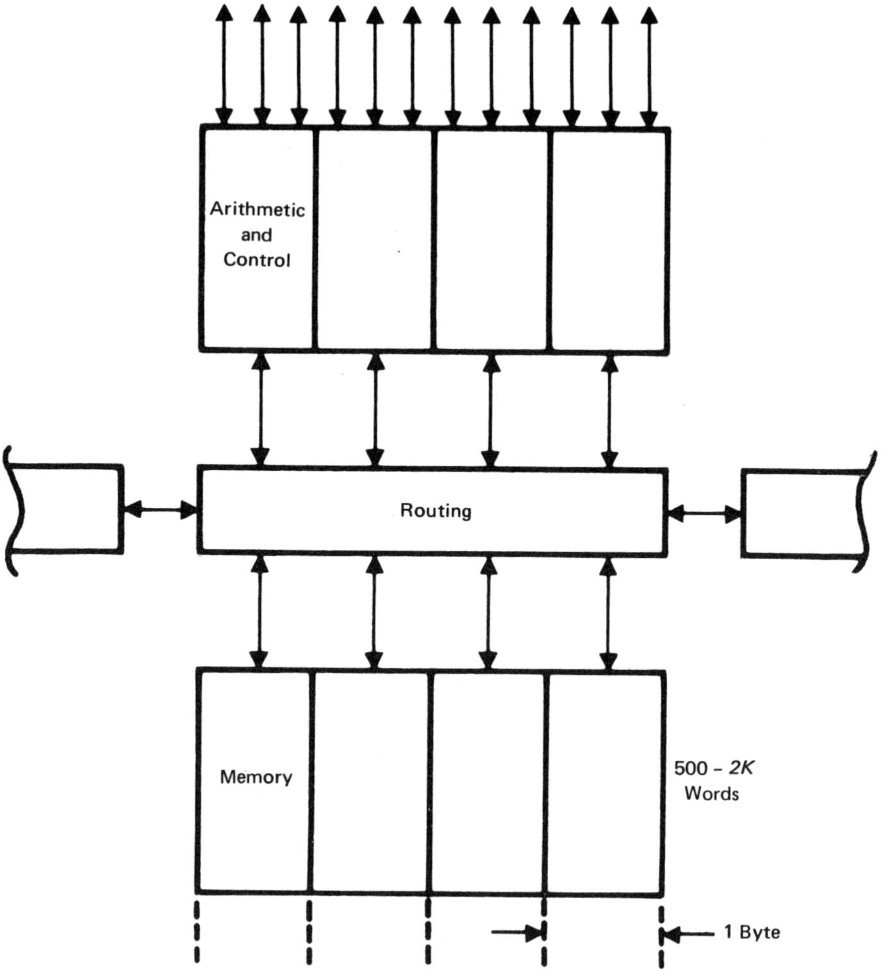

Fig. 1. Processing Element Detail

organization embodies those qualities needed to meet, as a minimum, 1975–85 naval airborne computer system requirements. The effectiveness of this organization is based, in part, on a functional distinction between sequentially organized problems, such as weapon delivery, navigation, communication, self-check, etc., and parallel organized problems, such as fast Fourier transformation, target signature recognition, and Kalman filtering. The sequential problems are handled by the *processing elements* (PE's) shown in the block diagram. Each PE

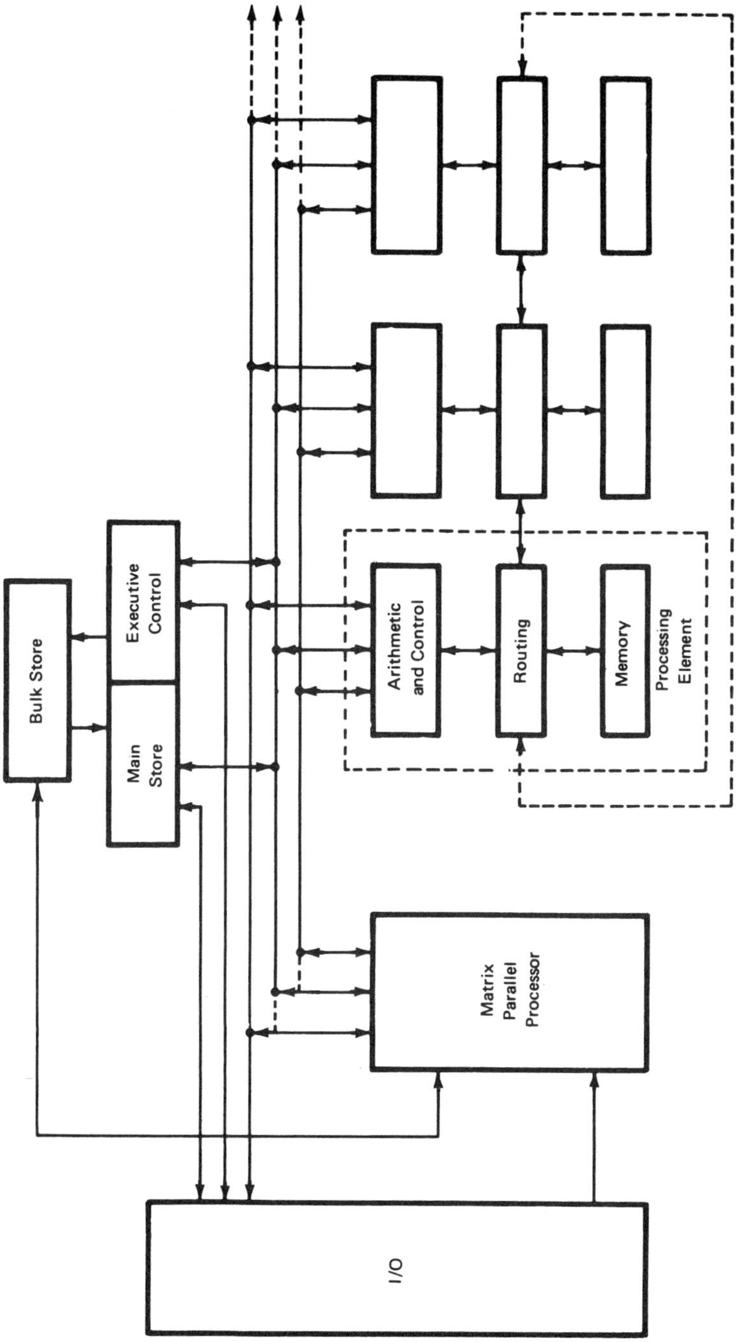

Fig. 2. Baseline Computer Organization

contains sufficient memory (approximately 500 to 2,000 words) to efficiently store and process whole problem macroroutines, or *program modules.* This block processing eliminates continual main memory access and, hence, most conflicts usually associated with multiprocessing. Parallel organized problems are handled by the programmable *matrix parallel processor.* This device consists of a combination of fast Fourier processor, associative processor, and associative or pseudo-associative memory. These three devices will be interconnected by switching logic controlled by the *master executive control,* (MEC), a processing element, or an internal microprogram controller, and will be both modular and programmable in nature.

The baseline performance characteristics of the AADC are a function of the size, speed, and complexity of the various byte-functional modules. If the following characteristics are assigned each major functional block:

main-store memory	:	250 *n*sec cycle time
processing element	:	250 *n*sec add time
		1 μsec multiply time
		(PE memory access time included)
number of PE's	:	5
operation mix	:	80% addition
		20% multiply

the basic operation rate of the multiprocessor section, taken alone, will be in excess of 10,000,000 OPS/sec. The operating rate of the matrix parallel processor will be typically on the order of $10^7 - 10^8$ OPS/sec.

Key to the successful operation of the baseline organization, is a master executive control capable of coordinating the powerful computational capability of the AADC. Figure 3 illustrates one scheme presently under consideration for the MEC. The scheme employs an associative memory to store all data pertaining to each program module (PM) stored in the *random access main-store memory* (RAMM), including, as a minimum:

1) address field
2) RAMM address
3) input/output instructions
4) special instructions
5) priority.

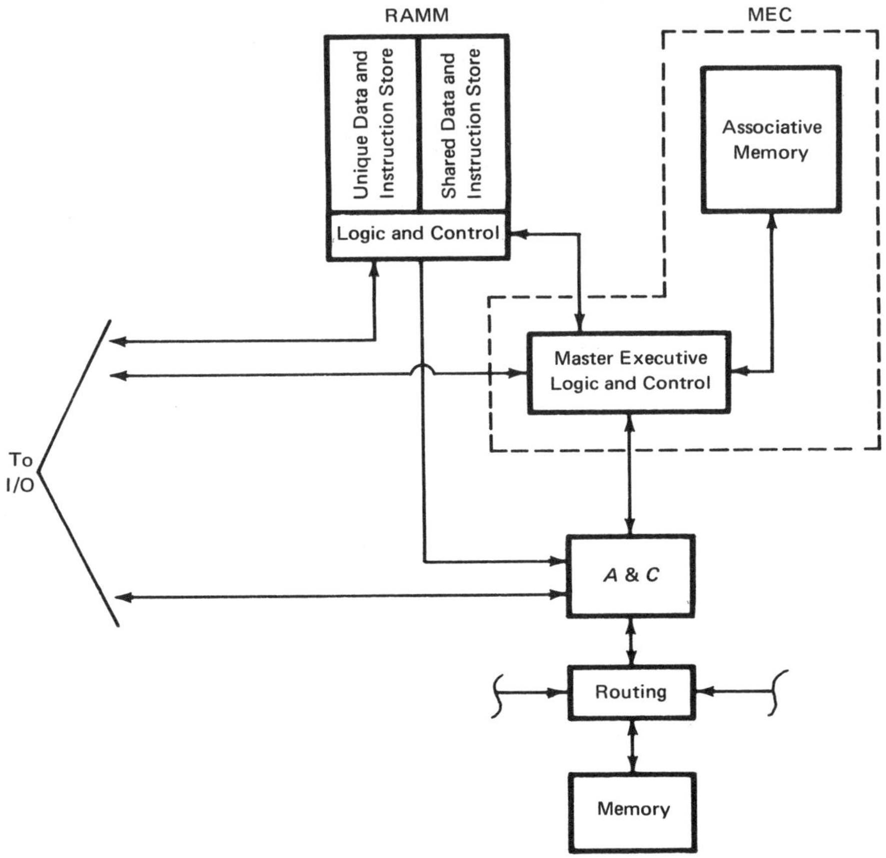

Fig. 3. Master Executive Control

The special instruction will relate each PM to other PM's in memory, as well as to the computer hardware, itself. The *executive program* consists of instructions used by the MEC to initiate system operation and to translate I/O and special instructions into sequences of logical operations. The executive program instruction flow would proceed, roughly, as described in Figs. 4 through 6. An associative memory is employed in lieu of a random access memory in order to reduce the processing time for this fairly involved executive program.

Operational program modules are stored in two areas in the RAMM: the first, entitled *unique data and instruction store* and the second, *shared data and instruction store.* Since memory access occurs singly, it

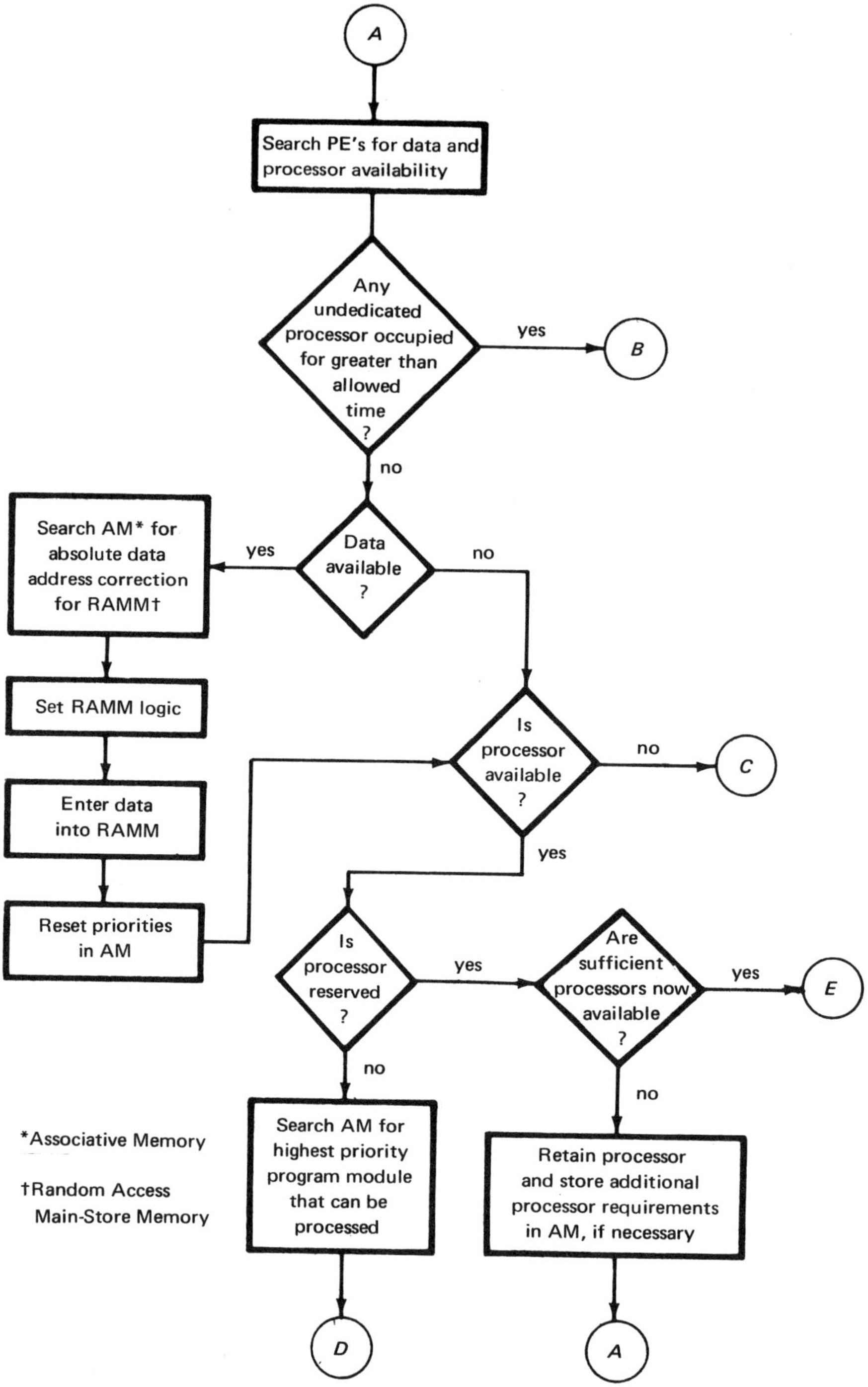

Fig. 4. Preliminary Executive Program Instruction Flow

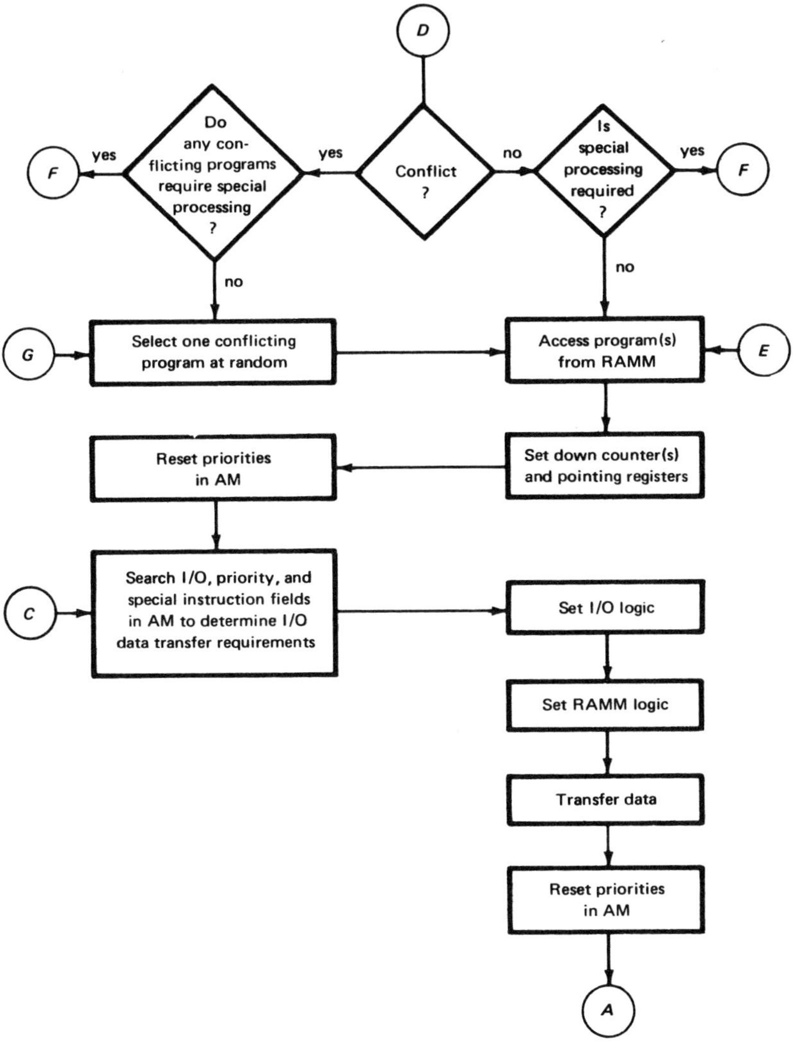

Fig. 4. Preliminary Executive Program Instruction Flow (cont'd)

is possible to factor out large portions of data common to various PM's. In general, a PM will be prefaced by a series of microprogram instructions that will be used to activate the microprogram area of a PE. The microprogram field will be followed by unique instructions and data throughout which the addresses of data and instructions stored in the *shared data* and *instruction store* will be interspersed. Upon PM access, these addresses will cause the RAMM control logic to insert the data and instructions stored in the designated areas into the unique data and instruction flow being bused to a PE memory.

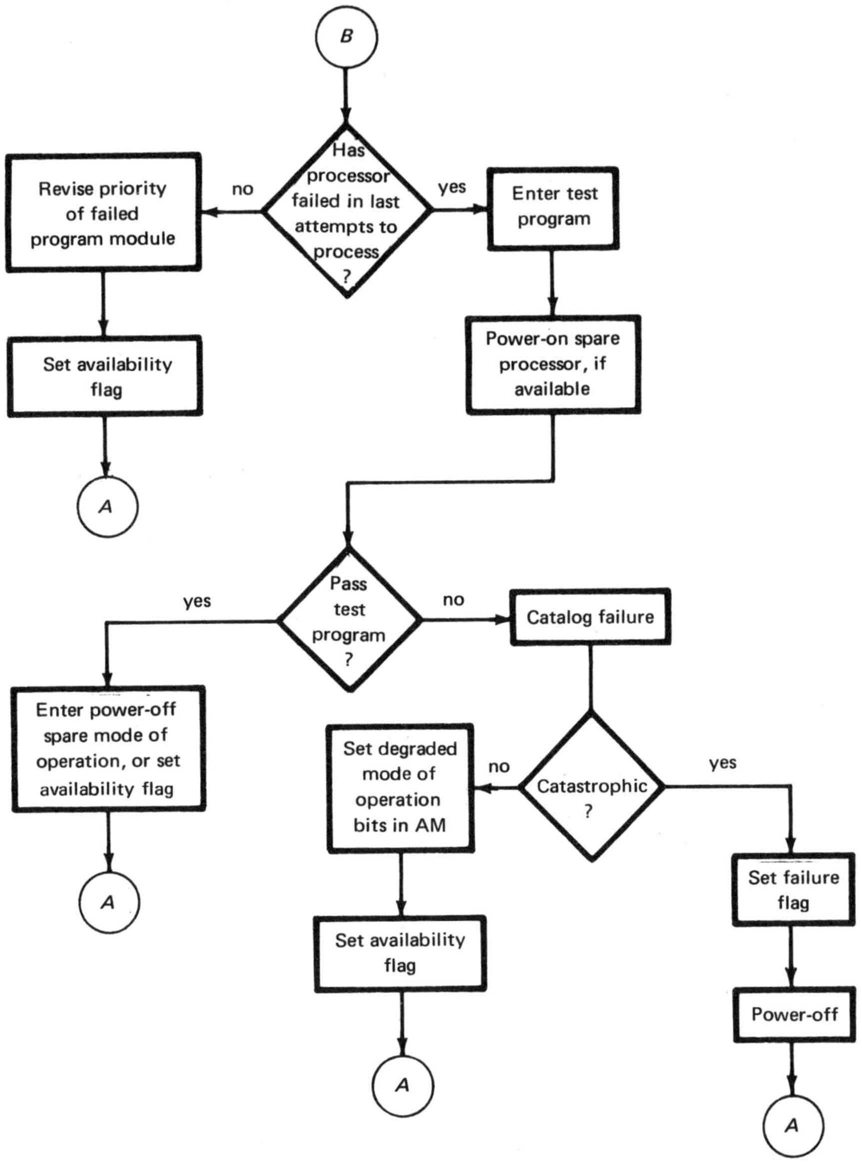

Fig. 5. Preliminary Failure Test Subroutine

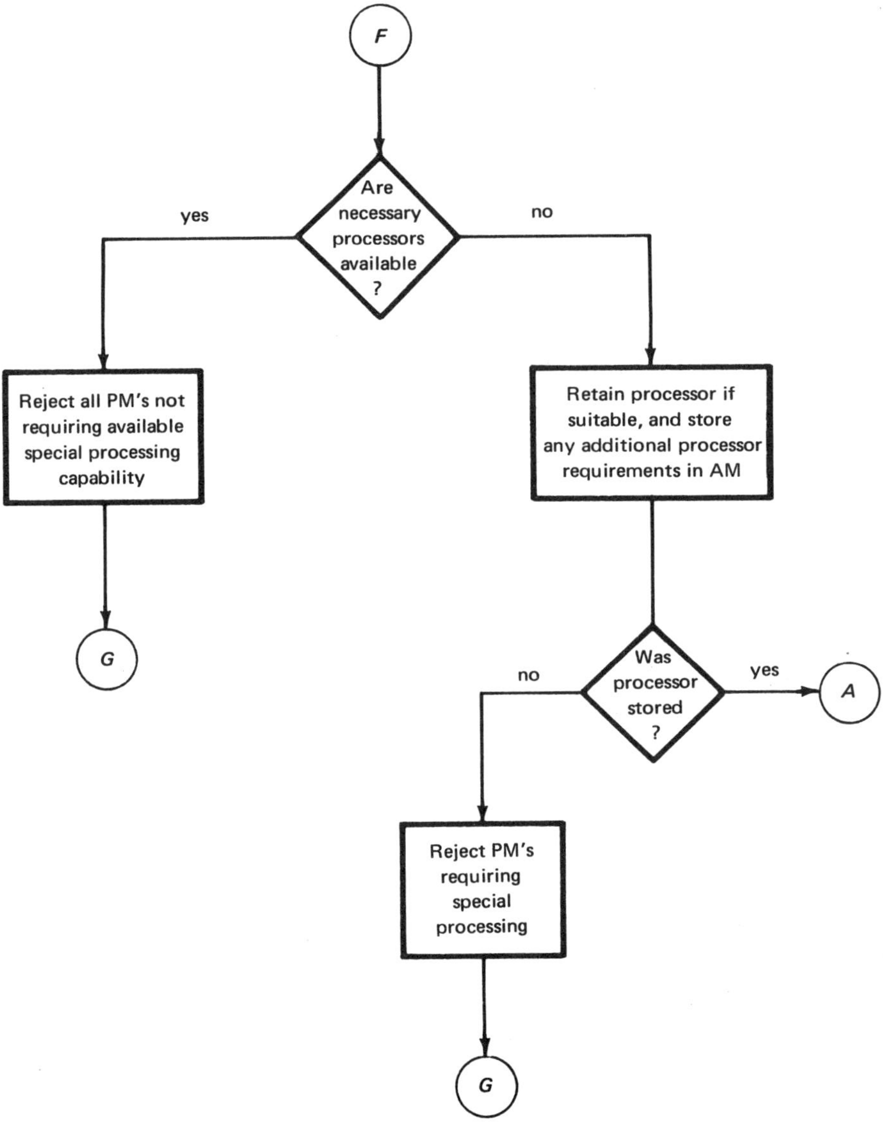

Fig. 6. Preliminary Special Requirements Subroutine

In addition to providing a high degree of hardware design flexibility, byte-functional modularity also permits the reallocation of processing resources via software control. The AADC, in particular, will utilize this capability to implement fixed point operation without paying the penalty of greater circuit complexity or greater processing time to perform the floating point function. This dual mode of operation is based on the fact that while, in general, data may be divided into bytes and then distributed among processor and memory modules, instructions must be replicated, *in toto,* in each module control area. For this reason, the capability exists to uniquely instruct each arithmetic and control module within any PE arithmetic and control block. If a PE consists of two or more bytes, then the instructions provided to the A & C modules can be divided in a manner such as to have one group of modules performing mantissa and the second, exponent computations. This technique will permit complete compatibility with fixed point operation in that the increased number of bits needed to retain precision and dynamic range for fixed point computation will come from bits used to perform the exponent function in the floating point mode of operation. No special hardware will be required to implement this dual mode of operation (Fig. 7).

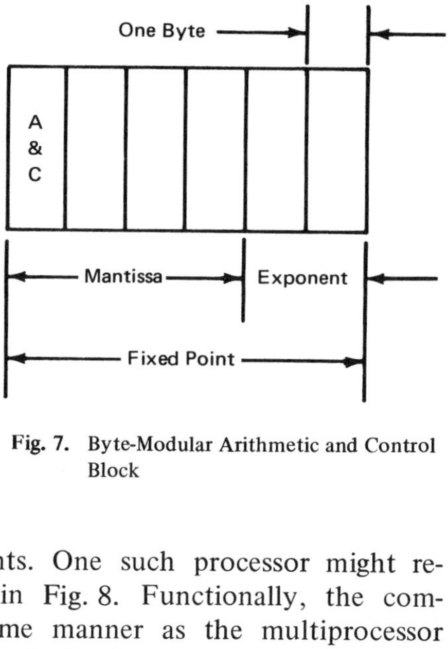

Fig. 7. Byte-Modular Arithmetic and Control Block

Taken one step further, the byte-functional module concept augers a computer system that will have the ability to efficiently accommodate various programs having different single precision word length requirements. One such processor might resemble the processor illustrated in Fig. 8. Functionally, the computer shown operates in the same manner as the multiprocessor portion of the baseline AADC. Physically, however, the A & C's and task memories have been integrated into a single, *monolithic multiprocessor.* Borrowing from the multiprogramming concept just described, it should be possible to allocate processors according to the processing requirements for each *program module.* It is therefore feasible to cause the monolithic multiprocessor to behave in the manner of M uniprocessors with M task memories, N combinations of uni- and multiprocessors with single or shared task memories, where $N \geq M$, up through a single uniprocessor, M bytes wide, functioning in either a fixed or floating point mode

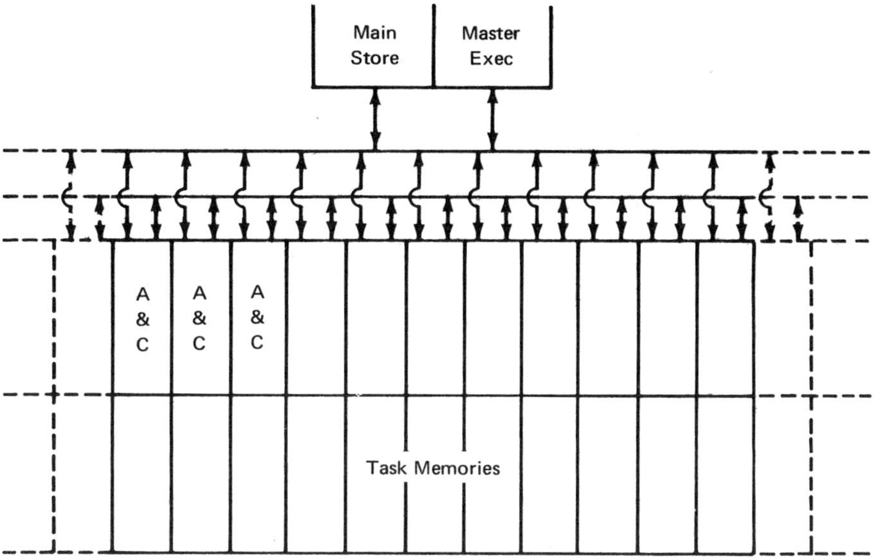

Fig. 8. Monolithic Multiprocessor

of operation. The basic advantages of this approach over the baseline AADC multiprocessor organization are:

1) processor word length as a function of specific program requirement.
2) elimination of routing block between PE, A & C and task memory.
3) utilization of all A & C's when sharing task memories.
4) Reduced module pin count.

While it is believed that the monolithic multiprocessor represents the most flexible application of byte-functional modularity, the executive control requirement for such a system may prove too involved for cost-effective implementation in the AADC. It should be noted, however, that the overall control problem can be greatly simplified by fixing the basic system word length prior to writing computer programs. If such is the case, monolithic multiprocessing may prove an attractive alternative to multiple processing elements. In any event, monolithic multiprocessing appears to be a reasonable design approach for the efficient utilization of LSI wafers with an excess of

10,000 gates. An LSI version of this processor might include several arithmetic and control units as well as a reasonable quantity of task memory modules.

CHAPTER 11

COMMUNICATION AND SEQUENTIAL PROBLEMS IN THE PARALLEL PROCESSOR

J. O. Campeau

Litton Industries
Woodland Hills, California

INTRODUCTION

Parallel processors offer a new approach to the handling of very large problems. A conventional single arithmetic unit machine cannot effectively come to grips with problems which require computation rates in excess of that permitted by its circuit speeds.[2] One approach to such large problems has been to develop computers with increased circuit speeds and more involved logical structures such as instruction look ahead, carry acceleration, etc., while still preserving the single arithmetic plant.

A different approach to the large problem consists of breaking it into a number of smaller problems. Now each smaller problem can be handled in parallel by its own processor. Individually the processor cells may have relatively modest circuit speeds and yet the array of such cells can have a powerful computation capability.

Some problems, by their very nature, are ideally suited to such parallel processing systems. For example, the type of problem originally proposed for the parallel processor, namely multivariable field problems, consists of a large number of local problems, each requiring only results from neighboring cells.

This work was supported in part by the United States Air Force Avionics Laboratory under Contracts No. F33615-68-C-1548 and F33615-67-C-1238.[1]

Such problems as pattern recognition and radar data processing, in which many targets must be processed simultaneously, also seem to fit the array nature of the parallel processor.

Parallel processors have properties which make them well suited to the special demands of *large scale integrated arrays* (LSI). This occurs since the nature of LSI requires a minimization of connections between arrays and the use of repetitive large logic blocks.

If each large logic block is a simple processor, it is evident that the number of signals leaving and entering will be less than would be the case if a "slice" were taken through the midst of a more complex single arithmetic plant.

In an earlier paper the organization of a parallel processor under development at Litton was shown to permit the use of entire LSI wafers in its construction.[3] Its cellular nature permits continued operation in spite of the defects normally present on the LSI wafer, without requiring a secondary metalization process. A different LSI wafer system does require a secondary metalization.[4]

The projected low manufacturing cost of the Litton parallel processor is based on the use of LSI wafers in its construction. These undiced MOS LSI wafers will each contain over 120,000 transistors.

The organization of the *block oriented computer* (BOC), under development at Litton Industries, features a system composed of large numbers of identical cells rather than a single cell as is found in the traditional Von Neumann computer. The term *cell* is here used to refer to an elementary processor, including arithmetic and some control functions. The term *cell* may also refer to a portion of a memory or of an input/output system.

The Litton parallel processor is similar to the ILLIAC IV, an experimental parallel processor being developed at the University of Illinois.[5] The organization of the BOC is shown in Fig. 1. It is modular and can be expanded by adding processor, input/output or memory modules.

A version of this computer is being specialized to a missile guidance and control application for the Air Force. Such a missile guidance and control problem is quite sequential in nautre, that is, final results tend to depend on earlier calculations rather than only relatively simple "local" calculations, as do problems that are inherently parallel in nature.

This paper will concern itself with problems encountered when the parallel processor is applied to such typical system applications as the guidance and control of a missile, which do not "fit" easily to the natural requirements of the parallel processor.

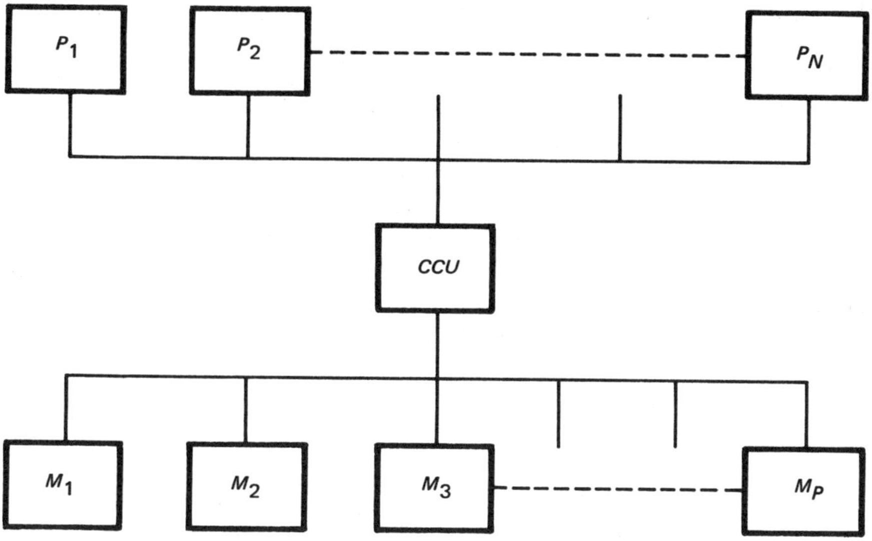

Fig. 1. Organization of Block-Oriented Computer

In this paper the terms *parallel processor* and *array processor* are used interchangeably.

PROBLEMS IN THE PARALLEL PROCESSOR

Communication

In the parallel, or array, processor, a multitude of simple arithmetic cells are used to perform the totality of computations required for the problem. In the type of problems originally proposed for the array processor, namely, multivariable field problems, a real increase in computation ability is achieved by this organization, due to its array mode of operation. Communication of information only occurs between adjacent cells, and control is global in nature (all cells do the same thing).

In a more sequential problem such as navigation and guidance, the array processor encounters difficulties. The communication problem assumes major proportions when such an array processor is configured to have a large number of computing cells (on the order of 100). For example, if the output of arithmetic unit number 10 is to be used by arithmetic units number 36, 40, 51, and 74, how would it be possible to communicate the result of this to the other cells?

If hard wiring is used, then the communication problem is solved in a most direct manner. However, the flexibility of the machine is greatly compromised by hard-wired buses.

In order to illustrate the principles involved, consider the multiplication of two numbers, *A* and *B*. This computation can be performed in a single processor cell, e.g. *AU*10. If the product *AB* is required in other *AU*'s, e.g., *AU*36, *AU*40, *AU*51, and *AU*74, how would this communication take place?

If hard wiring is used, then the output of *AU*10 gets to the other *AU*'s through a set of direct connections. At the end of the multiplication, the product would be ready for distribution. At this time it could be distributed on the direct connections, in a serial manner, to *AU*36, *AU*40, *AU*51, and *AU*74 (the example chosen).

Using hard wiring, more *AU*'s would be added if the problem to be solved grew in complexity with more multiplications, etc. If the problem became very large, however, the hardware requirements might become excessive.

Time Sharing

The possibility of time-sharing of individual arithmetic unit cells must be explored as one way to save hardware while processing a relatively large problem. This is done in the conventional single processor machine when one arithmetic unit does many different parts of the problem. The single *AU* is shared among all the different problem requirements in terms of +, −, ×, ÷, etc.

If it were possible to change the information flow paths in the array processor matrix of *N* cells, and if the processor performs a computation every *A* seconds while the requirements are for a complete solution every *B* seconds, then through time-sharing of the arithmetic cells, only *NA/B* would be required. One problem considered for example, only required ten solutions per second. With a machine able to iterate 400 times per second, only 10/400 or one-fortieth the number of cells were required.

To do this time-sharing in an array processor would mean mode changing; yet once a connection is hard-wired from point *A* to point *B*, no mere program branch is going to change it.

In the example presented above, if we attempt to use *AU*10 in a different computation, e.g., a division, and distribute the quotient to other *AU*'s, say *AU*6, *AU*23, *AU*42, etc., it is evident the hard-wired connections previously made cannot be rerouted without a physical change in the circuit. The previous connection from *AU*10 to *AU*36, *AU*51, and *AU*74 would in general have to be removed, and a new set of connections from the output of *AU*10 to the inputs of *AU*6, *AU*23, *AU*42, etc., would be made.

The above discussion illustrates the practical difficulties in time-sharing hard-wired array cells. A hard-wired array processor would be limited in the size of the problems it could handle, since it could not use a given *AU* for more than one problem.

From the above, hard wiring is seen to run contrary to the requirements for time-sharing. Time-sharing requires that the communication paths be changeable under the control of the program.

A system which would allow the paths to be changed under the control of the program can be designed for the array processor organization. A general, program-switched bus scheme may be employed. Yet if the switching logic and circuitry is not to become excessive, this system must be limited to the transfer of only a single number at a time between a sending and receiving cell. Such a low transfer rate can become the limiting factor in the processing capability of the parallel processor.

The array processor using traditional LSB first computing algorithms suffers a hardware-speed exchange problem. The results of the operations from an array of cells, available following a computing cycle, may be intercommunicated in a variety of word and bit serial/parallel combinations trading time for busing hardware.

The power increment algorithm described below permits the flexible high speed parallel communication of results in a relatively simple manner.

The Sequential Problem

In the normal flow of data, it may not be possible to begin a particular computation until the results of the previous computation have been completed. Problems may be classified as to the degree with which they have this *sequential* characteristic, that is, according to the number of independent parallel computations which can occur.

Qualitatively, a typical field problem such as the LaPlace equation, could be placed close to the *array* end, while a problem having many branch points involving only a single computation path would lie close to the *sequential* end.

Arithmetic computations even without the decision function also can cause the same sequential difficulty. For example, with a conventional multiplication process the product $(A \times B) \times (C \times D)$ could not be performed in an array processor during one compute cycle.

It would first be necessary to form $A \times B$ and $C \times D$ separately in one compute cycle. During the next compute cycle the "product of products" could be developed.

When the array processor is brought to bear on a problem predominately sequential in nature, it may be difficult if not impossible to use all of several hundred arithmetic units at any time. This results in a decrease in the effective computing ability of the array due to the idle arithmetic cells.

The organization of the Litton processor possesses unique properties that greatly enhance its capabilities when handling sequential problems.

COMPUTER ORGANIZATION

The Litton array processor is based upon the dinary power increment algorithm. This algorithm permits the design of an efficient communication system for the processor. In addition to this, the sequential problem is reduced by another property of the power increment algorithm which permits the simultaneous computation of multiple sums and products.

Figure 1 depicts the Litton parallel processor organization. It will be noted that memory cells and processing cells all connect together on a common bus. A single *central control unit* (CCU) serves to steer information between the memory cells and the processor cells. Not shown is a *matrix* bus system which operates between the processor cells to permit the parallel communication of results.

Figure 2 shows the BOC busing structure at a typical cell. The busing structure can be conveniently divided into two parts. One of the parts is concerned with the array processing mode (*matrix* mode) of operation of the system. The other bus system is concerned with the filling of arithmetic units with data from the memory (*scalar* mode).

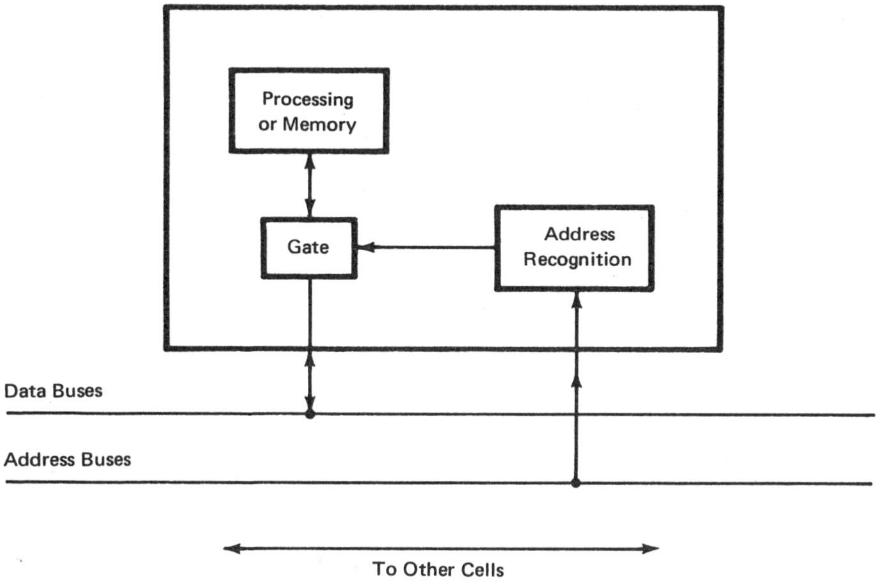

Fig. 2. Typical Cell Showing Bus System

Figure 3 shows a processor cell block diagram. The inputs and outputs from this cell are entities (called power increments) which in the ith of n minor iterations will have the value plus or minus 2^{-i} or 0 ($i = 1, \ldots, n$). This particular power increment coefficient may have one of three values in any particular word. Two binary bits are required for its representation. These bits of information are communicated between cells using what is called the Z communication bus system. Four such buses (each one a single metalized path) are available. A number communicated by power increments is said to be in "dynamic form." Conventional binary numbers (as in the Y register in Fig. 3) are said to be in *static* form.

The processor cell operates on incoming power increments and produces a power increment output. At the heart of the cell are two registers called, as in a DDA, the Y and the R registers. These registers are each 32-bits long and are added together serially. The addition is done at 1/2 clock speed so that a total of 64 clock times are required for each computer word of operation. At the end of each word time, a power increment output is developed.

The cell develops the power increment according to different rules depending upon whether the operation is (1) multiplication,

(2) division or square root, or (3) digital differential analyzer.

When operating to develop type (1) or (2) outputs mentioned above, the R register is caused to shift left with respect to the Y register. When operating to produce a DDA output, no relative shift occurs.

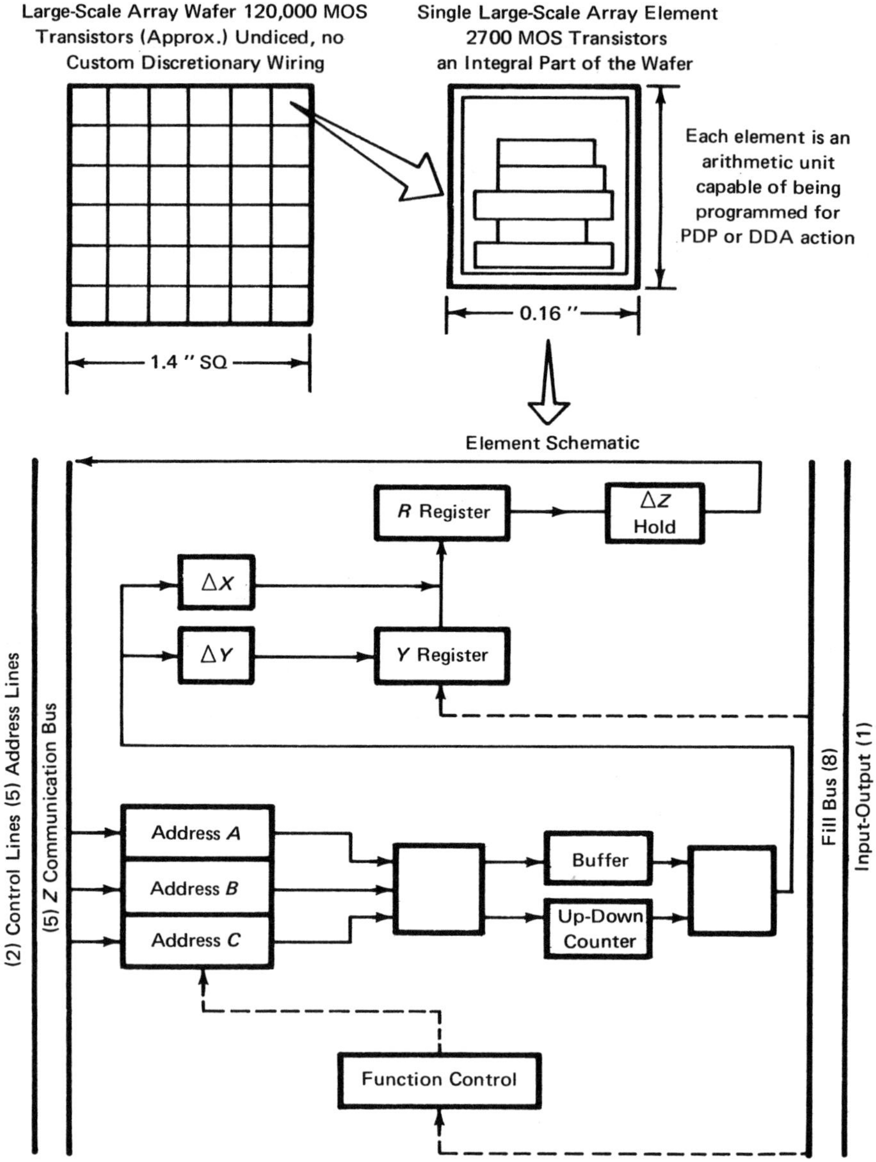

Fig. 3. Processor Cell

The power increment output of the cell is held in two storage bits called the ΔZ *hold*. This information is held until the proper time during the next word at which time it is placed on the Z communication bus line.

A group of 31 cells operates together to share a single communication bus line.

On the single bus line, the ith of 31 cells in the vector will occupy time position i ($i = 1, \ldots, 31$). The 32nd time position is used to represent the input zero.

One of the inputs to the arithmetic unit signals that its output (held in the ΔZ store) should be placed on the communication bus line. This select signal is also used to identify the cell which is to be filled from the memory. One control line common to all cells identifies whether the select refers to the communication bus lines or to the fill bus lines.

The fill bus lines connect to the memory through the central control. There are eight of these lines.

One final control line common to all cells is the end-of-word pulse which comes from the central control.

Inputs to the cell come from the communication bus system just described. Such inputs are extracted from the communication bus through three address systems called Address A, B, and C. These systems are filled with address data from the main memory, and select the information from the communication bus lines through a combination of space and time-wise selection.

Inputs are entered into a Buffer or an Up-Down Counter as shown in Fig. 3. At the end of each word time the information gathered is sent to two data stores called the ΔX and ΔY store.

The Function Control in the cell controls the operation directing whether multiplication, division, square root, etc., shall occur.

Array Processing with the Power Increment Algorithm

When all cells in a given subsection of the array are in compute this is said to be *array processing*. During array processing, the power increment algorithm has the property that computations proceed most significant bit first with partial results of the computation (addition, subtraction, multiplication, division or square root) available for further computation while the process itself is still underway.

By comparison, the conventional multiplier must complete its operation before the resultant product can be used elsewhere. In the Litton computer, the most significant bit of the product is available at the end of the first word of computation rather than requiring n words to obtain the total n-bit product. At the end of the nth word the n-most significant bits of the product will have been communicated throughout the array of processing cells and will have been absorbed into them in terms of performing multiple products, quotients, square roots, etc.

The BOC solves the communication problem with a minimum of hardware through use of a unique property possessed by the computing cells. Each cell produces the bits of a sum, difference, product, quotient or square root as it is being developed, rather than only at the end of the process. Each word time (equal to 32 clock times) one bit of the result is produced.

The bit representing the intermediate output of a cell requires a relatively small amount of time for serial transmission when compared to the 32 clock-time word length of the computing cells. It is therefore possible to time-share a single communication bus among a relatively large number of cells. Each cell has an output which occupies a definite time position on the bus. Other cells can then access the output of a relatively large number of cells simply by performing a time-gating operation on a single bus line.

The communication between cells is accomplished on a relatively small number of wires. Thus, for example, with 32 bit operands in a fully serial compute cycle of 32×32 bit times (e.g., multiplication), the outputs of 32 cells can be intercommunicated on a single bus line during the cycle itself.

Information paths between cells can be changed by simply entering an address into the receiving cell. This address will cause the receiving cell to take the data off a bus line at a different time. Extraction of data is merely a de-multiplexing of information using a time synchronized pulse.

In the BOC, a product of many terms such as $A \times B \times C \times \ldots$ can be performed during a single major computation cycle, due to the nature of the power increment process. While the products $A \times B$ and $C \times D$ are being developed, the product $(A \times B) \times (C \times D)$ may be developed at the same time.

The number of terms involved in any simultaneous multiplication is not limited in the power increment process. Similarly, sums of differences of products may be multiplied by other sums or differences of products at the same time these quantities are being developed.

A typical recursion formula used in the digital differential analyzer is presented to introduce the binary power increment computing principle

$$r_{k+1} = r_k + \left(y_k + \frac{\Delta y_k}{2}\right)\Delta x_k - \Delta z_k \quad (k = 1, \ldots, n) . \tag{1}$$

This is the formula for an interpolative trapezoidal DDA. Here, r_k is the number in the R-register at the kth step; y_k is the number in the Y-register. The quantities Δx_k and Δy_k are the inputs to the integrator, while Δz_k is the output. The initial quantities in the R and Y registers are r_1 and y_1. For the sake of simplicity in all the presentation to follow, scaling will not be considered. All variables will be considered to range from plus to minus 1. Under this simplification, if n bit accuracy is desired, the input and output increments have weight 2^{-n}.

As a side point, note that the number y_k in the above equation is built up by accumulating the Δy_k's according to the equation

$$y_k = \sum_{j=1}^{k-1} \Delta y_j + y_1 \quad (k = 1, \ldots, n) \tag{2}$$

In the digital differential analyzer a given increment always has the same value. In the Litton power increment method, on the other hand, the magnitude of the increments is allowed to vary in a systematic manner. The machine resets all its registers in general every nth iteration, a "major" iteration, where n bit accuracy is desired in the computation. Contrast this with the DDA where the registers are not reset during the computation. During the kth of n "minor" iterations all the increments in the power increment machine will have the value 2^{-k} .

This variable increment operation can be contrasted with the DDA where the size of the increments is constant at 2^{-n}. Over the n minor iterations involved in one complete major cycle of the machine, all the increments will have had the values $1/2, 1/4, \ldots, 1/2^n$. It can be shown that any number can be represented as the sum of such a series of increments. Thus for any terminating binary number z (where $-1 < z < 1$), there exists at least one set of ζ_k so that

$$\left.\begin{aligned} z &= \sum_{k=1}^{n} \zeta_k 2^{-k} \quad \text{where } \zeta_k = \pm 1,\ 0 \\ &= \sum_{k=1}^{n} \Delta z_k \quad \text{where } \Delta z_k = \pm 2^{-k},\ 0 \end{aligned}\right\} k = 1, \ldots, n . \tag{3}$$

Under this system, it is possible to build up any number in the machine to n bit accuracy in only n minor iterations. In the DDA a total of 2^n iterations would be required to start from initial conditions, in the worst case, and develop a quantity accurate to n bits.

It will be realized that the machine mechanization is not limited to a binary radix, although in the interest of brevity only the binary radix is considered here.

Make the following substitutions in eq. (1)

$$\left.\begin{aligned} 2^{-k}\,\xi_k &= \Delta x_k \\ 2^{-k}\,\zeta_k &= \Delta z_k \\ 2^{-k}\,\rho_{k+1} &= r_{k+1} \end{aligned}\right\} \quad k = 1, \ldots, n \, . \tag{4}$$

With these substitutions, eq. (1) takes

$$2^{-k}\rho_{k+1} = 2^{-k}2\rho_k + \left(y_k + \frac{\Delta y_k}{2}\right)2^{-k}\,\xi_k - 2^{-k}\,\zeta_k \, . \tag{5}$$

Now multiply through by 2^k

$$\rho_{k+1} = 2\rho_k + \left(y_k + \frac{\Delta y_k}{2}\right)\xi_k - \zeta_k \, . \tag{6}$$

Note that the equation for the Y-register has not been changed

$$y_k = \sum_{j=1}^{k-1} \Delta y_j + y_1$$

and that this implies that the summation must be performed taking into account the changing weight of the Δy power increments.

With this mechanization, even though the increments under consideration will build up numbers in the manner described in eq. (3), the recursion formula mechanized by the computation block is quite similar to the DDA formula given in eq. (1). The only difference is the factor of 2 associated with the quantity ρ_k.

In order to understand the method by which the output of a typical block, ζ_k, is developed at the kth of n steps, a typical minor iteration starts with ρ_k suitably scaled so that

$$-1/2 \leq \rho_k < 1/2 \qquad k = 1, \ldots, n \tag{7}$$

Then,

$$-1 \leq 2\rho_k < +1 \qquad k = 1, \ldots, n \tag{8}$$

Assume y_k is suitably scaled so that

$$-1/2 \leq y_k + \frac{\Delta y_k}{2} < +1/2 \qquad k = 1, \ldots, n\,. \tag{9}$$

Since the increments Δx, Δy, and Δz have the values

$$\Delta x_k,\ \Delta y_k,\ \Delta z_k = \pm 2^{-k}, 0 \qquad k = 1, \ldots, n\,. \tag{10}$$

It can be seen from eq. (1) that the quantities ξ_k and ζ_k will in general have the values

$$\xi_k,\ \zeta_k = \pm 1, 0 \qquad k = 1, \ldots, n\,. \tag{11}$$

Combining this fact with the bounds in eq. (9)

$$-1/2 \leq \left(y_k + \frac{\Delta y_k}{2}\right) \xi_k \leq +1/2 \quad k = 1, \ldots, n. \tag{12}$$

Further combining eqs. (8) and (12)

$$-3/2 \leq 2\rho_k + \left(y_k + \frac{\Delta y_k}{2}\right) \xi_k \leq +3/2 \qquad k = 1, \ldots, n. \tag{13}$$

The output of the block, ζk, is determined as follows: Set

$$2\rho_k + \left(y_k + \frac{\Delta y_k}{2}\right)\xi_k \equiv \sigma_k\,.$$

Then for:

$$\left.\begin{array}{r} \sigma_k \geq 1/2 \\ -1/2 \leq \sigma_k < 1/2 \\ \sigma_k < -1/2 \end{array}\right\} \text{overflow} \left\{\begin{array}{l} \zeta_k = +1 \\ \zeta_k = 0 \\ \zeta_k = -1 \end{array}\right\} \quad k = 1, \ldots, n\,. \tag{14}$$

From eqs. (13) and (14)

$$-1/2 \leq \rho_{k+1} = 2\rho_k + \left(y_k + \frac{\Delta y_k}{2}\right)\xi_k - \zeta_k < 1/2 \qquad k = 1, \ldots, n . \tag{15}$$

In other words, by mathematical induction, if the overflow logic described in eq. (14) is mechanized, the quantity ρ_k will always have the bounds given in eq. (7).

Although the particular cell configuration used in the Litton parallel processor only has two registers, an alternate configuration containing three registers will be employed to illustrate the operation of the arithmetic element by means of elementary examples. The cell may be represented symbolically by a block with two dynamic inputs and one dynamic output.

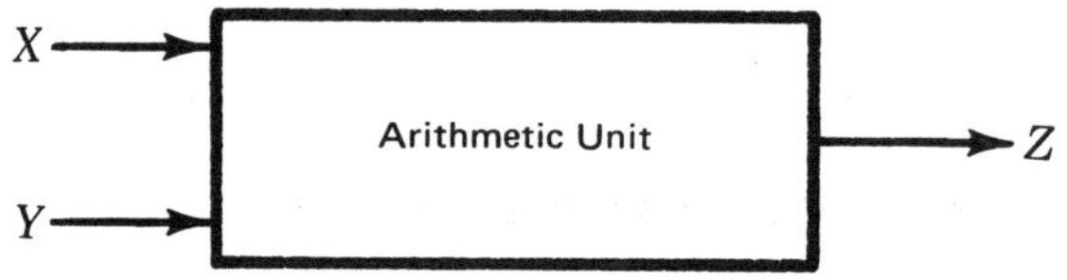

At each iteration the block takes in x and y and puts out z.

The quantities x, y, and z are called dynamic numbers. Dynamic numbers have only three values (+1, 0, −1) at any iteration, (i), and if weighted and accumulated, will yield the whole static number

$$X = \sum_{i=1}^{n} x_i 2^{-i}$$

$$Y = \sum_{i-1}^{n} y_i 2^{-i} \tag{16}$$

$$Z = \sum_{i=1}^{n} z_i 2^{-i}$$

The static accumulated quantity Z is the product of static quantities X and Y:

$$Z = XY$$

The block operates according to the equation

$$z_i = (X_i + 2^{-1}x_i)y_i + Y_i x_i + 2R_i - R_{i+1} \tag{17}$$

where X_i and Y_i are obtained from eq. (16) at iteration i.

The quantity z_i is part of the dynamic number and is the carry-out from the R register. It represents the overflow into the most significant half of the product. The quantity R_i is the least significant half of the product, and is a static number.

The block holds three static numbers, X, Y, and R, coded as

$$X_i\,(\text{coded}) = X_i\,(\text{true value}) + 1/4$$
$$Y_i\,(\text{coded}) = Y_i\,(\text{true value}) + 1/4$$
$$R_i\,(\text{coded}) = R_i\,(\text{true value}) + 1/2\,.$$

The quantities X and Y are generated according to eq. (13) and are bounded by 1/4 in magnitude

$$-1/4 \leq X, Y < 1/4, \tag{18}$$

while R has bounds

$$-1/2 \leq R < 1/2\,. \tag{19}$$

The input and output dynamic quantities x_i, y_i, and z_i of value +1, 0, or −1 are coded as

$$x, y, z\,(\text{coded value}) = x, y, z\,(\text{true value}) + 1\,. \tag{20}$$

A second mode of operation occurs when initial static constants are placed in the X and Y registers. In this event the numbers in X and Y are unaltered by the inputs, and eq. (17) assumes the reduced form

$$z_i = X_1 y_i + Y_1 x_i + 2R_i - R_{i+1}\,. \tag{21}$$

As an example of the system's operation, a static binary number will be converted to dynamic form. In order to do this the number is placed in the R register initially and the inputs are set to zero,

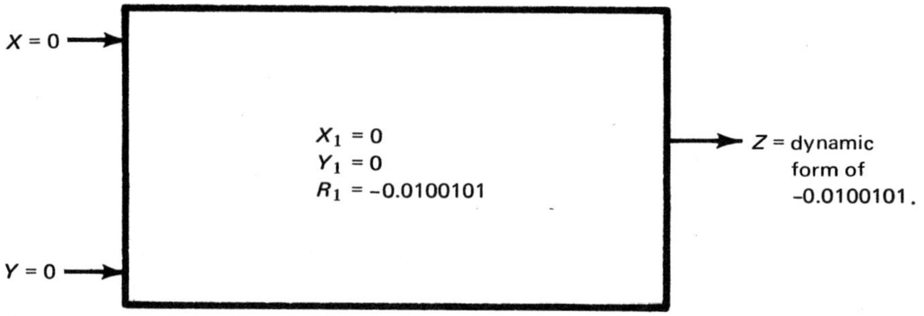

The static number -0.0100101 is coded according to eq. (17)

$$R_1 \text{ (coded)} = R_1 + 1/2 = 0.0011011 .$$

The coded forms of X_1 and Y_1, according to eq. (17), are

$$X_1 \text{ (coded)} = X_1 + 1/4 = 0.010000$$

$$Y_1 \text{ (coded)} = Y_1 + 1/4 = 0.010000 .$$

Since $X = x = Y = y = 0$, the term

$$\Delta = (X_1 + 2^{-i}x_i)y_i + Y_i x_i = 0.1000000,$$

coded according to eq. (17) where Δ is used as an abbreviation.

The development of the dynamic form of $R_1 = -0.0100101$ can best be summarized by Table 1.

Table 1
Dynamic Form Development

Iteration Number	R_i (true)	R_i (coded)	$2R_i$	$2R_i + \Delta$	$R_i + 1$	z_i	z_i (true)	$\sum_i z_i 2^{-i}$
1	-0.0100101	00.0011011	00.011011	00.111011	00.111011	00	-1	-0.1
2	+0.011011	00.111011	01.11011	10.01011	00.01011	10	+1	-0.01
3	-0.00101	00.01011	00.1011	01.0011	00.0011	01	0	-0.010
4	-0.0101	00.0011	00.011	00.111	00.111	00	-1	-0.0101
5	+0.011	00.111	01.11	10.01	00.01	10	+1	-0.01001
6	-0.01	00.01	00.1	01.0	00.00	01	0	-0.010010
7	-0.1	00.0	00.0	00.1	00.1	00	-1	-0.0100101
8	0	00.1						

The initial quantity, -0.0100101, is converted to dynamic form as shown in the column headed z_i(true). Notice that when the dynamic increments are weighted and accumulated in the column headed $\sum_i z_i 2^{-i}$, the original static quantity, -0.0100101, is recovered. This can be seen by comparing the last entry in the last column with the first entry in the first column.

To illustrate the development of Table 1, the first row will be considered. The coded form of $R_1 = -0.0100101$ is obtained by adding 0.1, according to eq. (14). The coded form is R_1 (coded) = R_1(true) + 1/2 = -0.0100101 + 0.1 = 0.0011011. This is the first entry in the second column headed R_i (coded). The next column is $2R_i$, and is obtained by a simple left shift. The quantity Δ, as mentioned above, is an abbreviated form of the X and Y quantities added to $2R_i$ according to eq. (22). The new value in R, namely $R_i + 1$, is the fractional part of $2R_i + \Delta = 0.111011$ so that z_i(coded) = 0. According to eq. (17), the true value of z_i is obtained from the coded value of subtracting unity. The true value of the first weighted increment output is z_i(true) = -1. This is shown in column 8 in the table.

The fractional part of $2R_i + \Delta$ is 0.111011, which becomes the new R number: $R_2 = 0.111011$. This number is the coded form of the positive quantity +0.011011 and represents the remainder at that point.

A second example will show the solution of three linear algebraic equations for three unknowns. The equations are

$$\left.\begin{aligned} 2X_1 + 0.106X_2 - 0.0613X_3 &= 0.212 \\ -0.0576X_1 + 2X_2 + 0.0921X_3 &= 0.327 \\ 0.212X_1 - 0.176X_2 + 2X_3 &= 0.197\ . \end{aligned}\right\} \tag{22}$$

The Litton computing technique requires three blocks for this computation connected together as shown in Fig. 4.

To demonstrate the process in operation, the initial constants must be converted to binary form. The binary equivalents are given as follows:

Block No. 1			
X	-0.106	-0.000110110010001	00.001001001101111
Y	0.0613	+0.000011111011001	00.010011111011001
R_1	0.212	+0.001101100100011	00.101101100100011
Block No. 2			
X	0.0576	+0.000011101011111	00.010011101011111
Y	-0.0921	-0.000101111001010	00.001010000110110
R_1	0.327	+0.010100111011011	00.110100111011011
Block No. 3			
X	-0.212	-0.001101100100011	00.000010011011101
Y	0.176	+0.001011010000111	00.011011010000111
R_1	0.197	+0.001100100110111	00.101100100110111

The coded forms of the 3 sets of X, Y, and R initial conditions are set into the 3 binary blocks and the system is run through 15 minor cycles. The coded contents of the R registers of the blocks and their weighted increment z overflows at each iteration are tabulated in tables 2, 3, and 4. The right-most column shows the development of the summed weighted-increment z overflows from each of the blocks. At the end of the 15th iteration, these numbers are:

Block 1:	$2X_1$ (true)	=	+0.001100110111110
	$2X_1$ (decimal)	=	+0.20111
	X_1	=	+0.100555
	X_1 (correct)	=	+0.10055857
Block 2:	$2X_2$ (true)	=	+0.010100101100101
	$2X_2$ (decimal)	=	+0.32339
	X_2	=	+0.161695
	X_2 (correct)	=	+0.1616957
Block 3:	$2X_3$ (true)	=	+0.001101000100001
	$2X_3$ (decimal)	=	+0.20413
	X_3	=	+0.102065
	X_3 (correct)	=	+0.10207002 .

The mathematically correct values for the three solutions are also listed above for comparison. The total number of multiplications and divisions needed to develop the values for X_1, X_2, and X_3 with conventional methods is 27. It is easy to see the economics of the three

Table 2
Block No. 1

x = 00.001001001101111
y = 00.010011111011001
R_i = 00.101101100100011

i	R_i	$2R_i$	$2R_i + \Delta_i$	z_i (true)	$2x_i = \Sigma z_i 2^{-i}$
1	00.101101100100011	01.011011001000110	01.111011001000110	0	0
2	00.111011001000110	01.110110010001100	10.001111011111011	+1	+0.01
3	00.001111011111011	00.011110111110110	01.001001101100000	0	+0.010
4	00.001001101100000	00.010011011000000	00.101100100101111	−1	+0.0011
5	00.101100100101111	01.011001001011110	01.111100000010110	0	+0.00110
6	00.111100000010110	01.111000000101100	10.011100000000101	+1	+0.001101
7	00.011100000000101	00.111000000001010	01.001101010100000	0	+0.0011010
8	00.001101010100000	00.011010101000000	01.000001011010001	0	+0.00110100
9	00.000001011010001	00.000010110100010	00.011100000010001	−1	+0.001100111
10	00.011100000010001	00.111000000100010	01.011000000100011	0	+0.0011001110
11	00.011000000100011	00.110000001000110	01.011010110110000	0	+0.00110011100
12	00.011010110110000	00.110101101100000	01.010101101100000	0	+0.001100111000
13	00.010101101100000	00.101011011000000	01.000100100101111	0	+0.0011001110000
14	00.000100100101111	00.001001001011110	00.101111111101111	−1	+0.00110011011111
15	00.101111111101111	01.011111111011110	01.111111111011110	0	+0.001100110111110

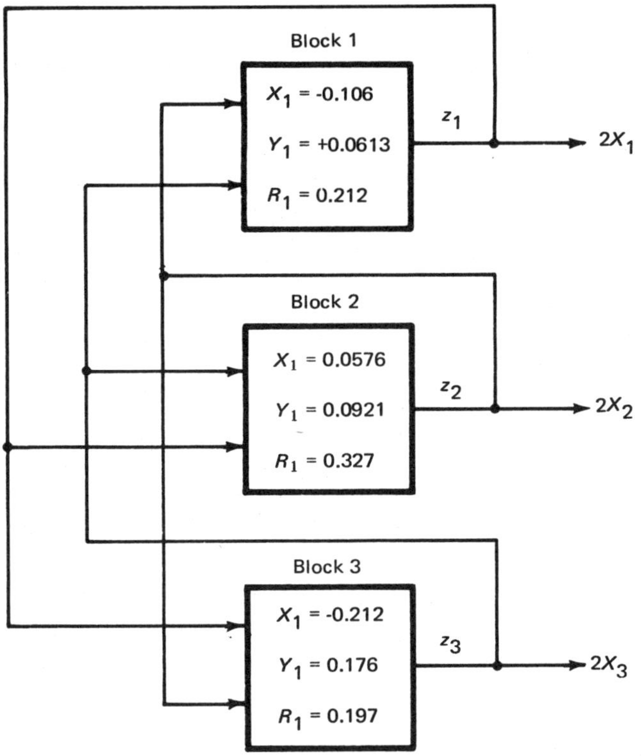

Fig. 4. Block Interconnections to Solve Three Linear Algebraic Equations in Three Unknowns

Table 3
Block No. 2

x = 00.010011101011111
y = 00.001010000110110
R_i = 00.110100111011011

i	R_i	$2R_i$	$2R_i + \Delta_i$	z_i (true)	$2x_2 = \Sigma z_i 2^{-i}$
1	00.110100111011011	01.101001110110110	10.001001110110110	+1	+0.1
2	00.001001110110110	00.010011101101100	00.110011101101100	-1	+0.01
3	00.110011101101100	01.100111011011000	10.000101001101101	+1	+0.011
4	00.000101001101101	00.001010011011010	00.101010011011010	-1	+0.0101
5	00.101010011011010	01.010100110110100	01.110111000011111	0	+0.01010
6	00.110111000011111	01.101110000111110	10.001000001110100	+1	+0.010101
7	00.001000001110100	00.010000011101000	00.111010000010001	-1	+0.0101001
8	00.111010000010001	01.110100000100010	10.010100000100010	+1	+0.01010011
9	00.010100000100010	00.101000001000100	01.001000001000100	0	+0.010100110
10	00.001000001000100	00.010000010001000	00.101100100101001	-1	+0.0101001011
11	00.101100100101001	01.011001001010010	01.110011010001000	0	+0.01010010110
12	00.110011010001000	01.100110100010000	10.000110100010000	+1	+0.010100101101
13	00.000110100010000	00.001101000100000	00.101101000100000	-1	+0.0101001011001
14	00.101101000100000	01.011010001000000	01.111010001000000	0	+0.01010010110010
15	00.111010001000000	01.110100010000000	10.010000100100001	+1	+0.010100101100101

Table 4
Block No. 3

x = 00.000010011011101
y = 00.011011010000111
R_i = 00.101100100110111

i	R_i	$2R_i$	$2R_i + \Delta_i$	z_i (true)	$2x_3 = \Sigma z_i 2^{-i}$
1	00.101100100110111	01.011001001101110	01.111001001101110	0	0
2	00.111001001101110	01.110010011011100	10.011101101100011	+1	+0.01
3	00.011101101100011	00.111011011000110	01.000010100011100	0	+0.010
4	00.000010100011100	00.000101000111000	00.110000010111111	−1	+0.0011
5	00.110000010111111	01.100000101111110	10.000011000011010	+1	+0.00111
6	00.000011000011010	00.000110000110100	00.100110000110100	−1	+0.001101
7	00.100110000110100	01.001100001101000	01.101001111001100	0	+0.0011010
8	00.101001111001100	01.010011110011000	01.101000100010001	0	+0.00110100
9	00.101000100010001	01.010001000100010	01.111100010101001	0	+0.001101000
10	00.111100010101001	01.111000101010010	10.100110001110101	+1	+0.0011010001
11	00.100110001110101	01.001100011101010	01.100001001100011	0	+0.00110100010
12	00.100001001100011	01.000010011000110	01.100010011000110	0	+0.001101000100
13	00.100010011000110	01.000100110001100	01.110000000010011	0	+0.0011010001000
14	00.110000000010011	01.100000000100110	01.110100110011111	0	+0.00110100010000
15	00.110100110011111	01.101001100111110	10.010111001100001	+1	+0.001101000100001

blocks connected together performing the same function. The dinary power increment process requires only 6 multiplications, since each block can be thought of as performing 2, simultaneously with the others.

References

1. J. O. Campeau *et al.*, "Final Technical Report, Advanced Air-To-Surface Missile Computer" (Air Force Avionics Laboratory-68-63, Vol. 1, AD839862 [July, 1968]).
2. A. W. Burks, H. H. Goldstine, and J. Von Neumann, "Preliminary Discussion of the Logical Design of an Electronic Computing Instrument" (June 28, 1946), *Datamation,* Part 1, (September, 1962), 36–41.
3. J. O. Campeau, "Array Processing and Large Scale Integration," *IEEE International Convention Proceedings,* March, 1969.
4. J. S. Kilby, "Device Fabrication" (International Solid State Circuits Conference, February 9, 1966).
5. D. Slotnick, "Unconventional Systems," *Proc. AFIPS Conference*, 30: (Spring, 1967), 477–81.

CHAPTER 12

A PIPELINE PUSH–DOWN STACK COMPUTER

Harold S. Stone

Stanford University
Palo Alto, California

Abstract

This paper describes a new computer organization which is a hybrid of the pipeline computer and the push-down stack computer. The central characteristic of the computer organization is a hardware translator that performs dynamic register allocation. Because dynamic allocation can eliminate register conflicts that would otherwise exist, the new computer organization is potentially capable of higher performance than conventional pipeline computers. The new computer can support an unusual form of multiprocessing, and can eliminate processing delays that occur in pipeline computers when conditional branches are executed.

INTRODUCTION

An important technique for the design of high performance computers is the technique that is known as *pipeline* design. The essential characteristic of pipeline computers is that they can support concurrent operations in the following sense. An operation can be initiated within a module before the completion of the preceding operation. The rate of computation is determined by the rate at which operations can be initiated rather than by the time required to perform

---

The research reported on this project was sponsored in part by the Air Force Cambridge Research Laboratories, Office of Aerospace Research, Bedford, Massachusetts, under Contracts AF 19(628)-5920, and AF 19(628)-2902, and in part by the Atomic Energy Commission under contract AT(04-3)-326, P. A. 23.

The author is on leave from Stanford Research Institute.

each operation. The IBM 360/91 and the Control Data 6600 are two examples of computers that make use of pipeline design.

Efficiency of pipeline computers depends strongly on the way that they are programmed. It is a nontrivial problem to write a machine language program that takes maximum advantage of the potential concurrency within a pipeline computer, and it is a far more difficult problem to construct a FORTRAN or ALGOL compiler that generates efficient programs for such a machine.[1,2] In this paper, we describe a pipeline machine organization that can be programmed very easily to achieve high efficiency computation. The computational portions of the processor are assumed to be similar to conventional pipeline computers. High efficiency is achieved by deferring part of the compilation process until execution. Instructions are coded in a polish string format, but as each instruction reaches the execution unit, it is recoded into a 3-address format. A translator module performs the recoding in such a way as to maintain high utilization of processor resources. Since register allocation is effectively performed dynamically during execution rather than fixed prior to execution, greater concurrency is possible than with a fixed allocation.

The second section of this paper, "A Model of a Pipeline Computer," gives a brief outline of the structure of a pipeline computer. The third section describes the structure of the pipeline push-down stack computer, including a description of the translation algorithm, and the fourth section contains a brief discussion of characteristics of the computer organization that support multiprocessing, executive control, and efficient memory access. A brief summary appears at the end of the paper.

A MODEL OF A PIPELINE COMPUTER

In the current computer vernacular, the term "pipeline" refers to the ability to execute two or more operations concurrently within a module when the interval between initiation of operations is smaller than the process time associated with each operation. A module that operates in pipeline mode can be executing one operation in an early process phase at the same time that it is executing another operation in a late process phase. The term *pipeline* is particularly descriptive because the rate of flow in a petroleum pipeline depends on the rate at which material is pumped into the pipeline, and does not depend directly on the time required to transfer material.

The particular type of pipeline operation described in this paper is a mode of execution of instructions in a central processor.

The computer model described below permits instructions to be issued from an execution station at a rate that is higher than the rate possible in a conventional computer. Several instructions may simultaneously be in process in this computer. The model is intended to exemplify the behavior of computers like the CDC 6600 and the IBM 360/91, but no attempt is made to characterize the actual physical design of such computers.

Figure 1 shows the pipeline computer organization. The processing portion of the computer is contained in a collection of registers, function units, and in a switching network that connects function units to registers. Each functional unit can perform an operation on one or two operands and return a single result. Operations may be fixed-point or floating-point arithmetic operations such as ADD, SUBTRACT, MULTIPLY, and DIVIDE; logical operations such as AND, OR, and EXCLUSIVE OR; magnitude comparisons such as GREATER THAN, EQUAL, and LESS THAN; field operations such as SHIFT, ISOLATE, and NORMALIZE; and other miscellaneous operations. We do not specify the operations of the function units themselves for there is considerable freedom in this choice. Each function unit might be capable of performing only one operation, any operation within a class of related operations, or any operation in the processor repertoire. Within the collection of function units, each function unit may be unique, may be one of a subset of identical units, or all function units may be identical. There is assumed to be at least one function unit that can perform each operation. Otherwise the capabilities of the individual units is a design parameter that can be specified on a cost-performance basis.

We assume that instructions specify one or two operand registers, an operation, and a result register. For example, typical instructions are $R_1 \leftarrow R_2 + R_3$, and $R_1 \leftarrow R_6/R_5$. Memory fetches and stores are coded in the format $R_1 \leftarrow X$ and $Y \leftarrow R_4$, respectively, where we use X and Y to represent memory addresses.

In addition to the processing hardware, the computer contains a unit called the *precedence memory,* an execution station, and an instruction queue. Instructions are processed in the following manner. Instructions are removed from the instruction queue one-at-a-time at the clock rate of the computer. The execution station passes each instruction to the precedence memory if it is not a conditional branch. If the execution station cannot identify the next instruction because of an uncomputed

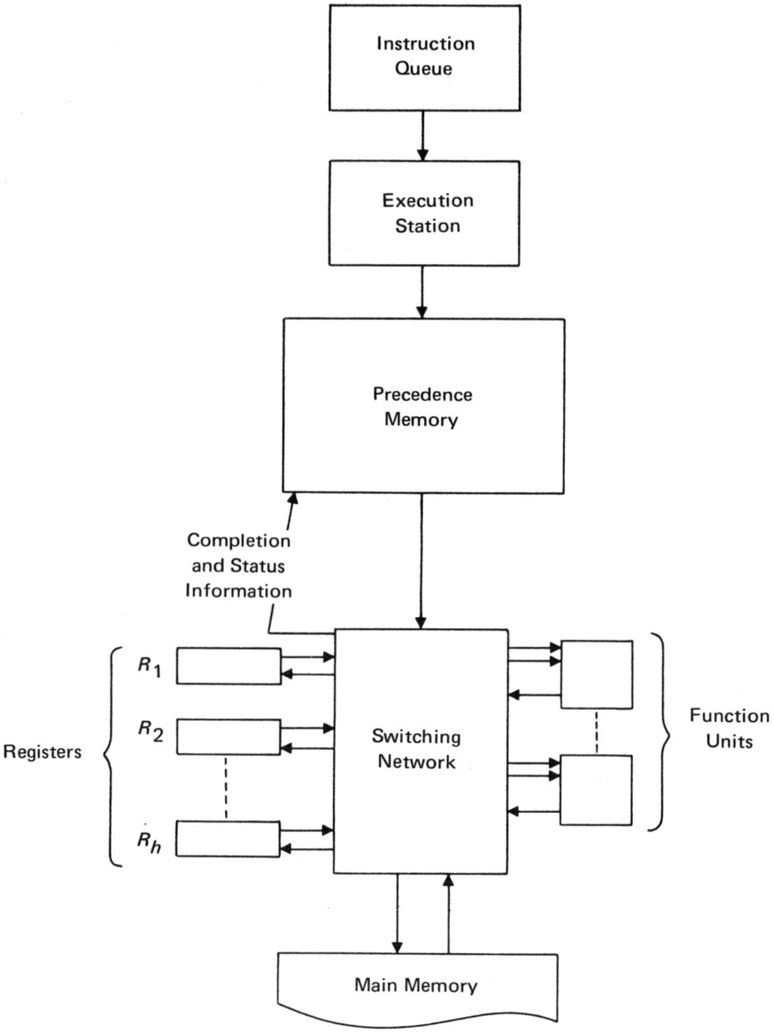

Fig. 1. A Model of a Pipeline Computer

branching condition, the execution station may pursue one of several courses of action in anticipation of the outcome of the conditional test. The precise strategy followed in this case is not an essential feature of the model.

The precedence memory is an active memory that defers or initiates instructions according to the following algorithm. As each instruction enters the precedence memory, the instruction is initiated if it can be initiated, otherwise it is deferred. Deferral of instructions can be caused by unavailability of one or both operands, function unit, or result register. With each deferred instruction the precedence memory records the status of items required for the instruction. Status information is updated as each instruction that has been initiated reaches completion. Status changes that permit any deferred instruction to be initiated are detected in the precedence memory, thereby causing initiation of such instructions. Information pertaining to completed instructions is purged from the memory to permit new instructions to be entered.

To illustrate the operation of the pipeline computer, consider the steps in the evaluation of the numerical assignment $A = B \times C + D \times E$. A typical instruction sequence for this calculation is:

1) $R_1 \leftarrow B$
2) $R_2 \leftarrow C$
3) $R_3 \leftarrow B \times C$
4) $R_4 \leftarrow D$
5) $R_5 \leftarrow E$
6) $R_6 \leftarrow R_4 \times R_5$
7) $R_7 \leftarrow R_6 + R_3$
8) $A \leftarrow R_7$.

All four memory fetches can be initiated concurrently because the instructions use different result registers. Instructions (3) and (6) can also be executed concurrently, but each of these instructions can be initiated only after both operands have been fetched from memory. Figure 2 shows the extent of concurrent execution for this example that might occur within a typical pipeline computer.

Further discussion of the characteristics and organization of pipeline computers can be found in Thornton, Chen, and Anderson *et al.*[3,4,5]

A PIPELINE PUSH-DOWN STACK COMPUTER

The model of a pipeline computer described in the previous section can achieve a high degree of concurrency during execution of instructions, but the degree of concurrency is sensitive to register assignments in the following sense. Two instructions that reference a common register cannot be executed concurrently if one uses the register for an operand and the other uses the register for a result. Hence, register allocation imposes sequential constraints on the execution of

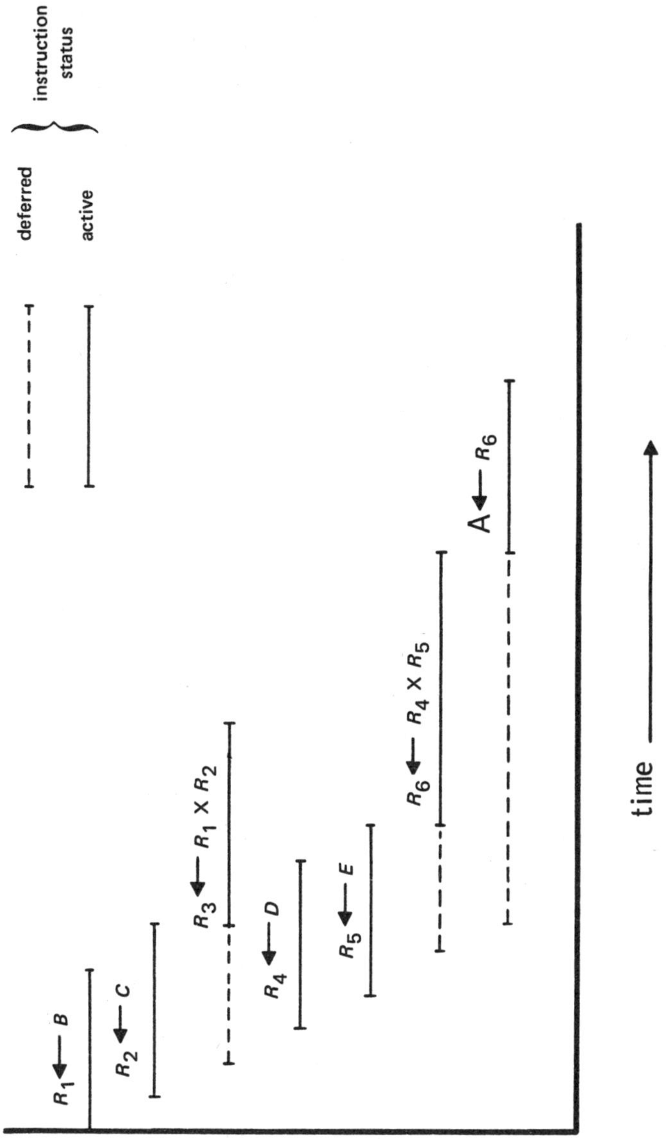

Fig. 2. Concurrent Execution of Instructions in a Pipeline Computer

instructions. "Good" register assignments are those which impose only the essential sequential constraints, and no more. In today's pipeline computers, register allocation is part of the compilation process, but it is the thesis of this paper that register allocation should occur dynamically. Register allocation that is fixed for an entire execution of a program is almost certainly suboptimal, because many segments of programs can be entered in different contexts, and a different optimal register assignment for the segment may exist for each context. The remainder of this section is devoted to the description of a machine organization in which register allocation occurs dynamically.

A machine organization which performs dynamic register allocation is shown in Fig. 3. We have added three modules to the pipeline computer. The module identified as a *translator* accepts instructions in a push-down stack format that will be described shortly. The output of the translator is a stream of instructions in the standard pipeline computer format. The "idle register queue" is an active memory that maintains a list of unassigned registers. The translator uses this list together with information in the "push-down stack" as it performs the task of register allocation. The salient feature of this machine organization is that register references in the input instruction stream are implicit rather than explicit. Explicit register references are compiled in instructions just prior to their actual execution, and the assignments are made on the basis of immediate availability of registers. Therefore, no instruction need be deferred unnecessarily because of register assignment conflicts.

The instructions at the input of the translator are in push-down stack format. That is, they operate on the top register or top two registers of a push-down stack. The instructions VALUE X and NAME X place the value and the memory address of symbolic variable *X* at the top of the stack, "pushing down" the previous contents of the stack. Arithmetic instructions ADD, SUB, MUL, and DIV use the top two registers of the stack as operands, compute an arithmetic result, and replace the top two registers with a single result. Arithmetic operations effectively "pop-up" the stack one time. The instruction STORE interprets the top element of the stack as a value and the next to top of the stack as an address. STORE causes the value at the top of the stack to be placed in the memory address given by the next to top element of the stack, and deletes both of the top two items from the stack. A complete instruction repertoire necessarily includes instructions for indexing, logical operations, conditional and unconditional branching, and subroutine calls. For the purposes of this paper, the seven instructions

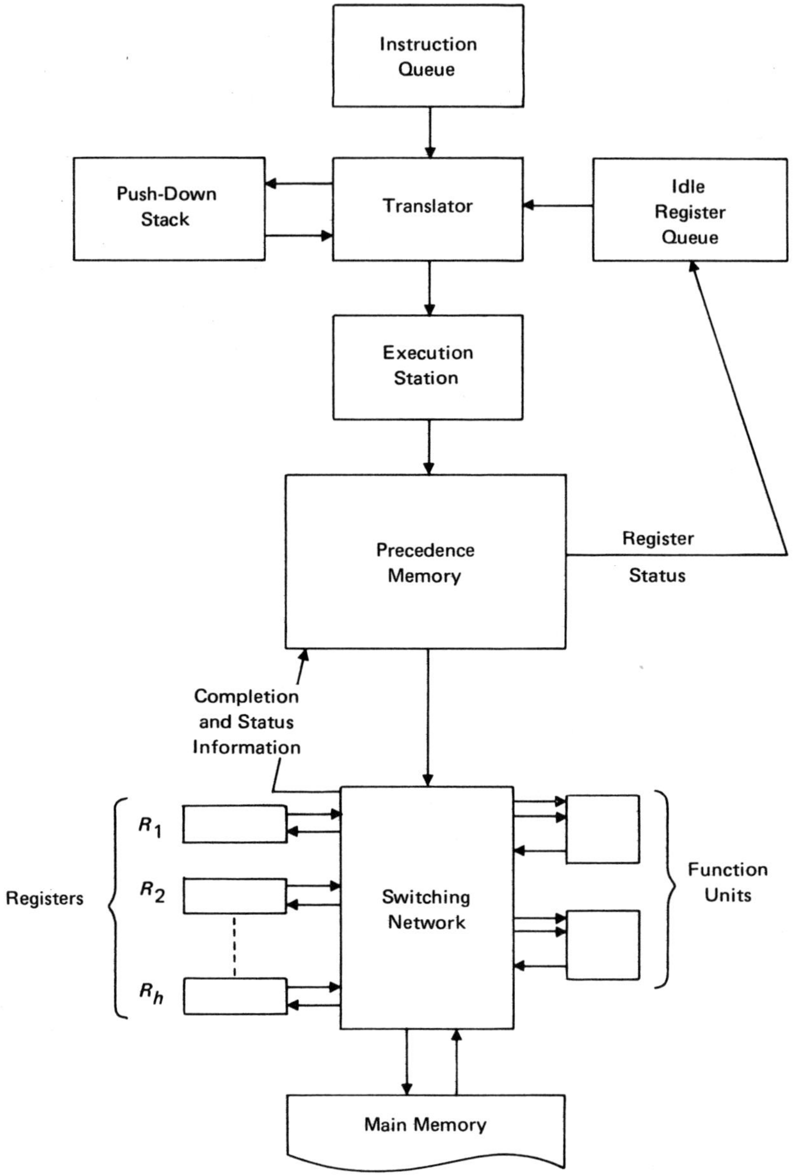

Fig. 3. A Model of a Pipeline Push-down Stack Computer

that have been described thus far are sufficient to illustrate the concept of the hardware translator. A complete description of a computer that uses push-down stack instructions can be found in Burroughs.[6]

The calculation $A = B \times C + D \times E$ given in the previous section can be written as a program for a push-down stack computer as follows:

1) NAME A
2) VALUE B
3) VALUE C
4) MUL
5) VALUE D
6) VALUE E
7) MUL
8) ADD
9) STORE.

A characteristic of the stack instruction set that is exemplified in the program above is that the program contains no explicit references to registers in a central processor other than the implicit references to the registers at the top of a push-down stack. When these instructions are executed in a push-down stack machine, the *implicit* references are turned into *explicit* references. It is not unusual, therefore, in stack organized computers to find that a particular instruction might refer to different machine registers at different execution times. Since register references in stack organized computers are not bound until the actual execution of instructions, instructions in push-down stack format are ideal for supporting dynamic register allocation.

Algorithms for translating a push-down stack format to a standard register format are not new. They have been incorporated in nearly all compilers. Compilers generally translate source text into an intermediate text known as a "polish string," which is roughly equivalent to a program in push-down stack format. The intermediate text is then translated into standard format by assigning registers to the instructions. See, for example, Horowitz *et al.*[7] For our model of a pipeline computer, we shall state an algorithm for translating the seven instructions VALUE, NAME, ADD, SUB, MUL, DIV, and STORE into a 3-address format. The algorithm can be generalized to a full stack instruction repertoire in a straightforward manner.

In the processor shown in Fig. 3, the idle register queue contains indices of unassigned registers. If the queue is empty when a request for a register is issued by the translator, the queue defers response until an idle register becomes available. (If a program issues a sufficiently long sequence of VALUE instructions, all registers can become unavailable simultaneously without a possibility of a register becoming available. We assume that programs never require more registers to be active simultaneously than the number of registers in the processor. The assumption is valid, in practice, for there exist several mechanisms for enforcing the assumption.)

Each cell of the push-down stack can contain a register index or a memory address. The stack itself is under control of the translator. The translation algorithm for our seven instruction repertoire is the following:

1) VALUE: The translator obtains a register from the idle register queue. Let this register be R_i. R_i is stacked on the push-down stack and the instruction VALUE X is translated to $R_i \leftarrow X$.
2) NAME: The instruction NAME X causes the address of X to be stacked. No translated instruction is produced.
3) ADD, SUB, MUL, DIV: Let the top two entries of the stack be the registers R_i and R_j. An idle register is obtained from the idle register queue. Let this register be R_k. Then the instruction $R_k \leftarrow R_i$ <op> R_j is produced where <op> is $+, -, \times$, or $/$. R_i and R_j are popped-up and R_k is pushed-down.
4) STORE: Let the top entry of the stack be the register R_i and the next to top entry be the memory address X. Then the translator produces the instruction $X \leftarrow R_i$ and the stack is popped-up twice.

To illustrate the algorithm, assume that the idle register queue initially holds the indices 1 through 7 in ascending order. Then if the translator operates on the program for the calculation of $A = B \times C + D \times E$ as given in this section in push-down stack format, the output of the translator will be the program for the same calculation in 3-address format that appears in the previous section.

The machine organization in Fig. 3 is a *hybrid* in the sense that it has characteristics of both the push-down stack computer and the pipeline computer. Since the translation process can operate concurrently with execution, and since register allocation is guaranteed to be optimal or near optimal, the new computer organization is potentially capable of higher performance than the pipeline computer. The new organization can also achieve higher performance than that obtainable in a conventional push-down stack computer. The latter claim is true because operations on the top of the stack impose sequential constraints on the execution of instructions in a push-down computer. It is not possible, in general, to execute push-down instructions concurrently without translation since the context of each instruction depends on the stack manipulation of the immediately preceding instructions. Any computer organization that can support concurrent execution of push-down stack instructions must simulate stack manipulation in some manner, and consequently must perform a process similar or equivalent to the translation process in the new computer organization.

There are several interesting facets of the new processor organization that deserve separate attention. We treat each of these in the next section. Further information on the organization and design of this pipeline push-down stack computer appears in Elspas *et al.*[8]

CHARACTERISTICS OF THE PIPELINE PUSH-DOWN STACK COMPUTER

In this section we describe how several conventional and advanced techniques of computer design are compatible with the organization of the pipeline push-down stack computer.

Multiprocessing

A multiprocessor can be constructed in a conventional way by replicating copies of the processor shown in Fig. 3. However, an unusual form of multiprocessing can also be supported by a single processor. Several independent instruction streams can share a single processor provided that each of the instruction streams has its own program counter, translator module, and push-down stack module. The number of registers and functional units must be chosen to satisfy the processing requirements of multiple instruction streams.

There is a potential advantage in operating one processor with multiple instruction streams instead of using multiple copies of a processor. Resources that are not in use temporarily by one process can be put to use by another process so that the multiple instruction stream processor can achieve higher utilization of its resources than a single instruction stream processor. If high resource utilization more than balances the performance degradation due to increased size and complexity, the multiple instruction stream processor is a potentially better vehicle for multiprocessing than a computer with multiple processors.

Execution of High Priority Processes

In many situations, high-priority processes can preempt a processor for brief periods of time. These processes may be associated with executive control, interrupt processing, or control of the external environment under a real-time deadline. In the conventional computer, a high-priority process either preempts an entire processor, or uses hardware that is private to the process. In the new computer organization, a high-priority process can share a processor with one or more low-priority processes.

A high-priority process can gain access to a processor as an independent instruction stream. To guarantee that the high-priority process obtains all of the registers and functional units that it requires, we postulate that a high-priority process can reserve resources for a short period of time. Low-priority processes can continue to operate, but with a smaller pool of registers and function units than would otherwise be available. Since the precedence memory and the idle register queue maintain a table of all available and unavailable resources, resources can be reserved or freed by making appropriate status changes in the resource table.

Look-aside Memory

The precedence memory can be used as a look-aside memory to eliminate unnecessary main memory references. Suppose, for example, that the instruction $X \leftarrow R_1$ has been deferred or is in the process of execution when the instruction $R_2 \leftarrow X$ enters the precedence memory. The precedence memory must be able to detect the fact that both instructions refer to the same memory address, X, and that the second instruction might have to be deferred until the completion of the first instruction. The precedence memory can eliminate a reference to main memory by changing the second instruction to the form $R_2 \leftarrow R_1$ since register R_1 has a copy of the contents of X. In this form, the instruction can be executed concurrently with the instruction $X \leftarrow R_1$.

Conditional Branches

A novel method exists for avoiding delays that are normally associated with conditional branches. The processor can proceed along both branches of a conditional branch until the correct branch is identified.

The following method is used to guarantee that computations on the two branches of a conditional branch cannot conflict with each other.

Computation proceeds normally until a conditional branch is reached. At that point, a second translator and push-down stack are brought into operation. The two translators then proceed concurrently along opposite paths of the conditional branch. Since register assignments are dynamic, the computations on the separate paths are guaranteed to use different sets of registers. Instructions generated by each translator must be identified with their respective translators so that all instructions associated with the incorrect branch can be purged from the precedence memory when the branching condition computation is completed. If either translator reaches a STORE instruction, it halts until the correct branch is identified since we cannot allow incorrect branches to modify main memory.

Translation into Two-address Instructions

In the pipeline processor, it may be advantageous to use a two-address format instead of the three-address format that has been assumed. Two-address instructions take the form $R_i \leftarrow R_i$ <op> R_j where the result register is the same as the first operand register. Two-address instructions impose a smaller storage requirement on the precedence memory than three-address instructions, and are also somewhat simpler to translate. In two-address processors, when a translator encounters the instruction ADD in the instruction stream, it can issue the translated version $R_i \leftarrow R_i + R_j$ without accessing the idle register queue. The indices i and j are assumed to be the top two indices on the push-down stack, and the stack is popped-up once. No new index need be rewritten in the stack. Translation into two-address format requires access to the idle register queue and requires that the contents of one register in the stack be modified.

Compiler Methods for Maximizing Parallelism

In spite of the fact that compilers need not perform register allocation for this computer organization, compilers can produce instruction sequences that take into account the parallelism available in the new processor organization. For example, the computation $A = B + C + D + E$ can be calculated as if it were $A = ([B + C] + D) + E$, or as if it were $A = (B + C) + (D + E)$. The latter is clearly more suitable for a pipeline computer because the terms $B + C$ and $D + E$ can be calculated concurrently. Algorithms for maximizing concurrency have been studied by several authors—Squire, Hellerman, Stone, and Baer *et al.*[9,10,11,12] The one-pass algorithm given by Stone is specifically designed for this machine organization, while the other algorithms are designed for more conventional pipeline computers.

Although there are several other minor points that bear discussion, the observations of this section cover the major characteristics of the new computer organization. One point that has purposely been avoided has been the organization of main memory. It should be clear that the processor can be used with any advanced memory organization of sufficient size and speed.

SUMMARY

The processor organization described in this paper is a hybrid of the push-down stack computer and the pipeline computer. Because it uses dynamic register allocation to minimize register conflicts, it is potentially capable of higher performance than that attainable by today's pipeline computers. The distinguishing characteristic of the new computer organization is a hardware translator that performs register allocation at execution time with the aid of algorithms that heretofore have been found in compilers.

A fundamental concept that underlies the processor organization reported here is that resource allocation of a general nature is inherently a dynamic process. In order to attain the highest achievable performance, we must expect to use hardware to perform dynamic resource allocation. Several questions pertaining to scheduling, memory allocation, and task partitioning are still open in this area, and must be resolved if we are to overcome the present limitations of high-speed computers.

References

1. R. W. Allard, K. A. Wolf, and R. A. Zemlin, "Some Effects of the 6600 Computer on Language Structures," *CACM,* 7: No. 2 (February, 1964), 112–19.
2. H. L. Nelson, "Program Optimizing Techniques for the CDC 6600 Central Processor," UCRL-12489 (University of Calif. Lawrence Rad. Lab. [1965]), also NASA N66-20536.

3. J. E. Thornton, "Parallel Operation in the Control Data 6600," *AFIPS FJCC Conf. Proc.*, 26: Part II (Washington, D. C.: Spartan Books, 1964), 33-40.
4. T. C. Chen, "The Overlap Design of the IBM System/360 Model 92 Central Processing Unit," *AFIPS FJCC Conf. Proc.*, 26: Part II (Washington, D. C.: Spartan Books, 1964), 73-80.
5. D. W. Anderson, F. J. Sparacio, and R. M. Tomasulo, "System/360 Model 91: Machine Philosophy and Instruction Handling," *IBMJR&D*, 11: No. 1 (January, 1967), 8-24.
6. *Burroughs B5500 Information Processing Systems Reference Manual* (Burroughs Corp., [Detroit, Mich., 1964]).
7. L. P. Horwitz *et al.*, "Index Register Allocation," *JACM*, 13: No. 1 (January, 1966), 43-61.
8. B. Elspas *et al.*, "Investigation of Propagation-limited Computer Network," AD-637-769 (Phase III Final Report, S.R.I. Project 4523, Air Force Cambridge Research Laboratories, Office of Aerospace Research [Bedford, Mass., June, 1966]).
9. J. S. Squire, "A Translation Algorithm for a Multiple Processor Computer," *Proc. 18th ACM Nat., Conf.* (Denver, Colorado, 1963).
10. H. Hellerman, "Parallel Processing of Algebraic Expressions," *IEEETEC*, EC-15: No. 1 (February, 1966), 82-90.
11. H. S. Stone, "One-pass Compilation of Arithmetic Expressions for a Parallel Processor," *CACM*, 10: No. 4 (April, 1967), 220-23.
12. J. L. Baer and D. P. Bovet, "Compilation of Arithmetic Expressions for Parallel Computations," *IFIPS Congress 68*, August, 1968, pp. B4-B10.

CHAPTER 13

A MULTIPLE INSTRUCTION STREAM PROCESSOR WITH SHARED RESOURCES

M. J. Flynn

Johns Hopkins University
Baltimore, Maryland

A. Podvin and K. Shimizu

International Business Machines Corporation, FSD
Los Angeles, California

Abstract

A parallel computer system organization is proposed consisting of 32 "skeleton" or virtual individual computers. Each virtual machine contains its own data and control registers, but lacks the more substantial execution facilities which, in turn, are shared by all machines. This sharing is done by closely synchronized time-phased switching. The execution resources, on the other hand, are heavily pipelined for maximum operational bandwidth. The pipelining factor for each of the execution functions is closely related to the synchronizing factor of the virtual machines. The resultant system avoids many of the contention problems associated with shared resource systems.

With an assumed circuit performance of 2 nanosecond/stage, a maximum system performance of 500 million instructions per second is projected.

A system simulator that interpretively executes test programs has also been developed to model the proposed system. The simulation results of two typical programs indicate that close to maximum performance can be achieved.

The authors are especially grateful to Mr. R. Rock and Mr. D. Bluett for the assistance in developing the simulator.

We are also indebted to Mr. B. O. Evans, President, Federal Systems Division, IBM, for his warm personal support of this study.

INTRODUCTION

While many problems exist which, because of their intrinsic spatial regularity, are well served by single instruction stream array processors, there is an even larger universe of operating environments where this is not true. These "environments" may be simply independent jobs or nonuniform segments of complex large programs. In an attempt to address this latter situation, the organization proposed in this paper was developed.

The traditional answer to the multiple simultaneous task problem is the *multiprocessor* organization. In this organization, n independent processors, each with identical physical arrangement, perform tasks (more or less independently). Only storage (at some level) is shared. The general criticism of this organization is that while the processor cost has risen n-fold, performance has risen by a lesser factor (due to interference).

As an alternative, resource shared organizations have been suggested. In these arrangements, various physical functions (adders, multipliers, etc.) are broken out and shared by "skeleton" or procedural machines. The most notable early effort along these lines was the GAMMA-60.[1] The potential problem with such sharing is that the switching overhead between virtual machine and resource militates against overall efficiency. One answer (as used in the adder of the peripheral processors of the CDC 6600) is to synchronously time-share the resources among the processors.[2] This arrangement was proposed on a more universal basis for time-sharing and independent job stream applications by the IMP proposal at Argonne National Laboratory.[3] The proposal described in this paper is an extension of the earlier proposals—to address the area of uniform and nonuniform segments of single large problems as well as the independent job stream case.

PREMISES AND OBJECTIVES OF THE PROPOSED SYSTEM

Technology

In the area of technology, the system would be envisioned as using the state of the art circuitry and storage. For circuitry, this is probably in the order of 2 nanoseconds per stage and corresponds to technology about twice that of System/360 Model 91. In the area of buffer storage, the availability of monolithic storage in small arrays (small in this case meaning from 128 words to 2,000 words) is assumed. Because of the nature of this system, it is relatively insensitive to main storage speed, and while this machine description centers on a main cycle of 300 nanoseconds, main storage as slow as 3/4 microseconds per cycle could be considered.

Performance

With the assumption of a logic technology (2 nanoseconds per stage), a peak performance in excess of 500 MIPs (million instructions per second) is expected.

Resource effectiveness

A serious attempt is made in this system to provide maximum computational performance per dollar invested in the system, especially in three areas:

1) *Technology:* To take advantage of the natural cost economies of monolithic circuitry, a high degree of replication of circuit arrays is needed. This is normally available in the execution area. The parallel organization extends replication to the instruction unit area.
2) *Execution areas:* The design includes a very high bandwidth execution unit, which is time-shared by 32 instruction units. By stressing bandwidth rather than latency, the burden of technology is lessened, while the instruction unit utilization of the execution area bandwidth assures high performance.
3) *Instruction area:* The use of parallel branching (forking) to processors gives the system an ability to adapt to the instruction unit resources.

MACHINE ORGANIZATION

It is well known that in classical high-speed organizations, the execution resources vastly exceed the requirements of the instruction stream; thus the IBM System/360 Model 91, had execution resources in excess of 70 MIPS (million instructions per second).[4] Thus, the floating-point adder could process an add or subtract operation every 60 nanoseconds (16 MIPS), the floating-point multiplier 6 MIPS, the variable field length/fixed point unit 16 MIPS, 16 MIPS for register loads, 16 MIPS for stores, etc. This potential was immediately restricted by the maximum instruction decode rate of 16 MIPS; and further, with an average incidence of conditional branch type instructions to 6 MIPS.

The reason for the discrepancy between 70 and 6 MIPS is twofold:

1) It is more efficient (in a performance-oriented sense) to build units specialized to a particular function rather than a universal execution unit. In the universal unit, both the increased size and control mitigate against performance.
2) A single stream of instructions as executed has heavy sequential dependency. Thus, for example, if P_a is the overall probability of incidence of a floating-point add, and P_b the probability of a Boolean instruction, and if i and $i + 1$ are two successive instructions in a stream, one would expect

$$(P_a)_{i+1} \mid (P_a)_i > P_a$$

and

$$(P_b)_{i+1} \mid (P_a)_i < P_b .$$

That is, the conditional probability of a floating-point add given an add is much higher than its normal incidence.

The organization proposed here is an attempt to develop a high-performance machine architecture by taking advantage of the specialized implementation of functional execution. In fact, within each execution unit the main emphasis will be on operative bandwidth (rate of execution), thru pipelining techniques, rather than operative latency (total time for one functional operation to complete).

To overcome the instruction stream dependency problem, a multiple instruction stream system with resource contention is proposed. The implicit assumption is that at time i the requests of n processors will be independent (or certainly more independent than the single stream processor).

Contention for, and sharing of, resources has a number of difficulties; the most notable being the size, extent, and performance of the communication system which links requestor to the resource. In order to eliminate the self-defeating bulk associated with a massive "cross-bar" switch, the proposed system uses a highly synchronized time-phased logical switching. Such a system uses the communication system at its maximum possible logical band width and avoids unnecessary logical bulk and repetition.

The multiple instruction stream proposed organization uses an execution area of very high bandwidth shared by 32 virtual machines. It is presumed that the latency in the execution area (that is, the total time needed to complete the operation from the presentation of the operands to the production of the result) is 64 nanoseconds. All algorithms, or execution units, are heavily staged or pipelined for maximum bandwidth. Sharing this execution unit are 4 time-phased arrays (or rings) of processors (Fig. 1); each ring contains 8 virtual processors. For the moment, each processor can be viewed as being the basic registers of a simple machine such as the IBM 7090. Each virtual processor is responsible for fetching its own operands, and preparing its own instructions. It does not execute the instruction (with the exception of load/store/branch), but rather requests the execution unit to do so. If the request is accepted, the result is returned 64 nanoseconds later. The total instruction preparation time is also 64 nanoseconds. Instruction preparation, and execution are overlapped. Thus, after a single processor prepares its instruction, generates its operands, and forwards them to the execution unit, it commences the preparation of the next instruction. All processors are closely time-synchronized, and no two processors within a ring are in the same phase of instruction preparation or execution. Each virtual machine is phased by 8 nanoseconds from either of its neighbors. As seen by the execution unit, each virtual machine has an 8-nanosecond slot in which it can request service on a request bus. In a separate accept bus, the virtual machine is informed 8 nanoseconds later whether or not its request was accepted. So far, we have described the action of one ring, its associated 8 processors, and their action on the execution area. There are 4 rings each with independent request and accept buses.

Four virtual machines, each from a different ring, contend for execution resources at a particular slot. This contention is time overlapped with operand fetch, etc. So long as requests are being made to separate resources, no problems ensue. If, however, two or more processors request a resource which can accept only one operation, a priority system must resolve the conflict. In this system, each ring has an assigned priority (0 through 3), and in conflicting cases, control passes to the highest priority requesting ring. Priority assignment rotates among the rings once every revolution (64 nanoseconds).

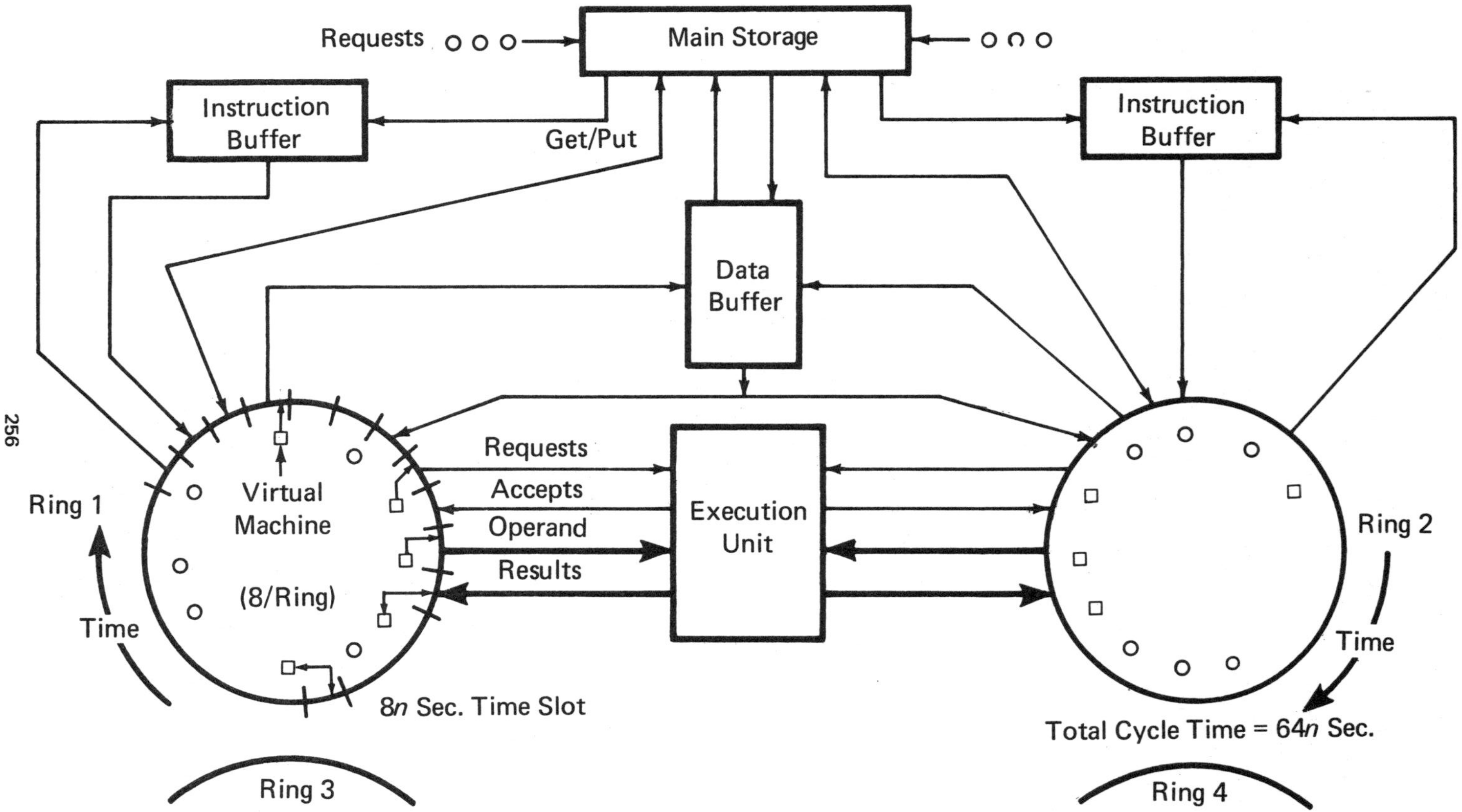

Fig. 1. The Idealized System

THE EXECUTION UNITS

The execution unit contains 7 independent, autonomous boxes or units: 2 floating-point add/subtract units, 1 floating-point multiply unit, 2 floating- and fixed-point divide units, and 2 "catch all" units responsible for fixed-point add/subtract, fixed-point multiply, Boolean and a shift or VFL (variable field length) operations. All operations except the divides have 64-nanosecond latency. In the case of some of the small operations such as Boolean and fixed point add, this will necessitate the addition of an appropriate delay.

In order to generate the high bandwidths required, the individual execution units are extensively staged (pipelined). The input staging area timing is necessarily uniform at some multiple of 8 nanoseconds. Note that the stages need not be completely embedded in the algorithm, but may be appended as high-speed dummy shifts. Figure 2 shows the allocation of the execution unit and the bandwidths required of each unit. In the floating-point area, two adders are used. They are staged (Fig. 3) at the output of the exponent difference, fraction justification, primary add, look-ahead add, exponent up-date, and post shift. Multiply is also naturally decomposable because of its tree structure, and the use of carry-save adder stages. The fixed-point add/subtract operations, together with the Boolean and shift operations, would not normally take a full latency period. Additional delay must be added in each of these areas to insure proper timing of the system. Divide (both fixed and floating point) is not a single latency operation, but requires four *revolutions,* or 4 × 64 nanoseconds for completion. In the case of a divide operation, a bit is set with the acceptance of the request by the divide unit in the virtual machine. This bit disables the virtual machine from initiating any new instructions. It is enabled one cycle before the quotient is scheduled to appear. All data paths are 48-bits wide, but result paths for floating-point operations are 96-bits wide. Instruction and data formats are shown in Fig. 4. The relative timing for machines in the use of the execution area is shown in Fig. 5.

INSTRUCTION UNIT

Each of the 32 instruction units (or virtual machines) is identical. They consist (as shown on Fig. 6) of a private data array of 128 words, the lower 4 of which words are used for 8 index registers. Sixteen base

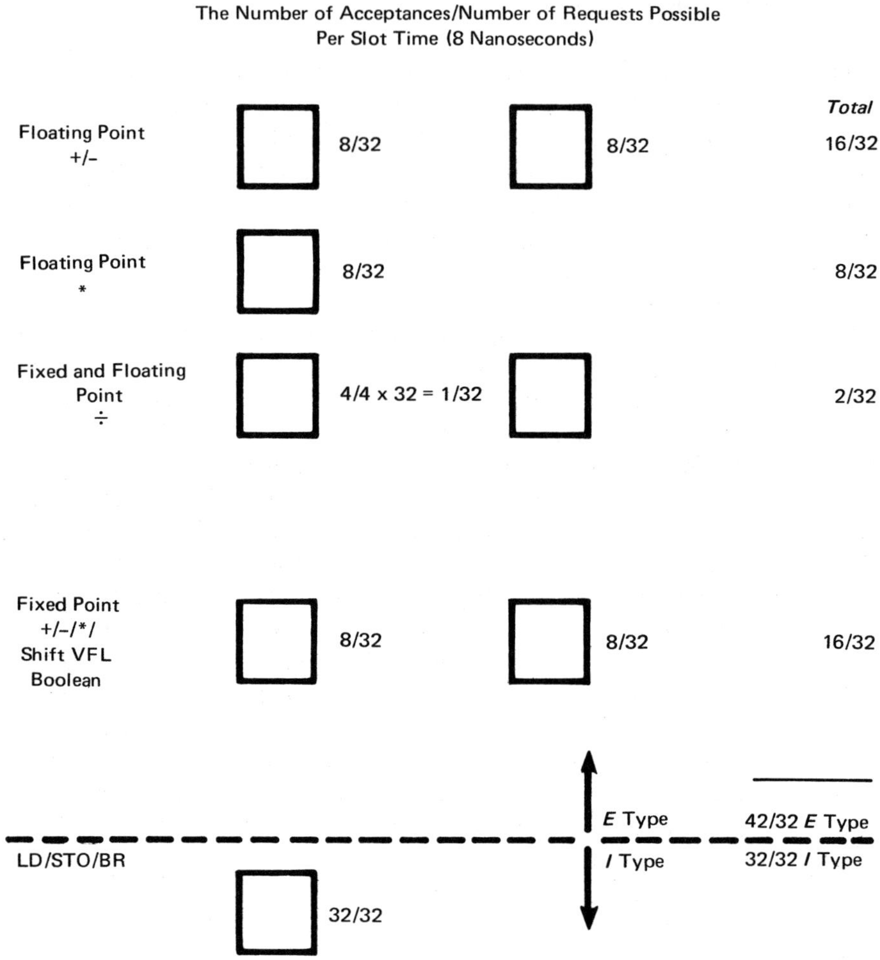

Fig. 2. Execution Area

registers per instruction unit are included in the system. An Accumulator–MQ is the sink for most operations. The high order bits of the Op Code in the instruction distinguish between the load-store-branch type instruction and the normal or executable instructions. The former instruction class is executed by the instruction unit itself. No request for execution unit service is made. If the high order bit of the Op Code is a "1" (indicating a non-LD/STO/BR), the next three bits in the Op Code will address 1 of 8 execution unit types, the remaining bits of the Op Code identify the suboperation, within that particular execution unit's capabilities, which is required to be performed. The 48-bit instruction format is used primarily for data transfers to and from main storage. Operation proceeds after a load from, or store to, main memory has been requested. Associated with every word in the private array is a load bit and a store bit. When the load bit is *on,* this implies that the word has been designated as a sink for information from main storage, and reference to such words as an operand will be denied until such time as it is turned *off* by completion signal from storage. If the store bit is *on* the contents of the particular word cannot be disturbed, and any subsequent store instructions into this particular location will be denied until the bit is turned *off*. The 16 base registers provide a relocation mechanism for main storage.

STORAGE

Storage Hierarchy Philosophy

Memory hierarchy considerations for a parallel processor involve two aspects. First, the parallel processing technique raises a need for sharing

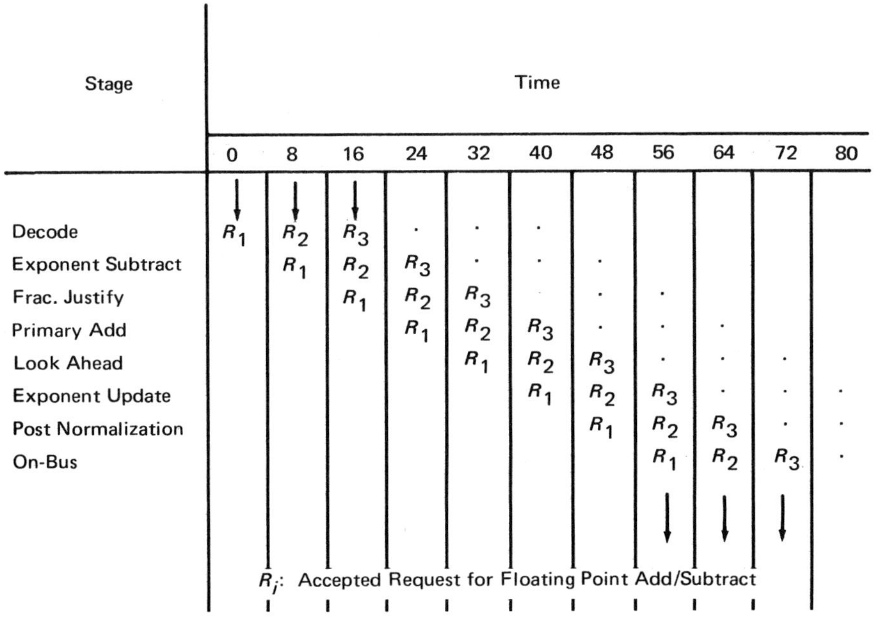

Fig. 3. Stages in Floating Point Add/Sub Unit

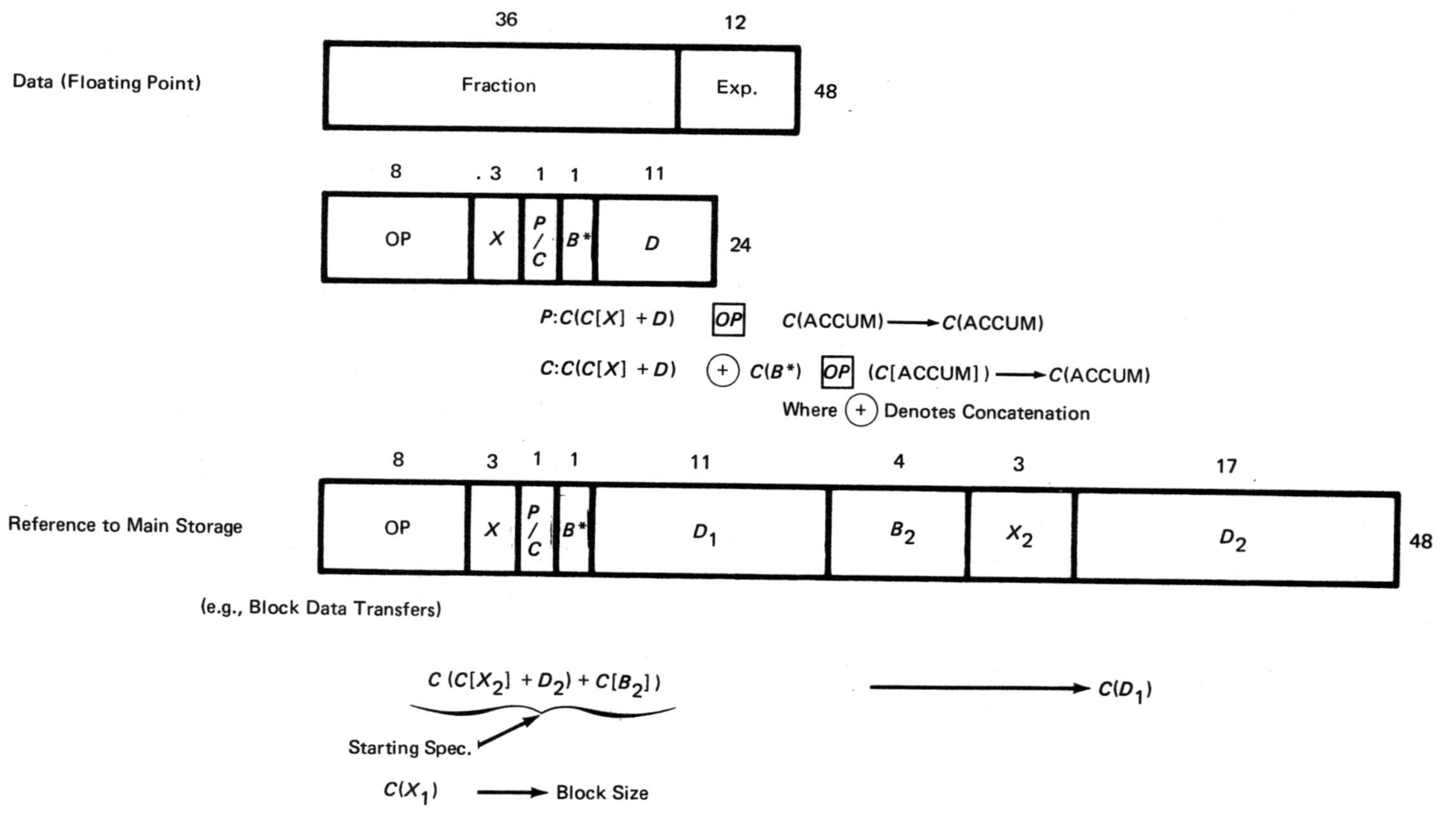

Fig. 4. Data and Instruction Format

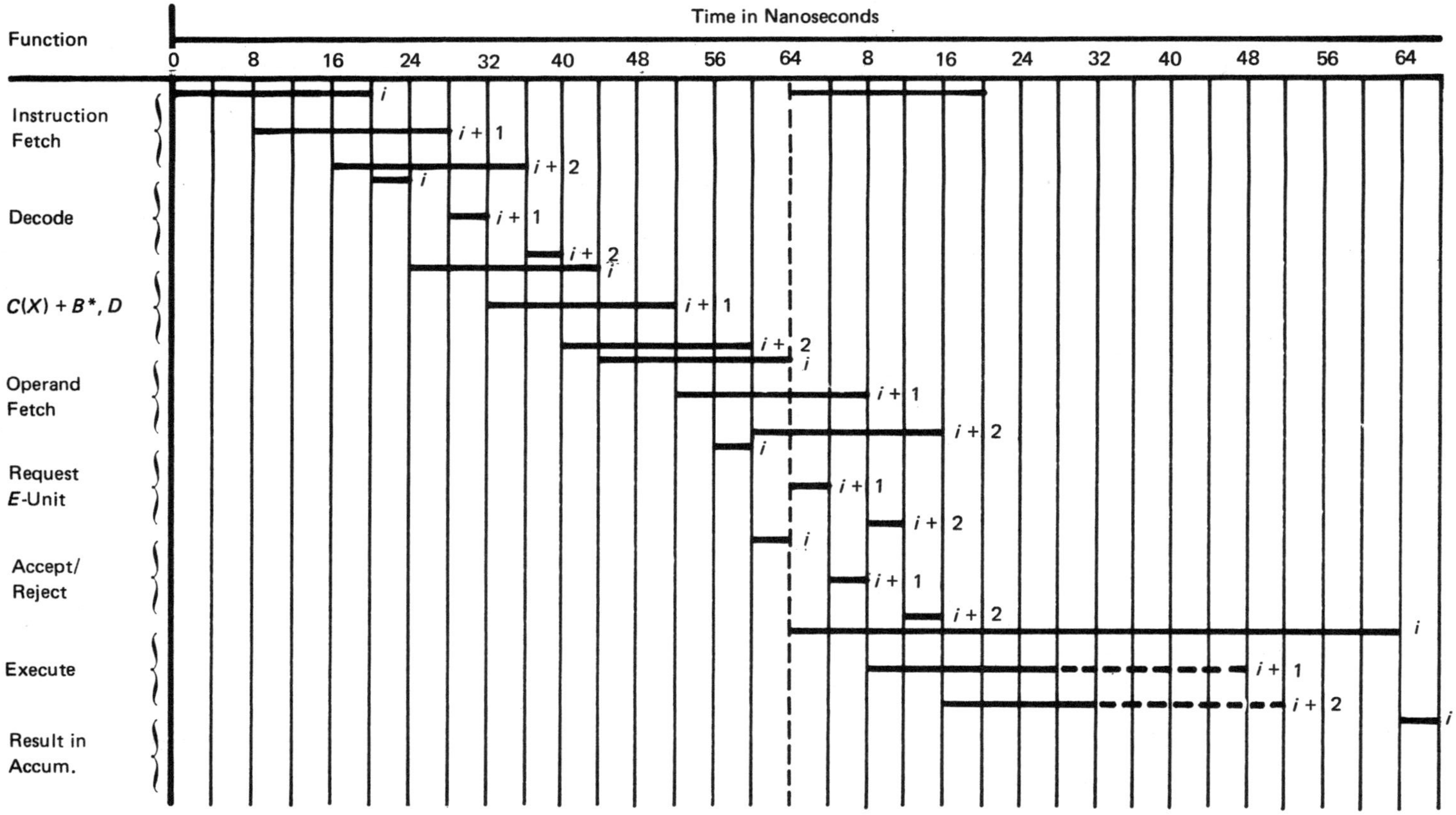

Fig. 5. Simplified Timing for Processor i, $i + 1$, and $i + 2$

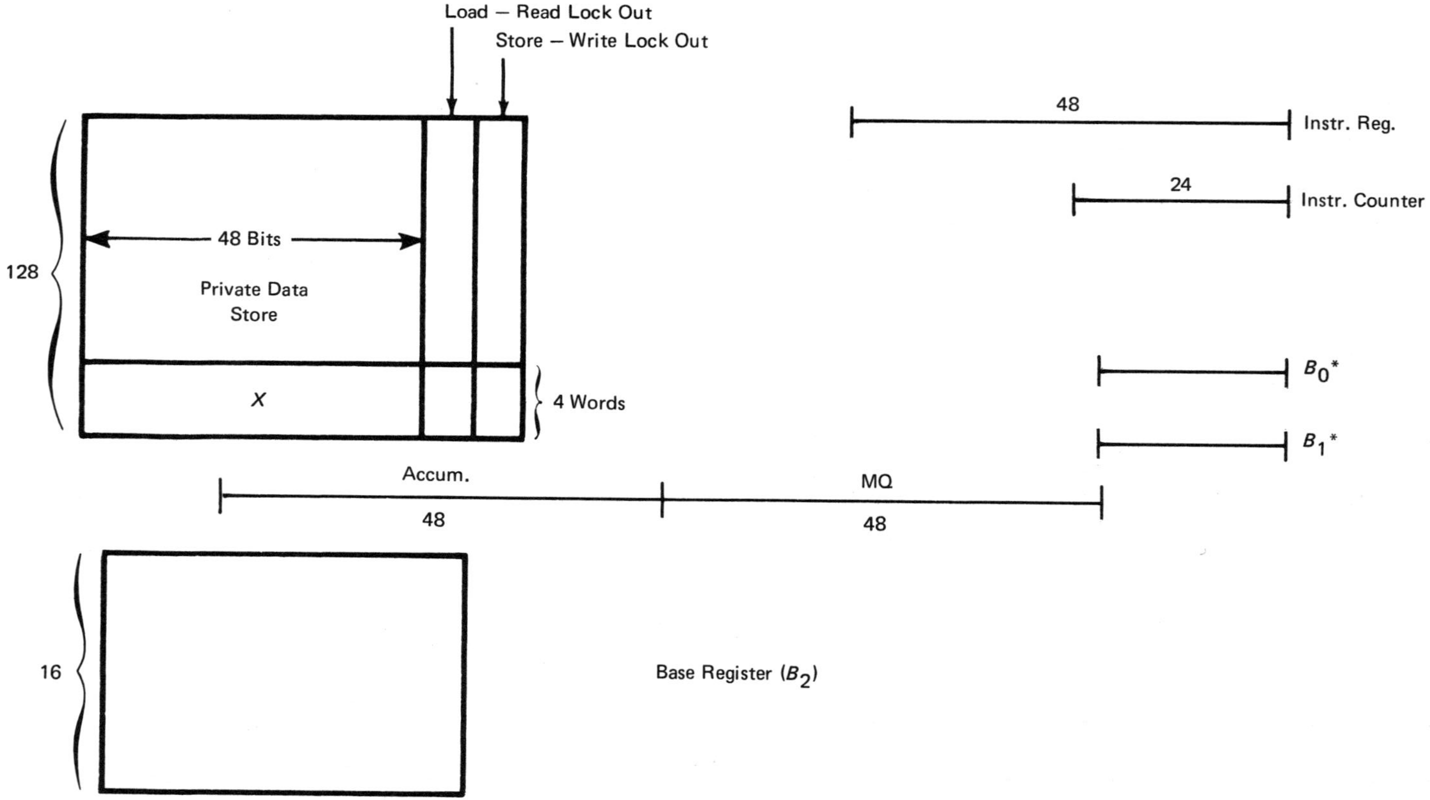

Fig. 6. Instruction Unit (Virtual Machine)

data, instructions, and control information, and for tasking and synchronization resulting in logical and hardware contention involving memory. This aspect of the memory hierarchy consideration uniquely applies to parallel processors, and, therefore, the objective is to minimize memory interference due to multiple accesses by a number of processors. The second aspect applies to any system, including uni- and multiprocessors, and the objective is to bring memory containing frequently used data and instructions closer to processors.

The first consideration results in a level of memory which is shared by all the processors. This level may be main memory or an intermediate buffer memory, depending on the remoteness (access time) of main storage. If an intermediate shared buffer arrangement is required, several possibilities arise: separate or similar treatment of data and instructions, and the implicit (addressed as main memory under hardware control) or explicit (directly addressed by programmer) use of the buffer (Fig. 7).

The second consideration establishes a requirement for a private buffer for each processor. The private buffer should be further divided

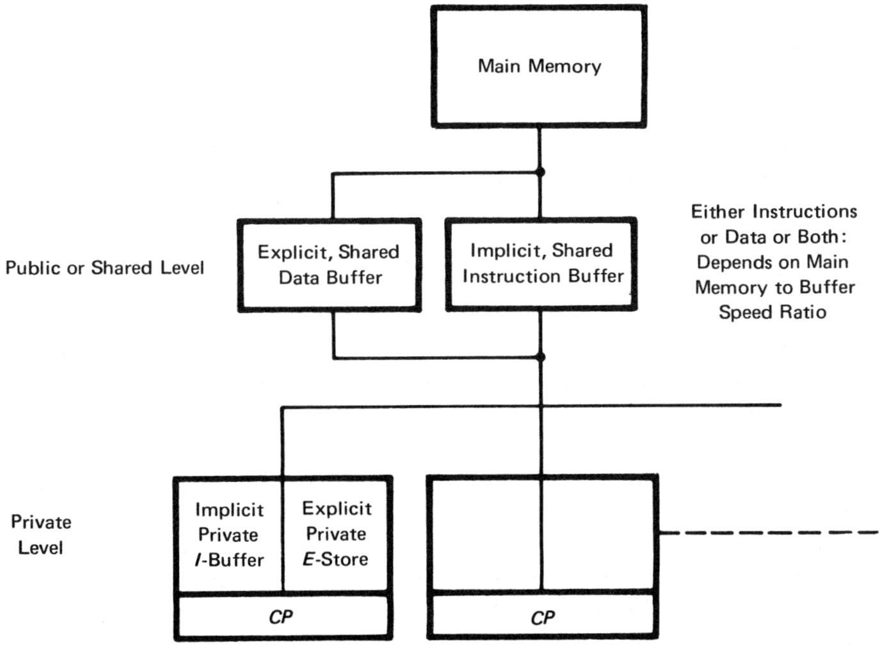

Fig. 7. A Storage Philosophy

into an explicitly addressed data store and implicitly addressed instruction buffer because data and instructions have their own unique characteristics.[5] Instructions are predominantly sequential and "read only," while data are referenced rather randomly and of "read/write" type. When an implicit data buffer is used, and the requested data is already in the buffer, the data would be loaded from the buffer. If not, a page containing the data would be fetched from main memory to the shared buffer first.

It is the philosophy of this system to treat *data* in an explicit fashion, that is, reference to an operand will be explicitly addressed to a buffer (either private or common buffer), or to main store. The private buffer is used when interstream communications are not necessary, or can be handled through common use of main storage if such reference is infrequent. The reference will be made to a public buffer when high performance interstream communications are required. The public buffer, of course, will be shared by all processors operating on that pool of data. Data transmission occurs between (1) main store and buffer store (private and common), and between (2) buffer store and working registers, i.e., accumulators and index registers. Transfers involving main store are of a block type, while transfers involving working registers are single or double word transfers.

Instruction addressing will be treated in an implicit fashion. The attempt here is to bring main storage images as close as possible to the processor, hence the individual processor will use ordinary memory addresses to fetch instructions. The target of the instruction fetch will be found either in the buffered "reflection" of main storage, or in main storage itself. The regularity of the instruction stream as well as the sharing of lines of code results in this different treatment of instruction and data buffering. Two levels of instructions buffering (one public and one private) are used, but both are implicitly addressed. The private instruction buffer is simply the single, small block of instructions in which a processor is presently operating. A branch or instruction counter overflow causes it to be refilled from the public buffer.

An interesting dilemma appears when the implementation of the public level of the instruction buffer is considered. Thus far, we have 32 virtual processors, each attempting to have equal and rapid accessing service to the public data buffer and (to a somewhat lesser degree with proper layout) the entry ports of the execution units. To attempt to also topologically locate an instruction buffer centrally, with respect to the requestors, would degrade performance all around. A better compromise appears to be the four-way segmentation of the public instruction buffer, so that each ring of (8) processors has its own buffer. This increases the complexity of processor-problem assignments, but the preservation of basic performance is insured.

While this description is based on the availability of 300 nsec main storage with access time at least 192 nsec, the following chart reflects the typical impact on the storage configuration of various main storage access times.

	Data Buffer		*Instruction Buffer*
Main storage access	explicit (public)	implicit (public)	implicit
up to 20 nsec	none	none	none
up to 64 nsec	useful (small)	none	simple private
*up to 192 nsec	necessary (1-2K words)	possible	common required
up to 512 nsec	necessary (2-8K words)	useful	full common required (more elaborate than described herein)

*this machine description

Storage Implementation

The treatment of storage will be broken down into three parts: main storage, and buffer storage for instructions and for data.

1) *Main Storage:* Fig. 8 shows a decomposition of the 24-bit address field when used to span a typical configuration of 512K, 48-bit words. The basic word size within the BOM (Basic Operational Module) or physical memory unit is 192-bits wide, or 4 full words wide (parity excluded). The lowest 3 bits of the address field refer to the half word within the physical memory word, 16 BOMs are interleaved on low order. Each BOM consists of 4,096 48-bit words or 1,024 physical memory words. The array of 16 BOMs interleaved on the low order 4 bits of a 192-bit physical word address represents 65K words of storage and is referred to as a *granule* of storage. The next 10 bits in the effective address format form the physical word address within a particular residue

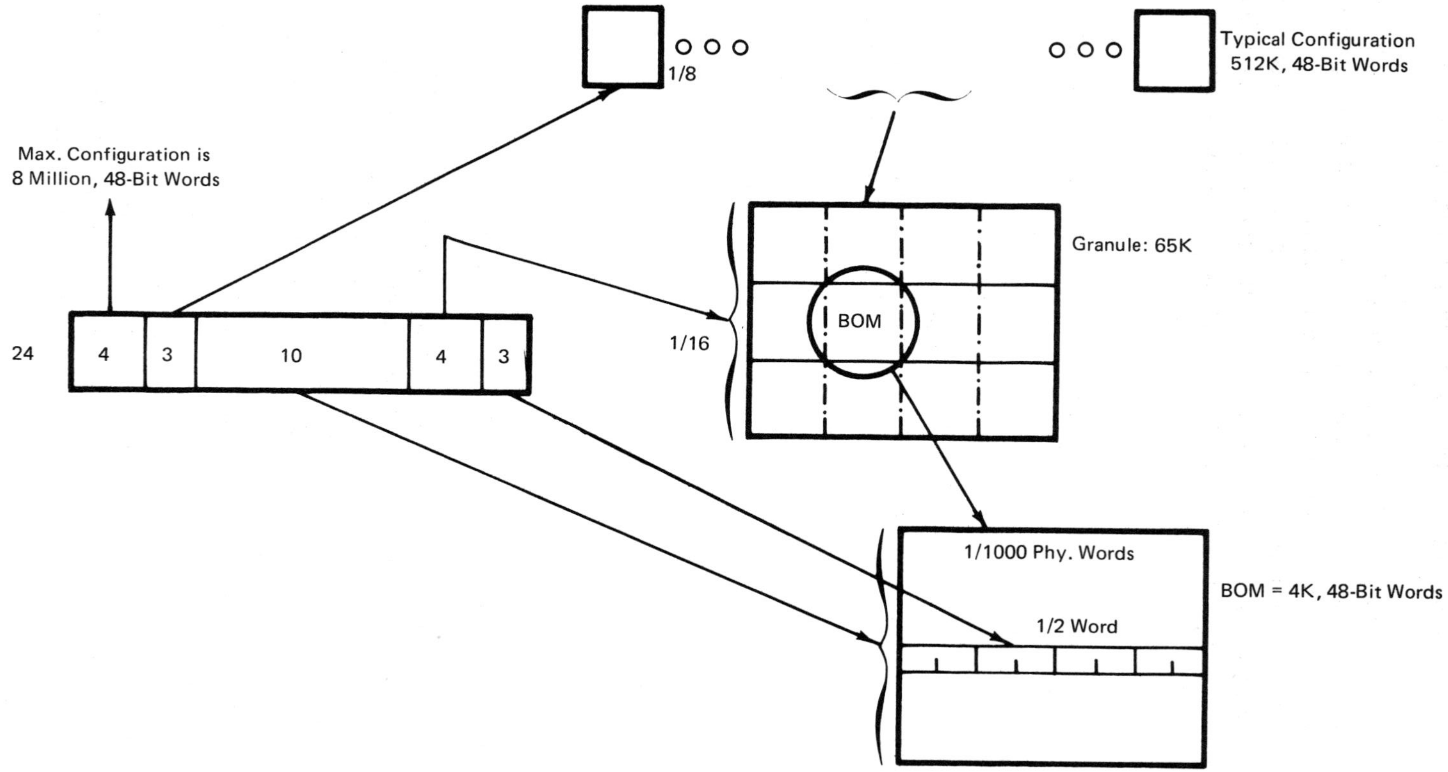

Fig. 8. Main Storage Organization

class of a granule. The higher 3 bits address 1 of 8 granules which comprise the complete system. Both high order and low order interleaving are used in the system. The high order interleaving provides a "fail soft" capability (by simply removing the defective granule of memory). The high order interleaving also provides separations between problem streams or tasks in which memory conflicts can be anticipated. Low order interleaving is used to improve bandwidth on block transfers. Main storage should be capable of providing 4 words of data every 20 nanoseconds per granule. Bus limitations may lower this somewhat.

2) *Instruction Buffer:* Four instruction buffers are provided per system, one on each ring. No communications are provided between buffers as the main function of the buffer is merely to make main storage more immediately available. The instruction buffer contains 2,048 instructions or 1,024 full words. It is arranged as shown on Fig. 9. With 64 X 32 instructions total, each of the 64 lines or reflections is a map of a 32-instruction block in main storage. A maximum of 64 reflections or blocks can be captured at any one time by the instruction buffer. Upon branching to a new location, the 32-instruction block which contains a target of the branch is entered into 1 of 64 positions of the buffer. The line address is entered also into the register associated with the individual processor instruction counter. Provided no overflow occurs out of bit 5 of the instruction counter, and provided further that no branch of the block occurs, the contents of its instruction register will be specified by the concatenation of the reflection address with the lower 5 bits of the instruction counter. When a branch or overflow occurs, reference is first made to 1 of 16 small associative storages. These associative storages are selected by the bits 6 through 9 in the instruction counter field. These bits specify the residue class of the target. When the appropriate class has been selected, an associative search is made with the query being the high order bits of the new instruction counter. If a match is detected, the contents associated with the match will be the line address which corresponds to that particular reflection. There are 16 associative memories, each unit contains 16 words, each word contains a 15-bit search field representing the high order 15 bits of the address and 6 bits which represent the target of the search; i.e., the address of the desired reflection.

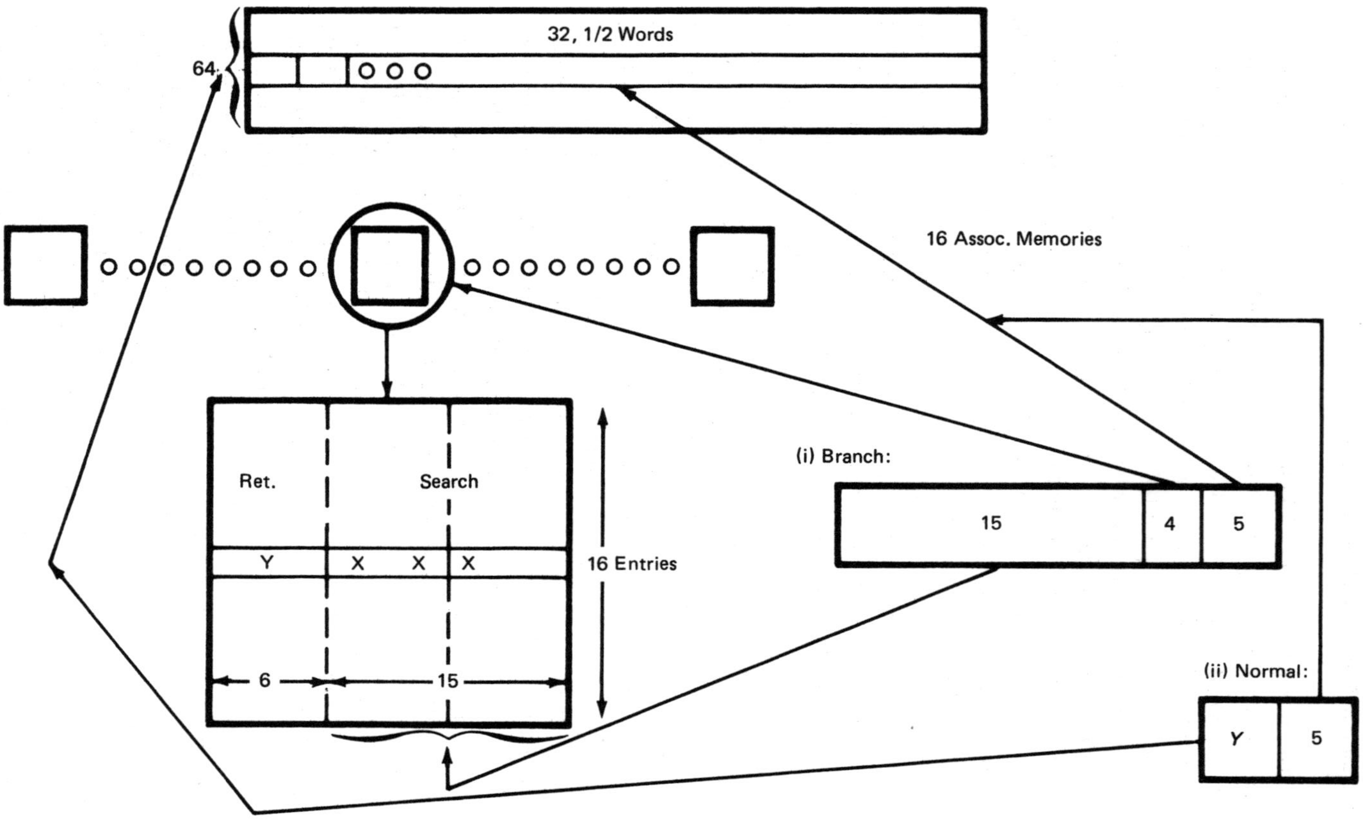

Fig. 9. Instruction Buffer

Stores into the instruction buffer are not permitted. Revisions must be made into main storage and recalled into the instruction stream.

3) *Data Buffer:* The direct accessing of operands from the data buffer is reserved to processors operating on frequently referenced common data, particularly that data associated with intertask communications. The data buffer is explicitly addressed. Addressing is done by using the private/common flag, setting it to 1. When the private/common flag indicates reference to the common storage, the *B* field is activated in the short word format and the contents of either base register zero or 1 will supply the high order bits for the displacement if the buffer storage is extended beyond 2,048 words.

Figure 10 illustrates the physical composition of the data buffer. A physical module consists of 128 words of monolithic storage. Access time is in the 20–24 nsec range, but the accessing mechanism is pipelined into three 8-nsec stages (decode, drive, sense). This

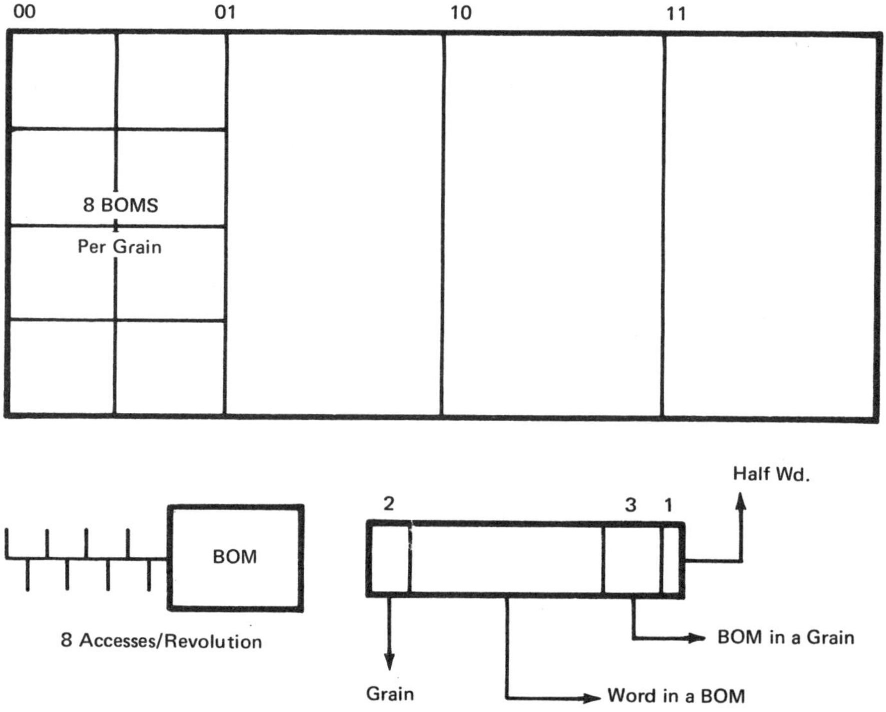

Fig. 10. Data Buffer

allows an accessing rate of 8 words/64 nsec. Eight modules (BOM) are low order interleaved to define a 1,024 word *grain* of buffer storage. High order interleaving of four grains completes the buffer.

So far as communications between streams is concerned, all status testing must be done by the individual programmer. This is accomplished by instructions similar to the IBM System/360, test and set.

CONTROL OF INSTRUCTION HANDLING RESOURCES

In order to provide resource effectiveness under various problem program environments, the assignment of virtual processors must be efficiently accomplished. A task itself is an independent sequence of instructions that forms a logical unit. If the task becomes too small, for example, single instructions, the overhead of assigning a task to an individual processor, etc., becomes excessive. Thus, ideally, the problem should be decomposed into as much "parallelism" or independent tasks as possible, the only proviso being that the tasks be of reasonable length.

A job always begins and ends with one processor (not necessarily the same). The program must have a mechanism for identifying multitask situations and also identifying the nature of the tasks. This mechanism is called the fork.[6] For every fork there must be a corresponding join instruction where all branches of the fork meet. With respect to a single fork, of course, there can be only a single join. The fork-join philosophy then generates the properties of being a well formed formula (just as right and left parentheses).

A "fork" (FORK α) is a request for execution of a set of tasks. The location of the fork status word is α. Depending upon the priorities involved, the fork will either wait to be executed or will cause the interruption of other tasks being executed in the processor.

SIMULATION OF SYSTEM

Simulator

A system simulator has been developed which allows the interpretive execution of programs on an instruction by instruction basis. This "Execution Simulator" was designed for flexibility in specifying simulation models and for the ease in altering existing models. The basis of the design is the centralization of resource description and flow control specifications in readily modifiable tables.

A master resource control table provides a ready means of assigning resources or resource pools to processors. All references to resources are made indirectly through a central resource table. Thus, widely different models may be constructed by means of input parameters.

In the Master Resource Table (Fig. 11) each processor (P_0, P_1, ... P_n) has the system resources allocated to it. These resources may be shared (as in the case of a floating-point add unit, F_0) or privately maintained (e.g., an instruction decoder, H_i). The processor may also make reference to the same resource in several different ways (the floating add unit, F_0, services both floating add and subtract instruction requests).

A digital simulation involves a numerical description of the "status" of the system resources to be simulated. This "status" is modified as "events" occur at various points in simulated time. Events may be impelled by forces outside the simulated resource or caused by preceding events associated with the resource. Flow control tables describe the actions to be carried out by the resources. The execution simulator provides a main timing routine to keep track of simulated time and invoke the various flow control tables in their proper sequence. Simulated time is advanced by variable increments rather than being broken into a sequence of fixed increments. Therefore, flow control tables may be invoked at any desired point in simulated time. When the execution of a particular flow control table is completed, simulated time is immediately advanced to the time of the next most imminent event.

In the Flow Control Table (Fig. 12), the actual operation of the resource subroutine is shown. The floating-point add routine, F_0, when called by processor P_i, calls the common priority routine, S_0, for conflict resolution. Control is then passed to the operative routine, S_1, which decodes the instruction into add, subtract, etc., classifications and performs the indicated operation. Upon completion, control passes to a common statistic gathering routine, S_2.

Simulator Results

Two test programs are presented here. The first (Fig. 13) is a pedegogic problem, heavily oriented toward the use of the execution

Resource \ Processor	P_0	P_1	P_2 ···
Processor	T_0	T_0	T_0
⋮			
FLT Add	F_0	F_0	F_0
FLT Sub	F_0	F_0	F_0
FLT Mpy	G_0	G_0	G_0
⋮			
Decoder	H_0	H_1	H_2
⋮			
Memory Mod. 1			
2			
⋮			
Memory Bus			
⋮			

Fig. 11. Master Resource Table

resources. The second problem (Fig. 16*a, b, c*) is an actual piece of weather modeling code.

In the simulator model (Fig. 13), the FORK instruction is implemented as a binary scalar (unlike any actual system). Thus, a processor begins the job and encounters a FORK DEX. This calls an additional processor, and each encounter FORK DEX 2. After the fifth FORK, 32

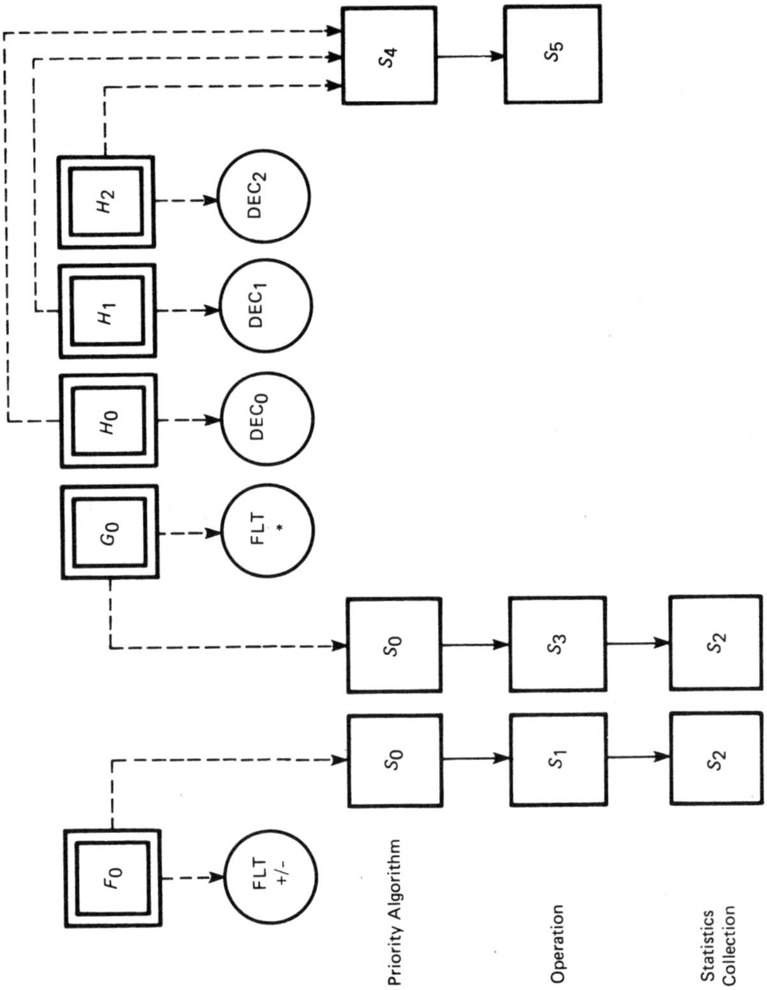

Fig. 12. Flow Control Table

```
INST                      Assembled Instruction Input for Execution Simulator

  LCC  Object Code        ADDR1    ADDR2    STMT    Source Statement


CC00C0                    462+Instrc     Start O Control Section for Input Instructions
                          463+*
CCCCCC CCC00000           464+IBEG       DC   A(O) For Length of Instructions

                          466  *
                          467  *
                          468  *
                          469  *
                          470            PRINT ON, NOGEN     Supress Instruction MACRO Expansions
                          471            FORK  DEX
                          480  DEX       FORK  DEX2
                          489  DEX2      FORK  DEX3
                          498  DEX3      FORK  DEX4
                          507  DEX4      FORK  TASKA
                          516  TASKA     LDX   X2,C10
                          526  TK2       FADD  A1, P1, A1                        INST4710
                          537            FSUB  A1, P1, A1                        INST4720
                          548            FMPY  A1, P1, A1                        INST4730
                          559            FDIV  A1, P1, A1                        INST4740
                          570            ADDD  A1, P1, A1                        INST4750
                          581            SUB   A1, P1, A1                        INST4760
                          592            MPY   A1, P1, A1                        INST4770
                          603            DIV   A1, P1, A1                        INST4780
                          614            SFX   C1, X2, X2                        INST4790
                          625            TXH   TK2, X2, C0                       INST4800
                          636            TEND  TASKA                             INST4810
                          645  *                                                 INST4820
                          646            Term               Cleanup              INST4830
                          683  *                                                 INST4840
```

Fig. 13. Test Case I

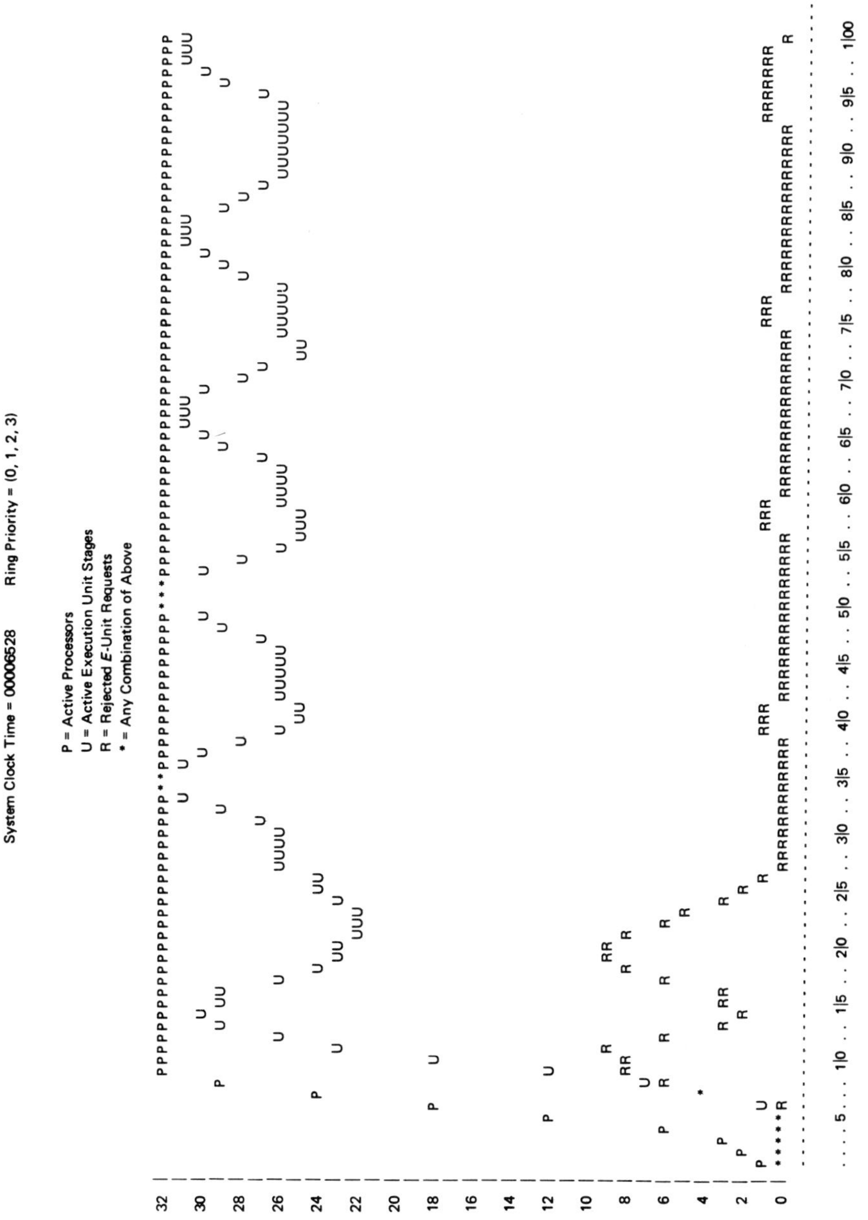

Fig. 14a. Execution Unit Utilization Test Case I

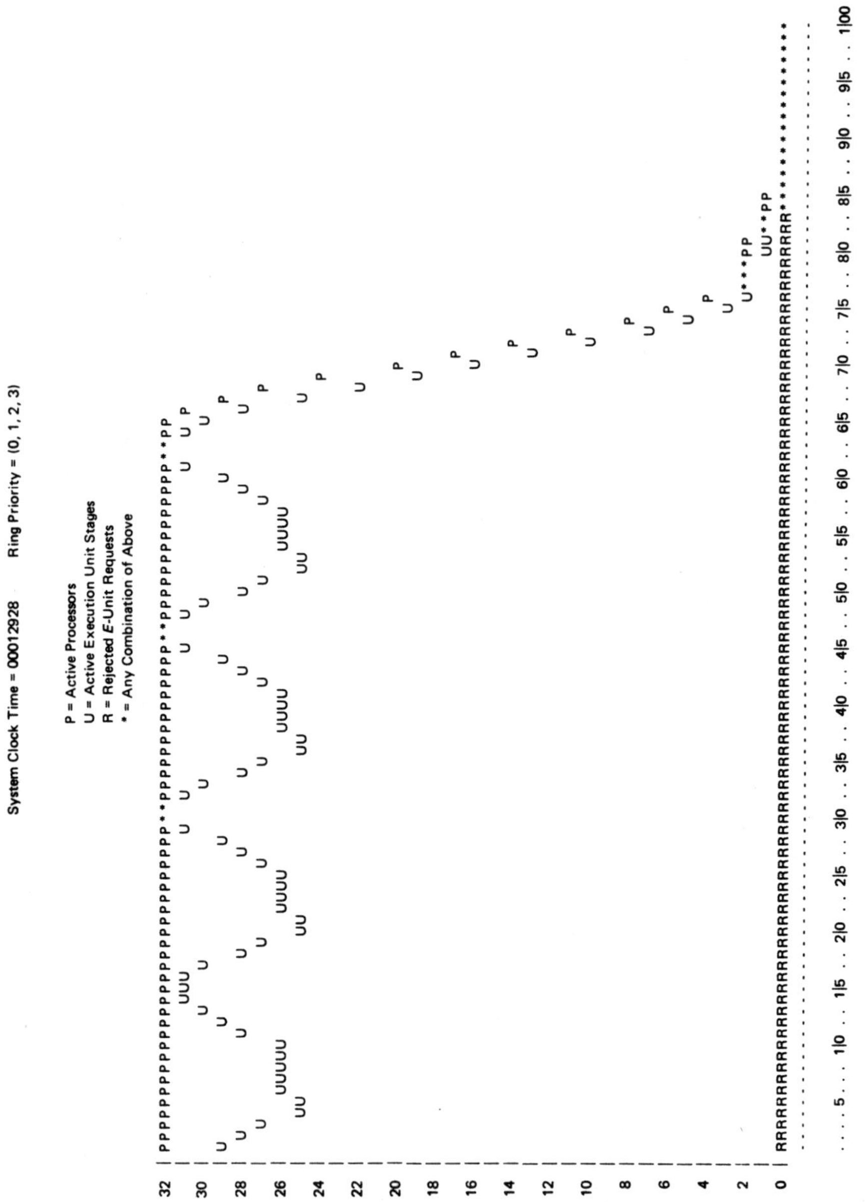

Fig. 14b. Execution Unit Utilization Test Case I

```
LCC  Object Code  ADDR1  ADDR2   STMT        Source Statement

     000000                      462+INSTRC       Start 0 Control Section for Input Instructions
                                 463+*
     000000 00000000             464+IBEG         DC   A(0) For Length of Instructions

                                 466  *
                                 467  *
                                 468  *
                                 469  *
                                 470              Print ON, NOGEN      Supress Instruction Macro Expansions
                                 471  *
                                 472  *    Instructions to Simulate "Subroutine CMP108"
                                 473  *
                                 474  *
                                 475  *           P r i v a t e  S t o r e  A s s i g n m e n t s
                                 476  *
                                 477  *       Variable                         Private Store Location
                                 478  *      ------                           ------------
                                 479  *        J                                    0 (XRO)
                                 480  *        IP1                                  1 (XR1)
                                 481  *        I                                    2 (XR2)
                                 482  *        IM                                   3 (XR3)
                                 483  *        IM1                                  4 (XR4)
                                 484  *        LOOP CNTRL                           5 (XR5)
                                 485  *        JM                                   6 (XR6)
                                 486  *        MRCH                                 7 (XR7)
                                 487  *
```

Fig. 15a. Test Case II

```
488 *       CONST  1                                    8
489 *       CONST  2                                    9
490 *       CONST  3                                   10
491 *       CONST  4                                   11
492 *       CONST  7                                   12
493 *       CONST 10                                   13
494 *       CONST  0.25                                14
495 *       CONST  0.5                                 15
496 *
497 *       ARRAY PU                                   20
498 *       ARRAY P                                    40
499 *       ARRAY PV                                   60
500 *       ARRAY DXU                                  80
501 *       ARRAY V                                   100
502 *            .TEMP.                                19
503 *
504 *
505 *       Get All Processors Going on Task
506 *
507 A              Fork  B
516 B              Fork  C
525 C              Fork  D
534 D              Fork  E
543 E              Fork  CMP108
552 *
553 CMP108         NOOP                                    Entry for Subroutine CMP108
561 *
562 *       Initialize Registers for Loop
563 *
564                LDX      X2, C5                         SET I TO 5
574                LDX      X3, C31                        SET IM TO 31
584                LDX      X6, C30                        SET JM TO 30
```

Fig. 15b. Test Case II

```
LCC Object Code  ADDR1  ADDR2    STMT         Source Statement

                                 594          LDX    X7, C2            SET MRCH TO 2—THEN 3, 4, 2, 3, 4, ETC
                                 604 *
                                 605          LDX    X0, C0            START J AT 0
                                 615          LDX    X5, C1            START LOOP CONTROL AT 1
                                 625 *
                                 626 *
                                 627 *    Instruction Counting Statements, i.e., CTR = CTR + 4, are not Simulated
                                 628 *
                                 629 LOOP     ATX    C1, X0, X0               J = J + 1
                                 640          TXH    SN10, X0, C2      IF (J.NE.2)
                                 651          TXLE   SN10, X0, C1                GO TO 10
                                 662          ATX    C1, X2, X1        IP1 = I + 1
                                 673          TXH    SKIP1, X2, X3     IF (I.EQ.IM) .... IP = 1
                                 684          SFX    C1, X3, X4        TEMP - X4 = IM - 1
                                 695          TXLE   SKIP1, X2, X4     IF .LE. IM-1, THEN .NE. IM
                                 706          LDX    X1, C1            IP1 = 1
                                 716 SKIP1    LDX    X4, X2            IM1 =
                                 726          SFX    C1, X4, X4          I - 1
                                 737          TXH    SN10, X2, C1      IF (.NOT.(I.EQ.1))
                                 748          TXLE   SN10, X2, C0                GO TO 10
                                 759          LDX    X4, X3            IM1 = IM
                                 769 *
                                 770 SN10     NOOP   ,                 WAS CTR = CTR + 4
                                 778          LOD    A1, X5            Simulate Saving 'CTR' Since
                                 788          STO    A1, P19                  It Does not Affect Flow
                                 798          TXH    SN2170, X0, X6    IF (J.EQ.JM) ... GO TO 2175—BRA If Higher
                                 809          SFX    C1, X6, X6        TEMP - JM = JM - 1
```

Fig. 15c. Test Case II

```
820             TXLE   SKIP2, X0, X6      COMPARING J TO JM-1
831             ATX    C1, X6, X6         RESTORING JM
842             BRA    SN2175             GO TO 2175 (J.EQ.JM)
851  *
852  SKIP2      ATX    C1, X6, X6         RESTORING JM (HERE IF J.NE.JM)
863  *
864  SN2170     LOD    A1, P40            P(J, I)
874             FADD   A1, P40, A1              + P(J, IP1)
885             FMPY   A1, P20, A1          ABOVE QUANTITY * PU(J, I)
896             STO    A1, P20            PU(J, I) = ABOVE RESULT
906  *
907  SN2175     NOOP   ,                  WAS CTR = CTR + 4
915             TXH    SN11, X7, C2       IF (.NOT. (MRCH.LE.2)) GO TO 11
926  *
927             LOD    A1, P100           V(J, I, L)
937             FADD   A1, P100, A1                + V(J, IM1, L)
948             STO    A1, P19            SAVE IN TEMP
958             LOD    A1, P40            P(J, I)
968             FADD   A1, P40, A1                 + P(J-1, I)
979             FMPY   A1, P19, A1        * (V(J, I, L) + V(J, IM1, L))
990             FMPY   A1, P80, A1        * DXU ( J)
1001            FMPY   A1, P14, A1        * 0.25
1012            STO    A1, P60            PV (J, I) = ABOVE PRODUCT
1022 *
1023 *
1024 SN11       NOOP   ,                  WAS CTR = CTR + 4
1032            TXH    SN12, X7, C3       IF (.NOT.(MRCH.EQ.3))
1043            TXLE   SN12, X7, C2                GO TO 12
1054 *
1055            LOD    A1, P40            P(J, I)
1065            FADD   A1, P40, A1              + P(J-1, I)
1076            FMPY   A1, P100, A1                   * V(J, I, L)
```

Fig. 15d. Test Case II

```
LCC  Object Code  ADDR1  ADDR2   STMT        Source Statement

                                 1087          FMPY    A1, P80, A1                                 * DXU (J)
                                 1098          FMPY    A1, P15, A1                                     * 0.5
                                 1109          STO     A1, P60             PV (J, I) = ABOVE PRODUCT
                                 1119  *
                                 1120  SN12    NOOP    ,                   WAS CTR = CTR + 4
                                 1128          TXH     SN13, X7, C4        IF (.NOT. (MRCH.EQ.4))
                                 1139          TXLE    SN13, X7, C3                GO TO 13
                                 1150  *
                                 1151          LOD     A1, P40             P (J, I)
                                 1161          FADD    A1, P40, A1               + P (J-1, I))
                                 1172          FMPY    A1, P100, A1                    * V (J, IM1, L)
                                 1183          FMPY    A1, P80, A1                           * DXU (J)
                                 1194          FMPY    A1, P15, A1                                 * 0.5
                                 1205          STO     A1, P60             PV (J, I) = ABOVE PRODUCT
                                 1215  *
                                 1216  SN13    NOOP    ,                   WAS CTR = CTR + 4
                                 1224          TXH     SN2181, X0, X6      IF (J.NE.JM) .... GO TO 2181
                                 1235          SFX     C1, X6, X6          TEMP - JM = JM - 1
                                 1246          TXLE    SKIP3, X0, X6       BRANCH IF J .LE. JM-1 ( .NE. JM )
                                 1257          ATX     C1, X6, X6          RESTORING JM ( J .EQ. JM )
                                 1268          BRA     SKIP4               BRANCH AROUND EQUAL CASE
                                 1277  *
                                 1278  SKIP3   ATX     C1, X6, X6          RESTORING JM - J .NE. JM
                                 1289          BRA     SN2181              GO TO 2181
                                 1298  *
                                 1299  SKIP4   ATX     C1, X2, X2          I = I + 1
                                 1310          LDX     X0, C1              J = 1
                                 1320  *
                                 1321  SN2181  NOOP    SN2181              CONTINUE
```

Fig. 15e. Test Case II

```
1330 *
1331           ATX    C1, X5, X5          LOOP COUNTER UP BY ONE
1342           TXH    TENDCMP, X5, C30         BRANCH TO TEND IF DONE
1353 *
1354           ATX    C1, X7, X7          MRCH = MRCH + 1
1365           TXLE   LOOP, X7, C4        BRANCH BACK IF MRCH STILL LE 4
1376 *
1377           LDX    X7, C2              RESET MRCH TO 2
1387           BRA    LOOP
1396 *
1397 *
1398 TENDCMP   TEND   CMP108
1407           TERM                       CLEANUP                      INST4830
1528 *                                                                 INST4840
1529 *                                                                 INST4850
```

Fig. 15f. Test Case II

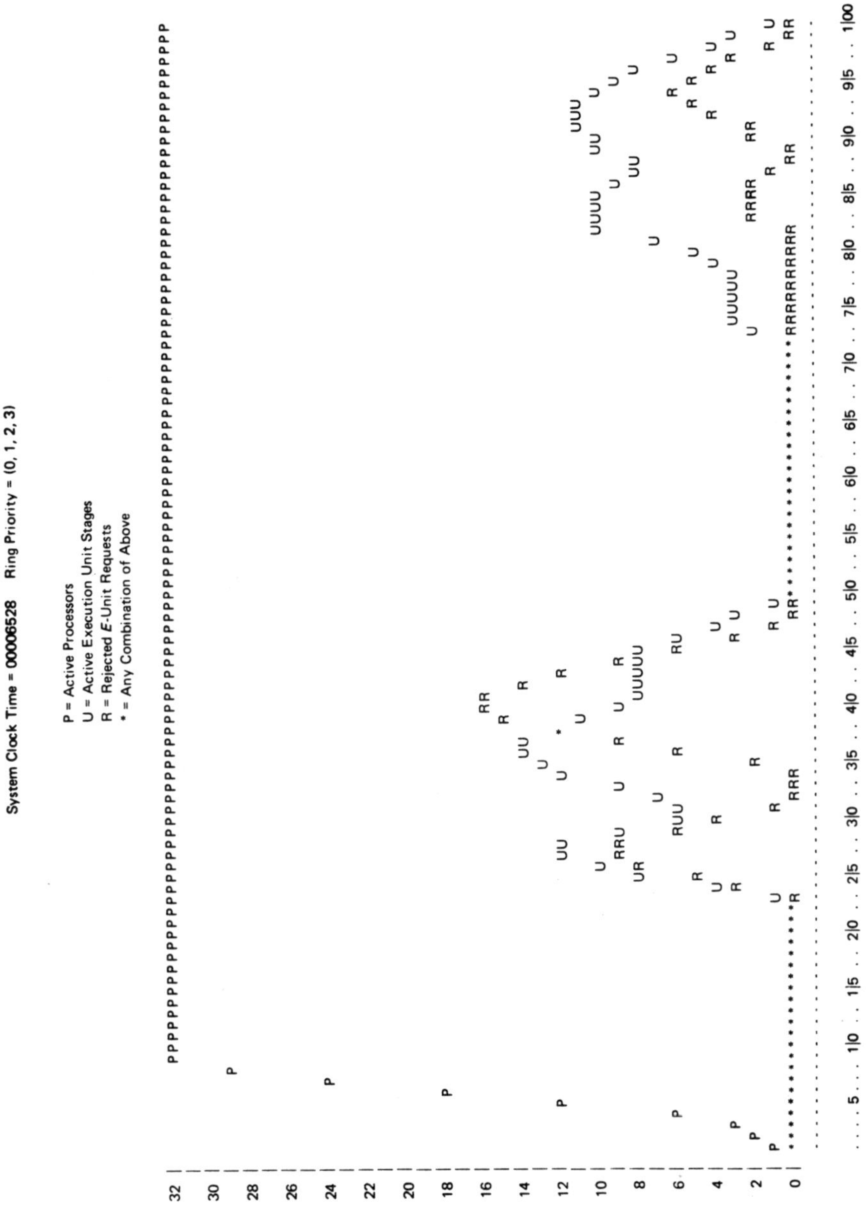

Fig. 16a. Execution Unit Utilization Test Case II

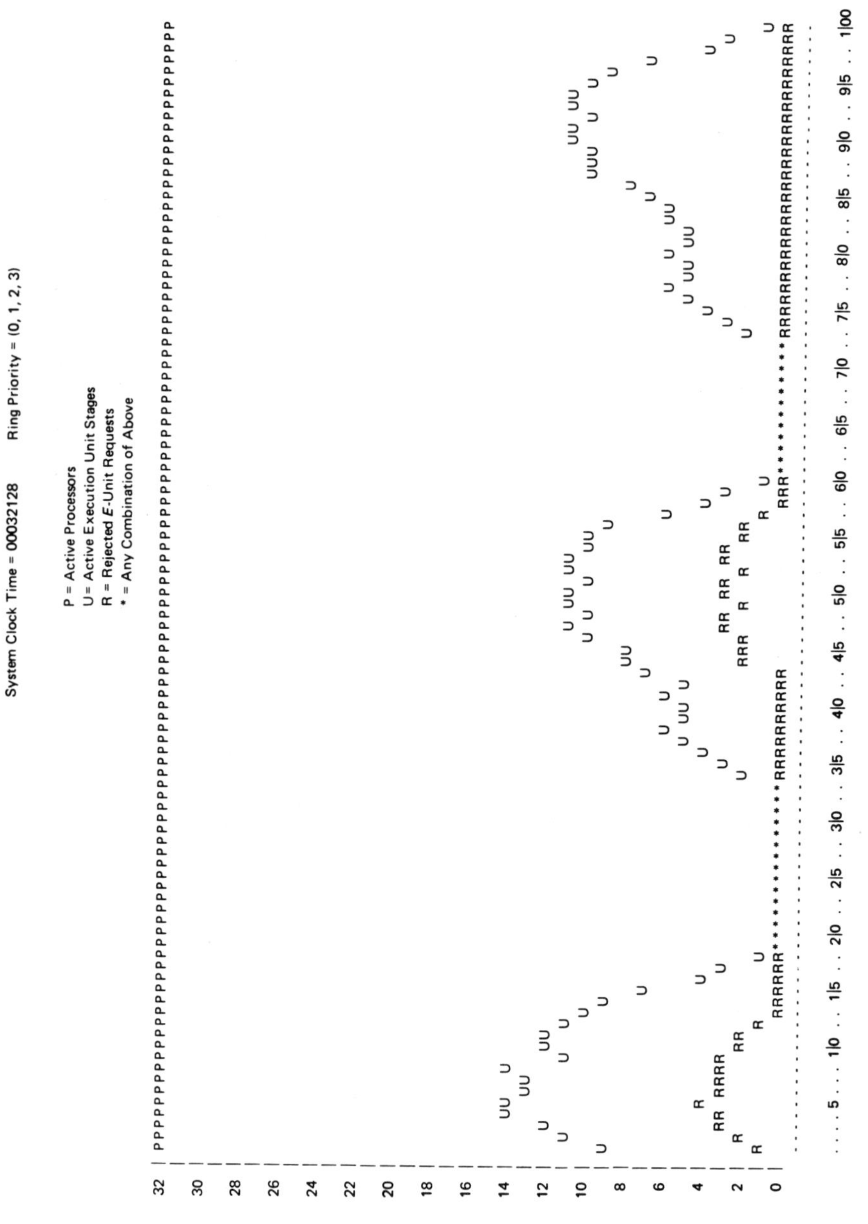

Fig. 16b. Execution Unit Utilization Test Case II

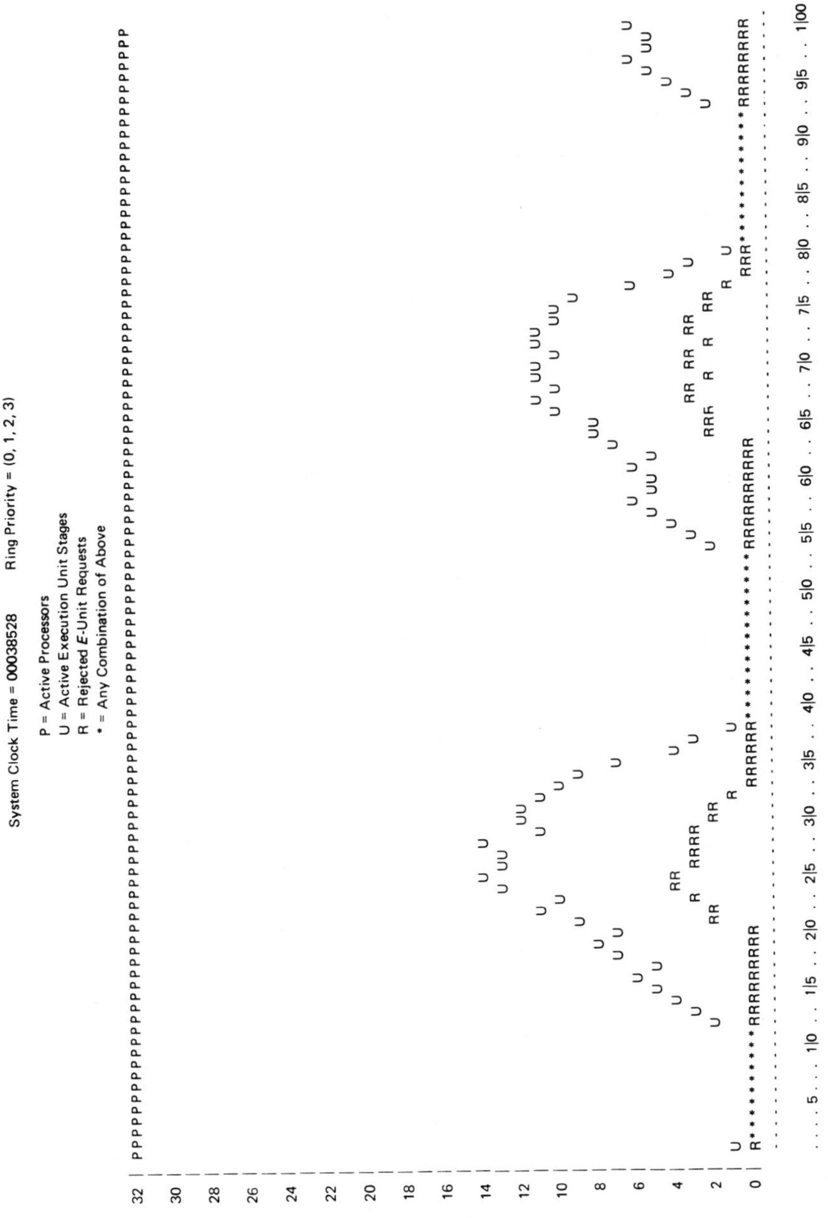

End of Simulation – Time Used = 0.0536 Hours, Performance = 6215 Instructions/Minute (Plot Mode)

Fig. 16c. Execution Unit Utilization Test Case II

processors have been called and each begins the TASKA routine. TASKA calls for 10 iterations of the subsequent code. LDX $X2$, $C10$ being interpreted as loading index register 2 with the constant, 10. The bulk of the remaining instructions are execution oriented. The triple address coding merely allows for simulation flexibility. For example, in the FADD $A1, P1, A1$, $A1$ is the single accumulator and $P1$ is a word in private storage.

The results (Fig. 14 *a, b*) show the start up of the processors (P), the early rejection (R) of execution unit requests, and the eventual elimination of conflict. As processors are rejected initially, they tend to lag unrejected units, and when they are "out of phase" there is no longer any resource contention.

The processors (P) are always busy after start-up and, due to the nature of the code, make substantial demands on the execution resources of the system (U). Notice that, as the rejection rate decreases, the utilization rate increases until both stabilize. The variation in execution utilization (U) is caused by the occurrence of the two privately processed instructions SFX and TXH.

The weather code (Fig. 15*a, b, c, d, e, f*) is probably much more typical code. The incidence of "procedural" instructions (load, store, branch, index operative) which are privately handled is much higher than in program 1. In addition, there is the typical clustering of resource requests (3 sequences of 3 floating-point multiples) which tax the system resources. The results (Fig. 16*a, b, c*) are somewhat similar to problem 1. Highest rejection rates are experienced initially; the acute initial utilization cycle (U) smoothes out as processors move out of conflicting patterns. The clumped multiples, however, prevent complete stabilization in a rejection free pattern.

References

1. P. Dreyfus, "Programming Design Features of the GAMMA 60 Computer," *Proc. EJCC,* December, 1958, pp. 174-81.
2. J. E. Thornton, "Parallel Operation in the Control Data 6600" *AFIPS Conference Proceedings,* 26: Part II, FJCC (1964), 33-41.
3. R. A. Ashenbrenner, M. J. Flynn, and G. A. Robinson, "Intrinsic Multiprocessing," *AFIPS Conference Proceedings,* 30: SJCC (1967), 81-6.
4. *IBM Journal of Research & Development,* Vol. II, No. I (January, 1967).
5. S. S. Sisson and M. J. Flynn, "Addressing Patterns and Memory Handling Algorithms," *AFIPS Conference Proceedings,* 33: Part II, FJCC (1968), 957-67.
6. M. E. Conway, "A Multiprocessor System Design," *AFIPS Conference Proceedings,* 24: FJCC, (1963), 139-46.

CHAPTER 14

ANALYSIS AND SYNTHESIS OF CONTROL MECHANISMS FOR PARALLEL PROCESSES

T. H. Bredt and E. J. McCluskey

Stanford University
Stanford, California

Abstract

The problem of control of concurrent processes has been studied extensively. In this paper, it is shown that flow table methods from sequential circuit theory can be applied to describe a control mechanism for such processes. This mechanism may be implemented either as a control circuit or as a control program. A further advantage is that there are no restrictions on the concurrent execution of any operations used in its implementation.

Flow table methods are also of value in the analysis of the system to ensure that it operates correctly. Using these methods, it is possible to consider the effects of both element and line delays in the system environment.

INTRODUCTION

This paper is concerned with the study of computer systems in which many processes may be running at the same time. A *process* is usually assumed to be a program which is being executed by a processor.[1] A

This work was supported in part by the Joint Services Electronic Programs U.S. Army, U. S. Navy, and U. S. Air Force under Contract Nonr-225(83), NR 373 360.

problem which occurs in such systems is the following: How can concurrent processes be controlled so that it is never possible for two or more processes to perform certain operations simultaneously? This problem is usually called the *mutual exclusion* or *interlock problem.* The exact nature of the operations performed is not important. They may represent the portion of a process which accesses a common data file, a common memory location, or a common system table. For the purposes of discussion, the special operations will be assumed to be contained in a portion of the process known as a *critical section.* Solutions have been proposed for this problem which depend on processes communicating via a common register or memory location.[2,3,4,5,6] Each process is given a special instruction to be executed whenever the process wishes to perform those operations which must be protected. An example of such an instruction is the *test and set* instruction for the IBM 360 series computers.[7] Using these instructions, it is possible to solve the mutual exclusion problem *if* it is not possible for two processes to execute these special instructions simultaneously. Thus, the problem of preventing two processes from entering critical sections simultaneously is reduced to the problem of preventing two hardware instructions from being executed simultaneously. The remainder of this paper discusses a new solution to the mutual exclusion problem which does not depend on the exclusive execution of any instructions used in its implementation. The solution has been developed by applying design procedures developed for sequential switching circuits to the design of control mechanisms of the type required for the mutual exclusion problem. The general goals of this research are the following:

1) to develop design techniques for parallel systems which can be applied in the design of both hardware and software
2) to state simply and precisely all assumptions made
3) to formally prove that, under these assumptions, a given problem has been solved.

If it is possible to produce such a proof, the computer system is said to operate *correctly* with respect to the problem posed.

The emphasis in this paper will be on the use of the mutual exclusion problem as an example to give an intuitive feel for the techniques proposed. The detailed explanation of the design procedures with complete proofs will be given in a future paper.

THE MUTUAL EXCLUSION PROBLEM

A precise statement of the mutual exclusion problem for two processes is given below.

Problem: (Mutual Exclusion Problem)

Given two concurrent processes, each containing a critical section, design a control mechanism such that the following two restrictions are satisfied:

Restriction 1:

At most one process may be in critical section at any instant.

Restriction 2:

If a process wants to enter its critical section, it must eventually be allowed to do so.

The first restriction was discussed in the introduction. The second restriction is necessary because in certain environments it might be possible for a process to want to enter its critical section but be blocked by other processes entering, leaving, and then reentering their critical sections. This possibility has been shown to exist in one proposed solution to the mutual exclusion problem.[8]

The environment we will consider is characterized by the following assumptions.

Assumption 1:

The speeds of the processes and the times for signals to propagate are finite, but unbounded.

Assumption 2:

There is no master clock for synchronizing the operation of the system.

The first assumption is desirable because the speeds of operation of computer components vary depending not only on the type of component but also on their age and many other factors. The second assumption is made because it is difficult to provide master clocks, particularly when components are separated by large distances.

We propose that processes which contain critical sections communicate not by accessing a common memory as in earlier solutions but rather through a common control mechanism developed especially for the mutual exclusion problem. When a process wants to enter its critical section, it must ask permission from the control and then wait until it is allowed to proceed. The control mechanism monitors the status of the processes and decides when it is possible for a process to enter its critical section. The communication between a process and the control is achieved by sending binary level signals over interconnecting lines. The specific configuration for a two process system is shown in Fig. 1. The variable associated with each line is known as a *control variable.*

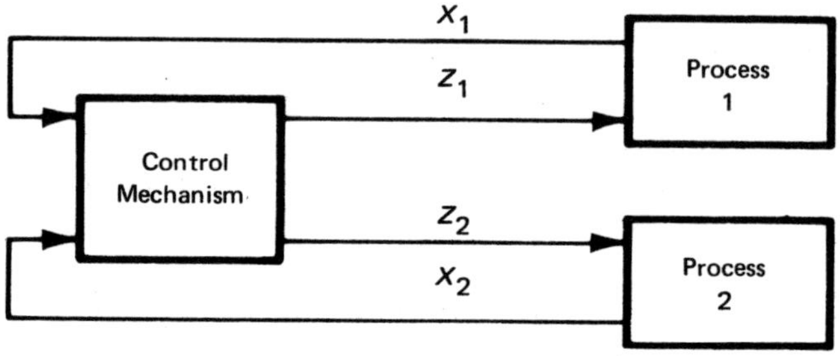

Fig. 1. System Configuration for the Mutual Exclusion Problem

The interpretation of the values of the control variables in Fig. 1 is given in Table 1. (CS*i* stands for the critical section of process *i*.)

Table 1
Interpretation of Control Variable Values

Values for Fig. 1 (i = 1,2)
$X_i = 1 \Rightarrow$ Process *i* wants to enter CS*i* *or* Process *i* is in CS*i*
$X_i = 0 \Rightarrow$ Process *i* does not want to enter CS*i* *and* process *i* is not in CS*i*
$Z_i = 1 \Rightarrow$ Process *i* may enter CS*i*
$Z_i = 0 \Rightarrow$ Process *i* may not enter CS*i*

By our first assumption, the times for signals to propagate along the lines and the speeds of the processes are unknown. One possible trouble condition caused by such delays is the following. Suppose process 1 (P1) is in its critical section (CS1), that is, $X_1 = 1$ and $Z_1 = 1$, and suppose process 2 wants to enter its critical section but is not enabled, $X_2 = 1$ and $Z_2 = 0$. Now suppose P1 leaves CS1 and sets X_1 to 0. This value change is propagated along the line to the control. The control sets the value of Z_1 to 0 and enables process 2 by setting Z_2 to 1. But, if process 1 loops and resets X_1 to 1 then examines Z_1 and finds it still has the value 1 (the 0 has not arrived yet), it is possible for both processes to be in their critical sections simultaneously. To prevent this possibility, we require that each process *Pi* test that $Z_i = 0$ before it attempts to set X_i to 1 indicating that it wants to enter its critical section. A typical process program showing the instructions which affect the control variables is given in Table 2. The *wait* statement in

Table 2
Program for Process 1

	$X_1 \leftarrow 0$
Start:	
	.
	.
	.
	Wait (Z_1 = 0, *L*)
L:	$X_1 \leftarrow 1$
	Wait (Z_1 = 1, CS1)
CS1:	Critical Section
	$X_1 \leftarrow 0$
	.
	.
	.
	Go to Start

this program is a special statement used to test the status of control variables. When this statement is executed, the process stops until the control variable has the value indicated; then, the process resumes by transferring to the specified location. This statement is similar to the *wait* statement in the PL/1 programming language. The assignment statements in the program such as $X_1 \leftarrow 0$ are not typical assignment statements but rather are used to change the value of the control variable on the appropriate interconnecting line.

In order to prove that systems such as this operate correctly, it is necessary to be able to describe precisely how each component of the system (process or control) changes control variable values and responds to changes in these values. To do this, we borrow the concept of a *flow table* used in switching theory for the design of sequential circuits.[9] A flow table can be used to describe not only how a circuit but also how a program interacts with its control variables. Using the process program given in Table 2 as an example, we obtain the flow table shown in Table 3. The input to the process is the variable Z_1 and

Table 3
Flow Table for Process 1

	Z_1 = 0	Z_1 = 1	X_1
A	B	(A)	0
B	(B)	A	1

it has two possible values 0 and 1. There are two corresponding columns in the flow table. Each row in the table corresponds to an internal state of the program. Associated with each row is a value of the output control variable X_1. Initially, Z_1 and X_1 have the value 0; the corresponding table entry is row A, column 0 $(A, 0)$. Since the program is running when in this state, it is *unstable*. The table entry is B indicating that eventually the program will reach internal state B where it will set X_1 to 1. Since the program waits in state B until Z_1 becomes equal to 1, the $B, 0$ entry is B. This entry is circled and is called a *stable state* since the program will take no further action until an input change occurs. When the value of Z_1 changes from 0 to 1, the program becomes unstable once more, enters its critical section, and then after an unspecified time returns to state A where X_1 is reset to 0. The program remains in this state until Z_1 becomes 0 enabling the process to attempt to enter its critical section once more. It is possible to give a normal procedure for constructing a flow table corresponding to a given program.

In addition to its role in the analysis of programs just described, the flow table can also be used in the synthesis of the control mechanism itself. The flow table that describes a solution for the mutual exclusion problem is given in Table 4. Two remarks must be made about this flow table. First, it allows simultaneous input changes to be described so that the objection to earlier solutions to the mutual exclusion problem has been overcome. Second, all possible input transitions must be considered explicitly. This is valuable in eliminating oversights in the design process. There are four internal states in this table which are used to remember which process had access last and which process is enabled

Table 4
Flow Table For Control Mechanism

		X_1X_2 00	01	11	10	Z_1Z_2
(2 last)	*A*	(*A*)	*B*	*C*	*C*	00
(2 gets)	*B*	*A*	(*B*)	(*B*)	*C*	01
(1 gets)	*C*	*D*	*B*	(*C*)	(*C*)	10
(1 last)	*D*	(*D*)	*B*	*B*	*C*	00

for access to its critical section. There are two input variables, X_1 and X_2, so there are four possible input states and four columns in the table.

Table 5
Control Program

A: $Z_1 \leftarrow 0$
$Z_2 \leftarrow 0$
Wait $(X_1X_2 = 01, B)$, $(X_1X_2 = 11, C)$, $(X_1X_2 = 10, C)$

B: $Z_1 \leftarrow 0$
$Z_2 \leftarrow 1$
Wait $(X_1X_2 = 00, A)$, $(X_1X_2 = 10, C)$

C: $Z_1 \leftarrow 1$
$Z_2 \leftarrow 0$
Wait $(X_1X_2 = 00, D)$, $(X_1X_2 = 01, B)$

D: $Z_1 \leftarrow 0$
$Z_2 \leftarrow 0$
Wait $(X_1X_2 = 01, B)$, $(X_1X_2 = 11, B)$, $(X_1X_2 = 10, C)$

Just as it is possible to analyze a program and produce its associated flow table, given a flow table such as shown in Table 4, we can define a *control program* to carry out the operations specified in the table. The program corresponding to the flow table of Table 4 is given in Table 5. The labels *A*, *B*, *C* and *D* correspond to the internal states *A*, *B*, *C*, and *D*. The *wait* statement is used to detect input transitions and to transfer to the appropriate next statement. Of course, it is also possible to use the flow table in the design of a circuit to perform the desired control function. It must be noted that in either implementation it must be the case that when the inputs are examined exactly one input state is valid.

We will now outline how one proves that this solution to the mutual exclusion problem is correct. The proof is performed in two steps. First, using the flow tables which describe how processes and control interact, a *control state transition diagram* is constructed. This diagram is a directed graph which represents for a given set of assumptions how the control variables change value. To determine if two processes can simultaneously be in critical sections, the diagram is examined to see if the $X_1 X_2 Z_1 Z_2 =$ 1111 state can be entered. A process can be excluded from its critical section when it desires to enter if certain cycles are present in the transition diagram. The second step in the proof involves showing that a particular circuit or program is an accurate representation of a given flow table. For circuits such questions are related to the presence or absence of hazards.[10]

CONCLUSION

We have shown how flow tables can be applied in the analysis and synthesis of control mechanisms for parallel systems. These mechanisms can be implemented as circuits or programs. A future and more complete paper on this subject will give details on the procedures for relating programs and flow tables, the modes of operation for the programs and circuits used in implementations, and an extension of the solution to handle more than two processes.

References

1. J. B. Dennis and E. C. Van Horn, "Programming Semantics for Multiprogrammed Computations," *Comm. ACM*, 9: (March, 1966), 143-55.
2. *Ibid.*
3. E. W. Dijkstra, "Solution of a Problem in Concurrent Programming Control," *Comm. ACM*, 8: (September, 1965), 569.

4. D. E. Knuth, "Additional Comments on a Problem in Concurrent Programming Control," *Comm. ACM,* 9: (May, 1966), 321-22.
5. E. W. Dijkstra, "Cooperating Sequential Processes," *EWD-123* (The Technological University, Eindhoven, The Netherlands [September, 1965]). Also, "The Structure of the 'THE' Multiprogramming System," *Comm. ACM,* 11: (May, 1968), 341-46.
6. B. W. Lampson, "A Scheduling Philosophy for Multiprocessing Systems," *Comm. ACM,* 11: (May, 1968). 347-60.
7. *IBM System 360 Principles of Operation.* File No. S360-01, Form A22-6821-49 (IBM Corporation [Poughkeepsie, N. Y., 1966]).
8. Knuth, *op. cit.*
9. E. J. McCluskey, *Introduction to the Theory of Switching Circuits* (New York: McGraw-Hill, 1965).
10. *Ibid.*

CHAPTER 15

ORTHOGONAL PROCESSING

William Shooman

Sanders Associates, Incorporated
Nashua, New Hampshire

Abstract

A machine is described that retains its general purpose word-serial-bit-parallel capabilities while attaining all the benefits of an associative processor in its word-parallel-bit-serial mode.

Representative computer operations are classified, and timing comparisons are made. For these operations it is shown that the Orthogonal computer is at least 100 times as fast as a conventional machine of the same memory speed.

Several applications are given, such as multiplication of matrix by vector, correlation, document search, and fast Fourier transform.

INTRODUCTION

The Orthogonal computer may be thought of as a conventional general purpose machine with additional capabilities provided by a modified memory and a second arithmetic unit. There are two operating modes, horizontal and vertical. The horizontal (conventional) mode, as usual, requires a separate instruction for each operation on a datum. By contrast, in the vertical mode many data are simultaneously processed one bit at a time by one instruction.

The important features include:

1) retention of the conventional general purpose capabilities
2) orders of magnitude speed gain

It is a pleasure to acknowledge my indebtedness:

To my colleagues J. E. Cook and H. S. Miller of Sanders Associates, who refused coauthorship of this paper because (I hopefully believe) of their too strict ethics. Each contributed significantly to ideas incorporated in this paper.

To R. L. Ward of Sanders Associates whose continued insistance that the time be taken to write this paper finally persevered.

To H. Campman of Sanders Associates for his continuing encouragement and support.

3) efficient memory and multiprecision utilization
4) the benefits of an associative memory at low cost
5) more efficient decision making via parallel testing
6) simple off the shelf circuitry
7) The vertical unit consists of 1,024 *identical* LSI chips, providing
    a) high system reliability at low cost
    b) Fail soft structure
    c) Easy maintainability
8) high data transfer rates from existing parallel devices
9) uncomplicated software requirements
10) simplified multiprogramming control.

DESCRIPTION

Figure 1 illustrates the major components of an Orthogonal computer and the data flow. That portion enclosed in dashed lines portrays a conventional general purpose computer with 32*K* words of random access memory, 32 bits to the word. The memory is partitioned into blocks, each consisting of 1,024 conventional words. A second addressing scheme provides the capability of reading and writing columns; a column consists of 1,024 bits, one bit from each of the 1,024 conventional words in that block, all at the same bit position. There are 1,024 vertical memory locations, comprising 32 columns in each of the 32 blocks (it is coincidence that the number of vertical memory locations equals the number of conventional words in a block). All registers in the vertical unit are also program addressable.

The function generator consists of 1,024 identical switching networks, each having the capability of generating any of the 16 Boolean functions of two variables, or the sum and carry bit of three variables. The two inputs for a Boolean function consist of any two vertical registers or any vertical register and any vertical memory location. Two of the three inputs to the full adder are the same as for the Boolean; the third input is one of the vertical carry registers. Any output of the function generator can be routed to any vertical location (i.e. vertical memory location or vertical register), with the exception that the carry output replaces the contents of the input carry register. In general, the same function is performed simultaneously at all 1,024 networks. Execution time in the function generator for either Boolean or add is one clock time (a 20 megahertz clock rate is the most likely candidate, with a one μs full memory cycle time).

The storage mask register, labeled *M,* automatically controls the flow of data from the vertical unit to memory. Only those bits (from among the 1,024 to be transferred to memory) whose positions correspond to 1 bits in *M* are actually transferred. Those memory bits whose relative positions correspond to 0 bits in *M* are unaltered. Each special purpose vertical register, when not being utilized for its specialty, is also a general purpose vertical register, with one slight exception.

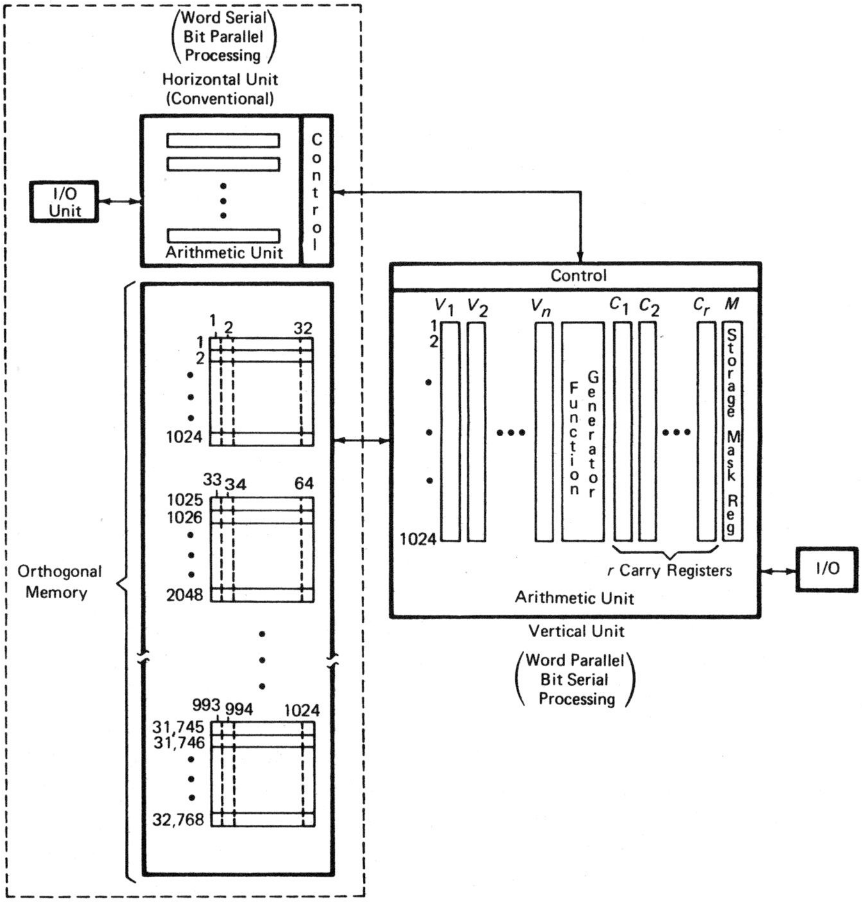

Fig. 1. Components of an Orthogonal Computer

SYSTEM

Overall system control resides in the conventional unit which interrogates the instructions. There are two classes of instructions: the conventional instructions, which refer to data operated on in the conventional unit, and the vertical unit instructions which refer to data operated on in the vertical unit (label these the VU instructions).* A VU instruction initiates transfer of control to the vertical unit.

A VU instruction occupies one, two, or three consecutive conventional memory locations and correspondingly contains one, two, or three vertical location addresses A_i. Each A_i refers to any vertical location, that is, any column in memory or any vertical register. Accompanying each A_i is a parameter P_i, which specifies the number of bits in each of the 1,024 data whose most significant bits are in A_i. The least significant bits of the 1,024 data are in $A_i + P_i$, so that $1 + P_i$ is the length of the data field.

A representative program in an Orthogonal computer would generally consist of a mix of conventional and VU instructions.

TIMING

VU instructions have been developed that perform (on many data simultaneously, one bit at a time) all the important instructions usual to conventional computers; for example, the arithmetic, Boolean, comparison, correlation, etc.

Figure 2 shows a classification of some computer operations of interest, each class represented by one of the curves. The horizontal coordinate marked along the bottom represents the number of bits in each of the operand data. The vertical coordinate uniquely selected by an intersection of one of the curves with a horizontal coordinate represents the speed increase of the indicated Orthogonal computer over a conventional computer of like memory speed, for the corresponding operations and data size. By way of example, the mark on the curve at the bottom shows an Orthogonal speed gain of about 140-1 for forming 1,024 18-bit sums. Similarly, the mark on the second curve from the top shows that we are about 500 times as fast in comparing a 15-bit constant to 1,024 15-bit data.

---

*Some instructions refer to data operated on in both units. These fit into the VU instruction class.

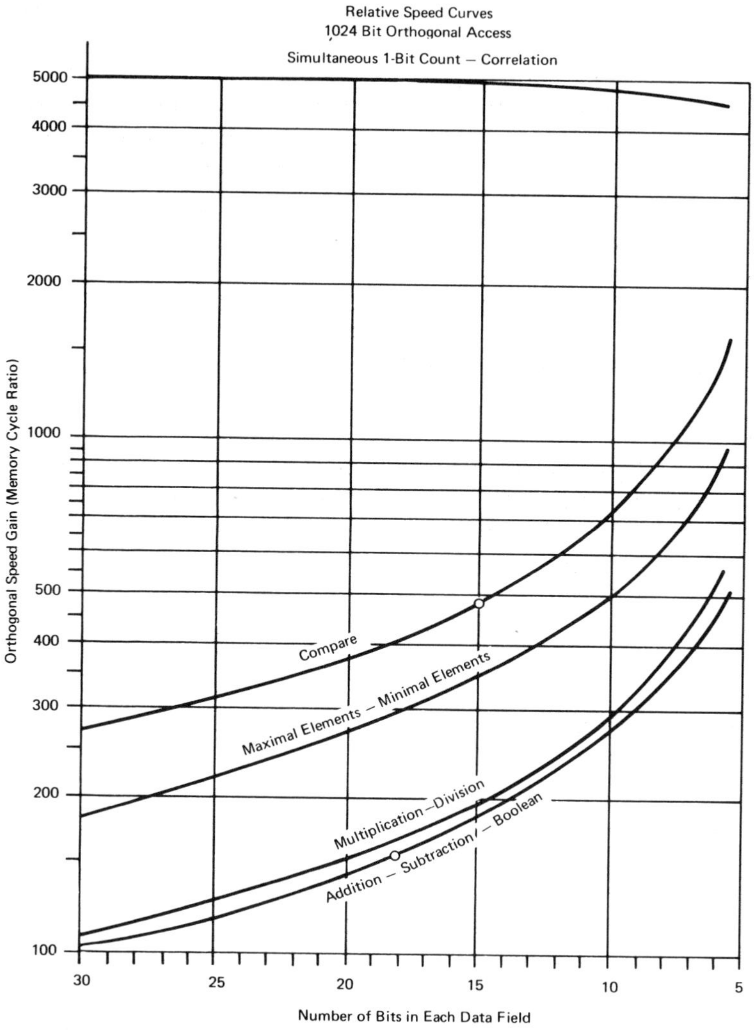

Fig. 2. Classification of Computer Operations

MEMORY AND MULTIPRECISION UTILIZATION

In a conventional machine, each data usually occupies an entire computer word, so that when data is smaller than the computer word, memory is wasted or time must be taken to unpack packed data. Since each VU instruction contains the dimensions of the associated operand data, and the data is handled one bit at a time, data fields can be packed with no loss in speed or generality.

Since vertical location 33 is *next to* 32, 65 is *next to* 64, and so on. It doesn't matter if a data field is in more than one block, so that multiprecision routines are not required in the Orthogonal computer in the vertical mode. If it takes N units of time to process a 32-bit data field vertically, it takes approximately $\frac{K}{32}N$ units of time to process a K bit data field vertically.*

ASSOCIATIVE MEMORY CAPABILITY

The VU instruction *compare* simultaneously compares 1,024 K-bit numbers to a K bit constant in K memory cycle times. The output of this instruction is a pair of vertical registers; label them $>$ and $=$. A bit in $>$ is a 1 bit if and only if the correspondingly positioned data is greater than the constant; analogously for the $=$ register. Since any of the Boolean functions can be performed on these registers, all the order relations are derived from the one instruction.

DECISION MAKING

Suppose it is required to perform computation T_1 on just those data within an Orthogonal block of memory that pass some specified test, while those that fail the test are to be processed by T_2. Since the same test is applied to all 1,024 data, the test can be performed simultaneously at all 1,024 levels in the vertical mode. The output of the test can be a vertical register whose 1 bits (0 bits) correspond in position to those data that pass (fail) the test. The contents of this register are transferred into the storage mask register, and T_1 is performed on all 1,024 data in the vertical mode. When result data are routed to memory only, those bits whose positions correspond to 1 bits in the mask are actually transferred. (See last paragraph under "Description.")

Get the ones complement of the mask into the mask and perform T_2 on all 1,024 data vertically. Since 0 bits in the new mask correspond in position to those data that passed the test, the results of T_1 will not be

*It is slightly less efficient for multiplication and division.

altered, and the results of T_2 on the data that failed the test are successfully routed to memory.

The Orthogonal speed increase in decision making is on the same order as that for the arithmetic operations.

OFF THE SHELF

A full scale Orthogonal computer is in design; all components are off the shelf discrete logic or MSI. The Orthogonal memory is now an in-house off the shelf item. An Orthogonal demonstrator has been implemented.

LSI

Redundancy

Figure 3 displays a portion (two of the flip-flops and the Boolean network) of the logic configuration at one-bit level of the vertical arithmetic unit. The vertical arithmetic unit consists of *one* logic configuration repeated 1,024 times. All the logic for one-bit level fits easily on one LSI chip, since this is the simple serial logic; so that the *entire* vertical arithmetic unit consists of 1,024 identical LSI chips. (Several LSI chip designs have been generated and trade-offs are being considered.)

Fail Soft

If one of the 1,024 logic levels fails, overall system performance is frequently only slightly degraded, in contrast to a conventional computer malfunction which usually affects all subsequent computations. This fail soft characteristic has special significance in a mission environment.

Simpler Diagnostic

A diagnostic routine could interrogate all 1,024 chips in parallel during processing time with *no* decrease in processing capability, since

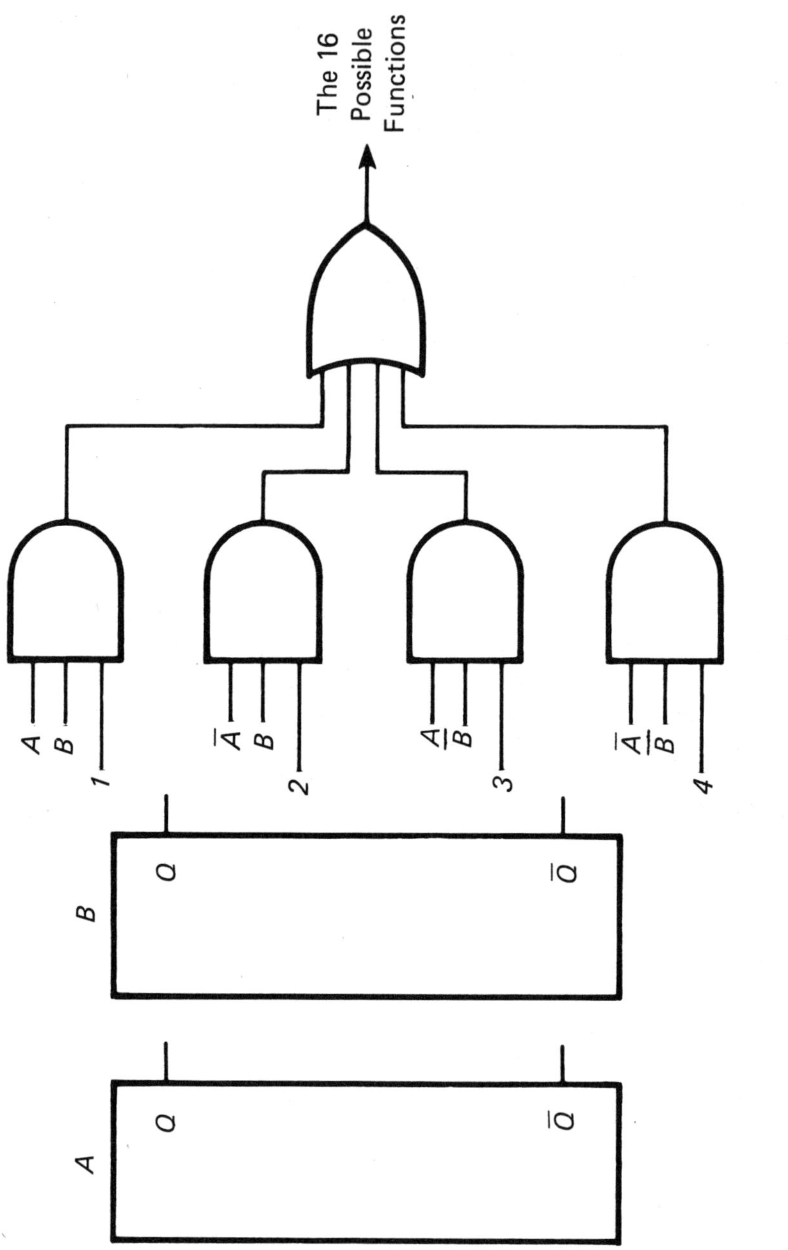

Fig. 3. Logic Configuration at One-bit Level

no VU instruction exhausts all the clock times available during execute time.

Maintainability (Healing)

Defective chips can be easily replaced in any environment since spare chips are light, compact, and inexpensive.

Automatic Healing

If it is required that the healing be automatic, two (and even more) complete sets of LSI chips can be implemented in such manner that failure at any operating chip causes automatic switching to another chip at the same level.

Majority Logic

If it is required that the probability of error be minimized, three complete sets of LSI chips can operate simultaneously using majority output. The probability of malfunction of two chips at the same level over a small time interval is extremely small.

PERIPHERALS

An example of the utility of the long vertical registers for I/O:

There is an off the shelf disk with 64 parallel read heads, 10^5 bits per track, totalling 6.4 megabits. At the rate of 1/3 μs for 64 bits, a 1,024 bit vertical register is loaded or unloaded in 5 1/3 μs. By paging two such registers, a data transfer rate equivalent to approximately 167 nanoseconds per 32-bit word is achieved in any Orthogonal memory with full memory cycle time $<$ 5 1/3 μs.

A denser and faster disk in development (twice the density and twice the speed) would provide a data transfer rate equivalent to approximately 42 nanoseconds per 32-bit word in any Orthogonal memory with full memory cycle time $<$ 4/3 μs.

If even higher transfer rates are required, disks with more than 64 parallel read heads are also in development.

None of the transfer rates discussed above depend on interleaved memory.

SOME ESPECIALLY IMPORTANT VU ALGORITHMS

Special Add

$$\sum_{i=1}^{2^r} \alpha_{i,j} (1 \leq j \leq 2^{x-r})$$

2^x equals the number of consecutive conventional memory locations in an Orthogonal block of memory. In the machine discussed above $2^x = 1{,}024$. Conceptually partition the Orthogonal block into 2^r subblocks, each subblock consisting of 2^{x-r} conventional locations. Each of the 2^{x-r} sums is composed of elements that are uniformly spaced 2^{x-r} locations apart. For K bit data *special add* takes $2K + r$ memory cycle times.

An important application of *special add* is vector times matrix multiplication. For this machine $x = 10$, $r = 5$ provides a 32 x 32 matrix. Store the data by column, that is $\alpha_{i,j}$ in location $2^{x-r}(j-1) + i$; $(1 \leq i \leq 2^r, 1 \leq j \leq 2^{x-r})$. Let the multiplying vector be labeled $X_1, X_2, \ldots, X_{32}$. During the VU multiply, the current multiplier bits of the vector are mapped into the following pattern of 1,024 bits

$$\overbrace{X_1, X_1, \ldots, X_1}^{32} \quad \overbrace{X_2, X_2, \ldots, X_2}^{32} \quad \ldots \quad \overbrace{X_{32}, X_{32}, \ldots, X_{32}}^{32}$$

into a vertical register (each X_j in this case symbolizes one bit rather than a K-bit word). The result of the VU multiply would then be

$$\alpha_{1,1}X_1, \alpha_{2,1}X_1, \ldots, \alpha_{32,1}X_1; \alpha_{1,2}X_2, \alpha_{2,2}X_2, \ldots, \alpha_{32,2}X_2; \ldots;$$

$$\alpha_{1,32}X_{32}, \alpha_{2,32}X_{32}, \ldots, \alpha_{32,32}X_{32}$$

in 1,024 consecutive conventional locations. Perform *special add* on these 1,024 data. As described above, each of the 32 sums is composed of elements spaced 32 locations apart; the sums are $\sum_{j=1}^{32} \alpha_{i,j} X_j \; (1 \leq i \leq 32)$, the desired vector times matrix.

Because of overlap of operations in the vertical arithmetic unit, the total time taken is approximately 65 memory cycle times for 20-bit data. A conventional machine would need to perform 1,024 multiplies and 1,024 additions, with an average of approximately 20,000 memory cycle times, a gain in speed of about 300 to 1.

Count Orthogonal

Given 1,024 data, α_i, in an Orthogonal block of memory. This instruction provides as output 1,024 numbers n_i such that n_i is the number of 1-bits in α_i. For K-bit data, the timing is K memory cycle times; that is, one memory cycle time per bit for all 1,024 counts.

An important application of count Orthogonal is correlation of various types. One example is to match a sample pattern to an existing library of patterns, where the probability of an exact match is extremely small. There is a VU instruction *coincidence* which records the number of bit coincidences of the sample with each of 1,024 patterns of the library. This instruction combines two VU instructions, the Boolean equal and count Orthogonal. The timing for coincidence on K-bit data is also K memory cycle times.

The VU instruction *compare* can then find all those patterns that exceed a specified number of coincidences. Another VU instruction *maximal elements* can find all those patterns that have the largest number of coincidences.

Another important example involves searching in an inverted file for the best documents with respect to a key word (or descriptor) request. Without going into any detail, the average search time for D descriptors is at the rate of approximately $D + 4$ milliseconds per 10^6 documents, implying the capability of responding to a 10 descriptor request in a 7 x 10^7 document file in one second.

In general, the Orthogonal computer is about 5,000 times as fast as a conventional computer for the count Orthogonal instruction.

FAST FOURIER TRANSFORM

The FFT is a natural application for the Orthogonal computer. The particular algorithm under consideration is *decimation in time*, using in-place calculations with normally ordered inputs. The usual technique is used, treating the even and odd numbered inputs as the real and imaginary parts of complex numbers.

Each input data consists of 10 bits including sign. 16 bits are retained during all computations. The Orthogonal computer can perform the transformation of 2,048 samples in less than one millisecond; real time processing at a two megahertz sampling rate.

A low cost special purpose Orthogonal FFT with a simplified vertical arithmetic unit of only 512 levels of logic can perform the FFT on 1,024 samples in less than one millisecond. This machine has no horizontal arithmetic unit. It has in its repertoire the usually associated functions such as IFFT, auto and cross correlation, cepstrum, etc.

Readings

1) G. Estrin, "Organization of Computer Systems–The Fixed plus Variable Structure Computer," *Proceedings of the WJCC*, 1960.
2. W. Shooman, "Parallel Computing With Vertical Data," *Proceedings of the EJCC*, 1960.
3. W. T. Cochran *et al*, "What Is the Fast Fourier Transform?" *Proceedings of the IEEE Trans. on A & E*, June, 1967.
4. H. S. Stone, "Associative Processing for General Purpose Computers Through the Use of Modified Memories," *Proceedings of the FJCC*, 1968.

PROGRAMMING AND SOFTWARE TECHNIQUES

CHAPTER 16

A MODEL FOR PARALLEL COMPUTATIONS

Duane A. Adams

United States Air Force Academy
Colorado Springs, Colorado

Abstract

The design of new generation computing equipment has generally been brought about by advances in the state-of-the-art of computer technology. Once the hardware has been designed, then begins the long and tedious process of determining how to efficiently use the hardware and how to design the appropriate software systems. Rarely has the organization of computations played a dominant role in the design of the hardware. With a parallel processor system, the designer is faced with an even greater problem than he has had in the past. If the system is not designed properly, it may not be possible to apply the greater computing power he has available to the problem at hand. Thus, the manner in which the computations are to be organized should have a strong effect on the design of the system.

In this paper, a model for parallel computations is presented in which the sequencing control is governed by the flow of the data. The directed graph is used to represent the computation. The nodes of the graph represent computation steps, and the edges represent transmission paths for data and control. An edge may be thought of as a queue of data produced by one node and waiting to be consumed by another. A computation step may be initiated whenever each edge directed into that node of the graph contains the amount of data required for the node to execute properly. The number of computation steps which may be executed at any given time is dynamically determined by the flow of data. Thus, unnecessary sequencing constraints may be eliminated. All programs expressed in this model are *determinate,* that is, the results of the computation do not depend on the number of processors used, their relative speeds, or the possible sequencing of computation steps among those nodes which are ready to be executed. An example of a graph program is presented, and some of the problems connected with the implementation of a system to execute graph programs are discussed.

INTRODUCTION

The general topic of parallel processing refers not only to parallel machines and the software systems which operate on them, but also to the organization of computations which are to be executed in parallel. The design of a parallel processor system plays a major role in determining how computations which execute on it are to be organized. To effectively use the ILLIAC IV, for example, the computations must be organized so that the same sequence of operations may be performed simultaneously on a number of sets of operands. Such a system limits the class of problems on which it may be used effectively. On the other hand, a parallel processor system in which the processors may function independently offers considerably more latitude in organizing the computations. One use of such a system would be the processing of several jobs simultaneously, but having only one processor assigned to a given job. The more interesting cases are those in which more than one processor has been assigned to a single job. Proposals have been made for allowing the programmer to specify processes which may be performed simultaneously with such constructs as the FORK and JOIN[1] and the AND.[2] Another approach to this problem has been to automatically detect the implicit parallelism in a program written in a language such as FORTRAN by examining the sequence of references to the variables in a program.[3] When two successive operations reference distinct variables, they may be performed simultaneously or in either order.

In this paper a computation model is described in which the sequencing control is governed by the flow of the data. The parallelism within a program is thus dynamically determined by the availability of the operands. The directed graph has been used to represent the computations. The nodes of the graph represent computation steps and the edges represent first-in first-out queues of data. During the execution of a graph program, a node may be executed whenever there is the required amount of data on each edge directed into that node. Thus more than one node may be executing simultaneously, and, there may be several executions of the same node simultaneously in progress. This model possesses the important property of *determinacy;* that is, the results of the computation do not depend on the number of processors used, their relative speeds, or the possible sequencing of computation steps among those nodes which are ready to be initiated. A model such as is described in this paper could provide insight into the design of a parallel processor system.

DESCRIPTION OF THE MODEL

The model described below is an attempt to provide a framework within which various classes of computations can be represented. The model has been developed so that computations represented within it will be determinate. The main emphasis has been on developing a model that can be programmed. This latter goal is reflected in our attempt to deal with data structures and a hierarchical description of the program. Data structures are treated with generality, and include the hardware defined structures such as bits and words, and the structures usually defined in a programming language such as arrays, strings, and lists. Our hierarchical program description is achieved by being able to treat each node of our graph as representing some operation—perhaps very complex—and also being able to represent as a graph the suboperations or instructions of which it is constructed. Thus a graph program will be defined recursively.

Elements of the Model

Our approach to defining a model for parallel computations is to provide a set of primitive elements (basic elements from which other structures can be constructed) and to define the rules for using these primitives. Various classes of computations can be described by varying the set of primitive elements. In the following definition, we shall use standard mathematical notation for sets and functions. Other terminology and notation used here frequently appears in the study of formal languages. In describing a data structure grammar, we shall use the following meta symbols:

1) $\rightarrow$ is used to separate the left and right hand sides of a production
2) * denotes zero or more repetitions.

Definition 1. Within the model M, a class of computations is defined by specifying the following:

1) A set of *data types* $d_1, \ldots, d_h$
The syntax of the data types is described by a grammar with vocabulary $V = V_N + V_T$, where $V_N = \{d_1, \ldots, d_n\}$, the non-terminal symbols, and $V_T = \{[,], Y_1, \ldots, Y_k\}$, the terminal symbols. The data types are described by productions of the form $d_i \to x$, with $d_i \epsilon V_N$, $d_i \neq x$, $x \neq \lambda$, the empty string, and either $x = \alpha_1\alpha_2 \ldots \alpha_n$ or $x = [\alpha_1\alpha_2 \ldots \alpha_n]$, where each α_j is one of
   a) an element of $V - \{[,]\}$ or
   b) an element of $V - \{[,]\}$ superscripted with * or an integer constant.
2) A set of *primitive nodes* $p_1, \ldots, p_{k-1}, p_k, \ldots, p_n$.
Each primitive node p_i, $1 \leq i \leq n$, has an ordered set of l inputs requiring data of types $d_{i1}, \ldots, d_{il}$, an ordered set of $m - l$ outputs producing data of types $d_{il+1}, \ldots, d_{im}$, and a single-valued function $f_i : (V_{i1}, \ldots, V_{il}) \to (v_{il+1}, \ldots, v_{im})$. In addition, for $k \leq i \leq n$, each input of p_i has a binary valued *input status* S_{ij}, $1 \leq j \leq l$, taking on values L and U, and p_i has a single-valued function g_i: $(S_{i1}, \ldots, S_{il}, V_{i1}, \ldots, V_{il}) \to (S'_{i1}, \ldots, S'_{il})$. Each V_{ij} in both f_i and g_i is either an element of type d_{ij}, or when $S_{ij} = L$, $V_{ij} = \varphi$, the null element. Neither f_i nor g_i is defined when all of the arguments $V_{i1}, \ldots, V_{i1}$ are the null element. Those nodes defined only by the function f will be called *r-nodes* and those nodes defined by both f and g will be called *s-nodes.*
3) *Directed edges.*
Each edge is of data type d, where $d\epsilon\{d_1, \ldots, d_n\}$, the set of data types.

Edges The edges in the model M serve as temporary storage areas for data; they are first-in first-out queues of data. The parameter d identifies the structure of the elements found on the queue, and the data structure grammar describes the composition of these elements. Thus, depending on the data type of an edge, the elements on that edge may be very complex structures. Each element on a queue is implicitly separated from the remaining elements; that is, each element that is added to the queue may later be unambiguously removed from the queue. This implicit separation of elements on a queue is a generalization of the implicit separation of words in a computer memory. The length of each queue is assumed to be unbounded.

Data types As stated in Definition 1, the primitive nodes are defined to operate on specific types of data, and each of the edges is defined to hold data of a particular type. For each data type $d_i \epsilon V_N$, the grammar states precisely what strings and ordered sets of terminal symbols are elements of that type. The productions which define the data types fall into two distinct categories:

1) productions of the form $d_i \rightarrow \alpha_1 \alpha_2 \ldots \alpha_n$
2) productions of the form $d_i \rightarrow [\alpha_1 \alpha_2 \ldots \alpha_n]$.

In the first case, an element of type d_i is the concatenation of the elements which correspond to the α's. This is the usual sense in which productions are used in the description of formal languages. In the second case, an element of type d_i is an ordered set of the elements $\alpha_1, \alpha_2, \ldots, \alpha_n$, enclosed in brackets. The brackets are used both as terminal symbols which enclose the remaining elements and as an indication that the enclosed elements form an ordered set. By treating the elements between the brackets as an ordered set, there is an implicit separation of elements. This is meant to correspond to the notion of a queue of elements on an edge. A symbol superscripted by $*$ denotes that zero or more copies of that symbol may appear in the productions, and a symbol superscripted by an integer constant denotes that a fixed number of copies of that symbol appear in the production.

A production of the form $d_j \rightarrow [d_i^*]$ would be interpreted to mean that each element of type d_j is a queue of elements, where each element on the queue is of type d_i. The queue may be of arbitrary length, including the queue with no elements at all. The ability to treat a queue of elements as a single element on another queue serves as a naming convention. It allows us to do what is done in ALGOL when a name is declared to be, say, an array of some given size. In procedure calls, we can then pass a structure as a single element to the procedure, just as in ALGOL the name of a structure is used in the procedure call. Another reason for treating a queue of elements as a single element on another queue is that the only storage in our model is in the form of first-in first-out queues of data stored on the edges between nodes; there is no random access storage. Thus, new structures cannot be formed by creating pointers to substructures, as is done in the implementation of LISP, for example. However, the use of pointers would be a natural way to simulate these data structures on a computer with a random access memory.

The set V_T of terminal symbols always includes the brackets, [and], although they need not be used. A special use of the brackets in connection with procedures will be discussed later. For the remaining symbols, $Y_1, \ldots, Y_k$, we might choose 0 and 1. These terminal symbols could be used to represent the data in a binary storage device. On the other hand, we may choose as terminal symbols the alphabet, decimal digits and perhaps some special symbols such as parentheses. This might be an appropriate set if one is concerned with symbolic manipulation.

In the following example of a set of data types $V_T = \{[,],0,1\}$ and $V_N = \{d_1, d_2, d_3, d_4\}$. The productions are given by

$$d_1 \to 0|1$$

$$d_2 \to d_1^{32}$$

$$d_3 \to [d_2 d_2^*]$$

$$d_4 \to d_2|[d_4^*] \ .$$

The vertical bar used above is a meta symbol meaning *or*. In this language each element on a queue of type d_1 would be either a 0 or a 1. An element on a queue of type d_2 would be any string consisting of 32 zeros and ones, hence a 32-bit word. The elements on a queue of type d_3 would be 1-dimensional arrays of 32-bit words, each array containing at least one element. Data type d_4 is defined recursively and allows tree-structured data to be represented. Each of the branches eventually leads to either an ordered set of 32-bit words or the empty set. Some of the allowed structures are

$$[\,],\ [[\,]],\ [d_2 d_2]\ , \text{ and } [d_2 [d_2 d_2][\,]]\ ,$$

where in this example each d_2 represents a 32-bit word.

Note that there is a difference between the production $d_i \to [d_j]$ where $d_j \to \alpha_1 \ldots \alpha_n$ and the production $d_i \to [\alpha_1 \ldots \alpha_n]$, even though the same elements appear in each case. In the first case, a single element appears between the brackets; in the second it is an ordered set of n elements. When $n = 1$, the two productions would be the same. The production $d_i \to [d_j^*]$ is well defined, but the production $d_i \to d_j^*$ is not, since it would admit the empty string λ as an element of type d_i. When we speak of the null element φ being placed on an edge, it is to be interpreted that no output is placed on the edge, φ is distinct from the empty set [ ].

Certain nonterminal symbols may be introduced as intermediate symbols in creating other structures. Besides being used to represent a rather general quantity such as the 32-bit word defined above, the data types may be used for such purposes as distinguishing between integers and floating point numbers. The use of structured data in programs within the model will be explained after the notions of a graph procedure and the execution of a graph program have been explained.

Primitive Nodes The primitive nodes represent the "building blocks" from which the programs must be constructed. Each of the primitive nodes has a particular well defined function associated with it. The particular choice of primitive nodes determines the class of computations, which can be represented within the model, and the ease with which such computations can be programmed. The functions performed by the primitive nodes might range from logical operations performed on bits, to such complex operations as matrix multiplication.

Each input status setting for an s-node may be either *locked* or *unlocked*, denoted by L and U, respectively. When the jth input is in locked status, then the argument V_j is implicitly defined to be φ for both functions f and g. That is, any values on a locked edge remain on the edge and are not used in the computation until that edge later becomes unlocked. The g function associated with an s-node allows the status settings of that node to be changed as a function of the current settings and the input arguments to the node. The s-nodes allow data to be selected from mutually exclusive sources in a determinate manner, thus avoiding race conditions.

The function f associated with each primitive node is used in a very general sense. The domain of the function has been defined to permit the null element within the l-tuple of arguments for f, as long as the entire l-tuple does not consist of null elements. The null element is *only* allowed as an argument of f when the corresponding status setting is locked. Thus, for an r-node where there are no status settings, the null element is not allowed as an argument of f. The function f may be a conditional function which produces output on only some or possibly none of its edges. The null element φ will indicate that no output has been placed on an edge. More than one element may be placed on an edge as output. For a given ordered set of inputs, both f and g have uniquely defined outputs.

The following example of an s-node illustrates some of the ideas which we have been discussing. Suppose the node has three inputs and one output, and its function is to accept a Boolean signal on its first input, and depending on the value of the Boolean, accept the next element of data from either the second or third input and place it on the output. Suppose that the second and third inputs and the output are all of type word. The f and g functions for this node are

g:	$(U,L,L,T,\varphi,\varphi)$	$\rightarrow$	(L,U,L)	f:	(T,φ,φ)	$\rightarrow$	(φ)
	$(U,L,L,F,\varphi,\varphi)$	$\rightarrow$	(L,L,U)		(F,φ,φ)	$\rightarrow$	(φ)
	$(L,U,L,\varphi,V,\varphi)$	$\rightarrow$	(U,L,L)		(φ,V,φ)	$\rightarrow$	(V)
	$(L,L,U,\varphi,\varphi,V)$	$\rightarrow$	(U,L,L)		(φ,φ,V)	$\rightarrow$	(V)

Note that not all combinations of status settings have been taken into account in the above functions. Those not listed are not meaningful for the above problem. For completeness of definition, however, we will adopt the convention that for the combinations of status settings not listed, the g function locks all inputs and the f function produces all null outputs.

Composition of Graph Programs

Definition 2. A *graph procedure* G consists of a node set $\eta = \{N_1, \ldots, N_l\}$ and an edge set $\mathcal{E} = \{e_1, \ldots, e_m\}$, where each edge $e_i \epsilon \mathcal{E}$ must satisfy one of the following three conditions:

1) be directed from the output of a node $N_j \epsilon \eta$ to the input of a node $N_k \epsilon \eta$
2) only be directed to the input of a node $N_k \epsilon \eta$; the ordered set of such edges is called the *input set* of G, and is nonnull
3) only be directed from the output of a node $N_j \epsilon \eta$. This ordered set of edges is called the *output set* of G, and is nonnull.

Some $e_i \epsilon \mathcal{E}$ must be connected to each input and each output of the nodes in η. Each edge must be data type compatible with the nodes onto which it is connected. The nodes of η may be either primitive nodes or procedure nodes.

Definition 3. A *procedure node* P is a node whose operation is described by a graph procedure G. The ordered set of inputs and the ordered set of outputs of the procedure node must agree in number and data type with the input set of G and the output set of G, respectively.

Definition 4. An edge e of data type d_i is *compatible* with the node or pair of nodes onto which it is connected when the following conditions are satisfied:

1) if e is directed into a node, then that input into which it is directed must be of type d_i
2) if e is directed from a primitive node, that output from which it is

directed must be of type d_i

3) if e is directed from a procedure node, that output from which it is directed may either be of type d_i or, if there is a production in the data type grammar of the form $d_i \rightarrow [d_j^*]$, that output may be of type d_j .

Definition 5. A *graph program* $\mathcal{P}$ is an ordered pair (G,X), where G is a graph procedure and X is the set of all graph procedures such that:

1) if P is a procedure node in G, and Y is the graph procedure defining P, then $Y \in X$
2) if $Y \in X$, P is a procedure node in Y, and W is the graph procedure defining P, then $W \in X$
3) X is defined only by (1) and (2), with no graph procedure appearing more than once in X.

The following discussion gives an informal definition of a graph program execution. The initial data for the program is placed on the appropriate edges. During the execution of a graph program, a node in the graph may be executed whenever there is at least one element of data on each of its unlocked input edges. When the node executes, it removes one element of data from each unlocked input, performs the indicated computation, and places the results on the appropriate edges leading out from the node. For an s-node, the status settings may also be changed. When no further nodes can be executed, the computation is complete, and the results will be left on the edges designated as the output set of G.

The definition of a graph procedure allows us to write recursively defined programs. A procedure node may be defined by the graph procedure within which it appears. Thus it will not always be possible to simply replace each procedure node in a graph program by its defining graph procedure, since the process may be nonterminating. When a procedure node is ready to be executed, a copy of the defining graph procedure is created. This is similar to the ALGOL 60 copy rule for procedures. The arguments of the procedure are placed on the edges of the graph procedure which are designated as the input set, and the graph procedure is then allowed to perform the indicated computation. When the execution of the procedure has been completed, the results on the edges of the output set are transferred onto the corresponding edges directed out from the procedure node.

The definition of a graph program makes no reference to reserved storage locations, and, in fact, there are none. The data is dynamically structured during the execution of a program, as is done in LISP. There are two basic methods for structuring data in the model, whereas in LISP there is only one. In LISP, the *cons* operation forms a new S-expression from two smaller S-expressions. The operation is sequential, and to form a list, for example, requires the repeated consing of elements onto the partially formed structure. In the graph model, the first method of structuring data is to have primitive nodes which accept an operand on each of their input edges, and output the structured result. A primitive node might be included whose operation is identical to the *cons* operation in LISP. More generality is allowed in the graph model since the number of operands need not be two, other operations besides the *cons* may be used, and it is possible to define structures other than S-expressions.

The second method for structuring data is referred to as *bracketing,* and is used only in connection with procedures. The elements on an edge in the output set of a graph procedure may logically form a unit of data, and we would like to treat these elements as a unit rather than as the individual elements on the queue. For example, a procedure may produce a sequence of words which we wish to consider as a vector, or a sequence of elements which forms a list. For determinancy reasons, it is not possible to have a primitive node remove a variable number of elements from a queue and form the desired structure. Bracketing can occur only after the procedure has finished executing, and thus all of the elements which are to be placed in the structure have been formed.

Whether or not the elements on an edge are to be bracketed is determined by the data type of the edge in the output set and the data type of the edge leading out from the procedure node. We note first of all that when a procedure terminates, the data is transferred from the edges of the graph procedure to the edges directed out from the procedure node. For each edge in the output set of the graph procedure, the corresponding edge directed out from the procedure node may either be the same type or it may be a different data type; and for these two alternatives, the data types may or may not be related by a production of the form $d_i \rightarrow [d_j^*]$. We consider these four cases. For an edge of type d_j in the output set of the graph procedure with corresponding edge of type d_i directed from the procedure node:

1) $d_i \neq d_j$ *and* $d_i \not\rightarrow [d_j^*]$ —this combination is not allowed by the definition of compatible
2) $d_i \neq d_j$ *and* $d_i \rightarrow [d_j^*]$ —bracketing occurs
3) $d_i = d_j$ *and* $d_i \not\rightarrow [d_j^*]$ —*no* bracketing occurs
4) $d_i = d_j$ *and* $d_i \rightarrow [d_j^*]$ —bracketing occurs.

Only in case (4) was there a choice of bracketing or not bracketing, and we chose to bracket the output since the operation of removing unwanted brackets is easier than having to sequentially form the structured data. An alternative to (4) would be to provide some indication in the graph that the output is to be bracketed.

The input arguments to a procedure node are simply passed to the defining graph procedure. There is no operation which can automatically remove the brackets from an input element. If the correct operation of the graph procedure requires that brackets be removed, then this must be done by a primitive node within the graph procedure. The consequences of this requirement is that any attempt to remove brackets from unbracketed data will cause an error within a primitive node and not an error in the initiation of a procedure node.

By requiring that the data type of an edge be compatible with the data type of the nodes onto which it is connected, we are insuring that the nodes always receive the type of data for which they are defined.

Notation for Graph Programs

The definitions which we have given thus far have been mathematical in nature. The objects defined, however, are meant to have an intuitive appeal. Furthermore, we need to have a notation for representing the programs in our model.

A node is represented by a circle.

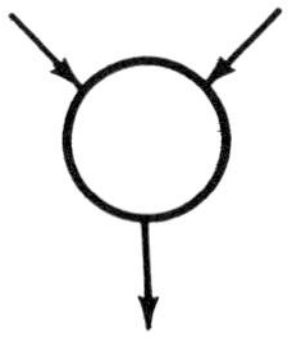

When discussing the properties of a node, the inputs and outputs will be designated by directed line segments.

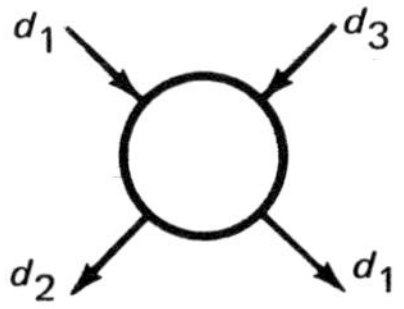

The data type will be indicated beside each directed line segment.

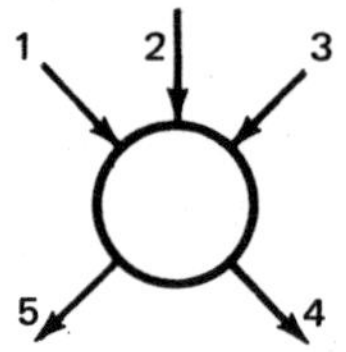

We shall assume that the inputs and the outputs each appear in consecutive order, and that they are ordered in a clockwise direction starting with the first input.

There must be at least one input and one output for each node, thus this ordering is unique. When the edges are not explicitly ordered, it is assumed this convention is being used.

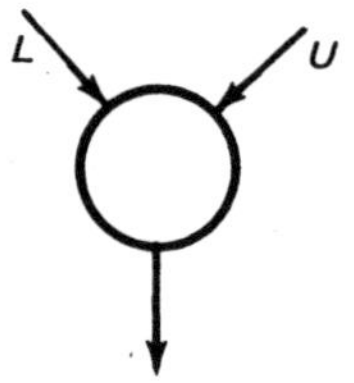

The initial status settings for an *s*-node will be denoted by writing an *L* or a *U* beside the corresponding input.

One drawback of the graph model as a means for expressing programs is that there is no convenient way for representing the data. The Karp and Miller model used an edge parameter to show the amount of data initially placed on each edge, but there was no means of indicating the values of the operands.[4]

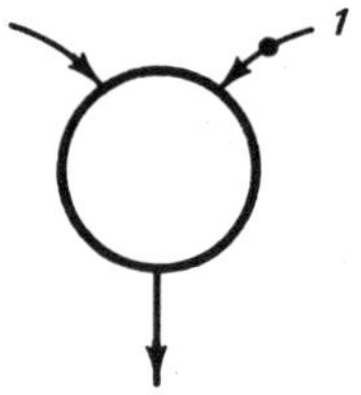

When an algorithm can be better understood by knowing certain values, especially constants, we shall denote it like this.

The interpretation here is that a 1 is initially placed on the edge directed into input 2.

We will identify the nodes in a graph program, both primitive and procedure, by placing an identifier within the node. For the procedure node, this identifier will be the same as the name of the defining graph procedure. The identifiers may be alphabetic strings, numbers, or mathematical symbols denoting the operations to be performed.

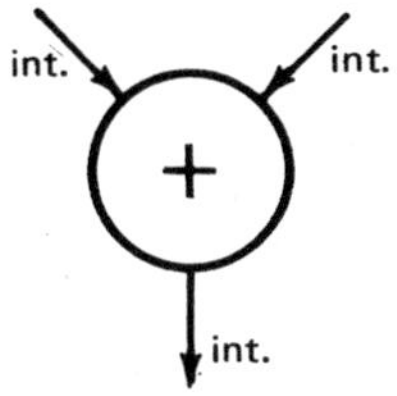

This node might be used to denote a primitive node which performs the addition of two integers.

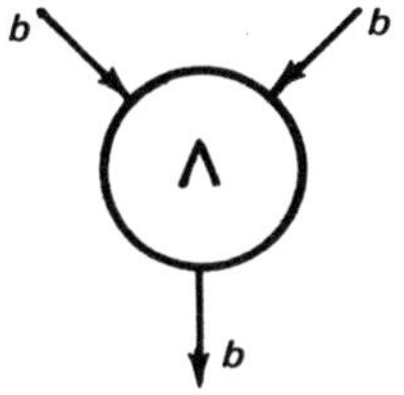

This primitive node might be used to denote the logical AND of two bits.

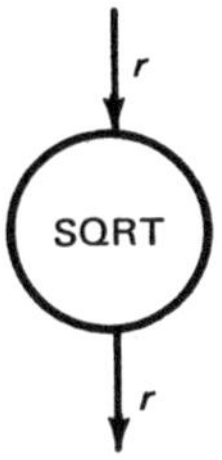

This may be a procedure node, where SQRT is the name of a graph procedure which calculates the square root of a real number.

Here it is assumed that b, int., and r have all been precisely defined by a data type grammar, and that each primitive node has also been given a precise definition.

There is really no theoretical limitation as to the kind of functions the primitive nodes may be defined to perform. A primitive node is primitive in the sense that we are not interested in its structure below a certain level. This does not mean that it cannot be described in terms of even more basic functions by another graph program with perhaps a different set of primitive nodes. This raises the question as to how we should describe the semantics for the primitive nodes. At present we have not selected a specific set of nodes and data types from which all others can be constructed. Even if we had, it might be very laborious to define our semantics from such primitives. We shall, in general, use conditional expressions to define the semantics for the primitive nodes, but we may use tables or verbal descriptions, if appropriate. In defining the g function for an s-node we shall use the notation

$$g\text{: ULL} \to \text{LUL}$$

which means that the status settings for edges 1, 2, and 3 initially are unlocked-locked-locked, and are reset to locked-unlocked-and locked, respectively, when the node executes. If there is a condition governing what the new setting will be, that condition will be specified as

$$g\text{: ULL} \rightarrow \text{LUL if } V_1 = T$$

is the same as

$$g\text{: }(\text{U,L,L,T},\varphi,\varphi) \rightarrow (\text{L,U,L})\ .$$

GRAPH PROGRAM EXAMPLE

A number of algorithms have been expressed as graph programs, including algorithms for Gaussian elimination, sorting, algebraic simplification, and numerical integration. We shall use the numerical integration example to illustrate the concept of a graph program and also to show one method of organizing computations for parallel execution. The adaptive trapezoid method is used. There is less parallelism in this method than in the adaptive Simpson method, but the algorithm is somewhat easier to understand. An approximation to the integral is obtained using the trapezoid method with only one subdivision. The interval is then subdivided and the trapezoid method used to recompute the integral which now has two subdivisions. If the two approximations are sufficiently close, the process terminates; otherwise the procedure is recursively called for each of the subdivisions.

Data types The data types for this example are defined using the terminal symbols 0 and 1.

$$b \rightarrow 0|1$$

$$\text{exp} \rightarrow b^8$$

$$\text{fraction} \rightarrow b^{40}$$

$$fp \rightarrow \text{exp fraction}$$

$$p \rightarrow [fp\ fp]$$

The above data types are given the following interpretation: *b* is used to denote both a bit and a Boolean variable where we equate 1 with *true* and 0 with *false*; *fp* denotes a 48-bit floating point number having an 8-bit exponent and a 40-bit mantissa; *p* is a pair of floating point numbers, actually an array of length 2.

Primitive nodes The primitive nodes are defined by stating what is produced at each output in terms of the values received at the unlocked inputs. The result at each output will be denoted by a small v subscripted by the output number, and the argument at each unlocked input will be denoted by a capital V subscripted by the input number. Thus $v_2 \leftarrow V_1$ should be read "the value resulting at output 2 is the value from input 1." The symbol φ is used to denote that no output was placed on a given edge. ALGOL arithmetic expressions are used wherever possible to describe the semantics of the nodes, although verbal descriptions are occasionally used.

The primitive nodes may be classified according to those (1) which perform control functions, (2) which perform data structuring operations, (3) which are predicates, and (4) which are logical and arithmetic operations. Only those nodes used in this example are given; other examples have required the definition of other functions. Certain of the functions to be defined are applicable to all data types while some functions are defined only for particular data types. We shall use the word *all* to denote an input or output which is defined for all possible data types. When more than one input into a node or output from a node has *all* as its type, then each such input or output must be of the same type whenever that node is used in a graph procedure.

The following 3 nodes are used only to control the flow of data.

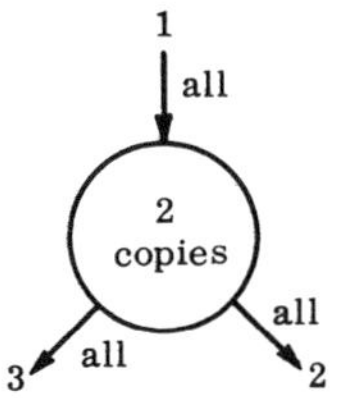

f: $v_2 \leftarrow V_1$;
$v_3 \leftarrow V_1$;

A node which creates 3 copies of an operand is also used.

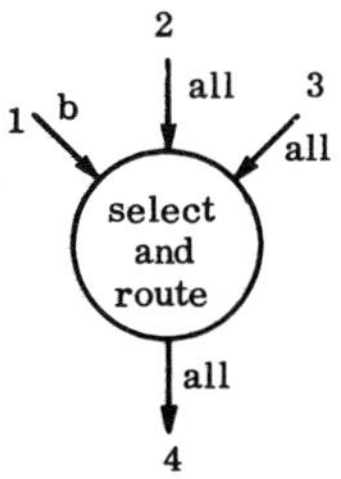

g: ULL→LUL <u>if</u> V_1 = <u>true</u>	f:	$v_4 \leftarrow \varphi$;
ULL→LLU <u>if</u> V_1 = <u>false</u>		$v_4 \leftarrow \varphi$;
LUL→ULL <u>for</u> <u>all</u> V_2		$v_4 \leftarrow V_2$;
LLU→ULL <u>for</u> <u>all</u> V_3		$v_4 \leftarrow V_3$;

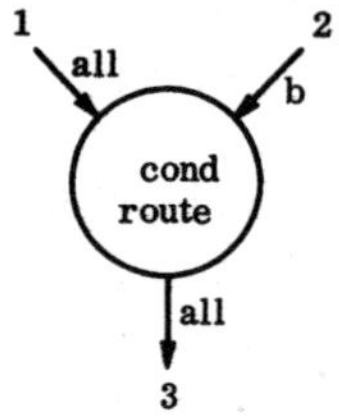

f: $v_3 \leftarrow$ <u>if</u> V_2 = <u>true</u> <u>then</u> V_1 <u>else</u> φ;

The following node is the only predicate node used in this example.

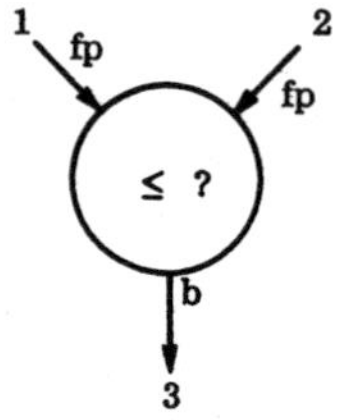

f: $v_3 \leftarrow$ <u>if</u> $V_1 \leq V_2$ <u>then</u> <u>true</u> <u>else</u> <u>false</u>;

There are 2 data structuring nodes used in this example.

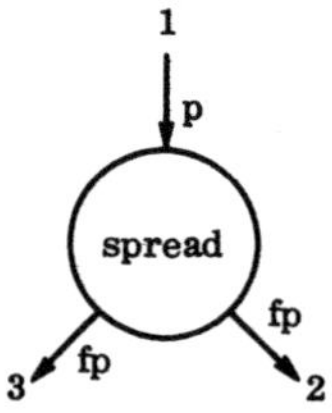

$v_3 \leftarrow$ the first element in V_1;
$v_2 \leftarrow$ the second element in V_1;

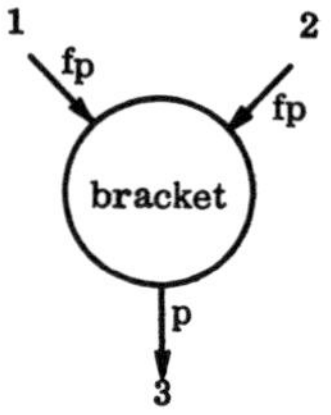

$v_3 \leftarrow [V_1 \ V_2]$;

The arithmetic operations used are

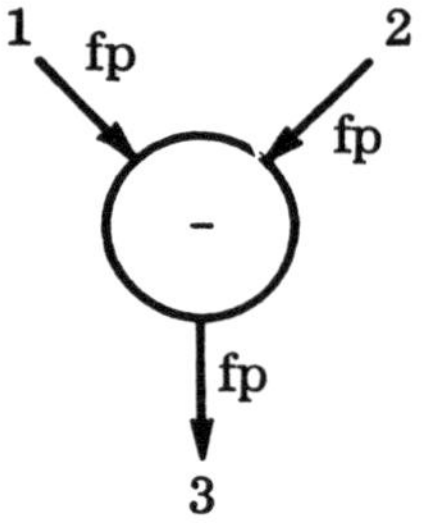

f: $v_3 \leftarrow V_1 - V_2$;

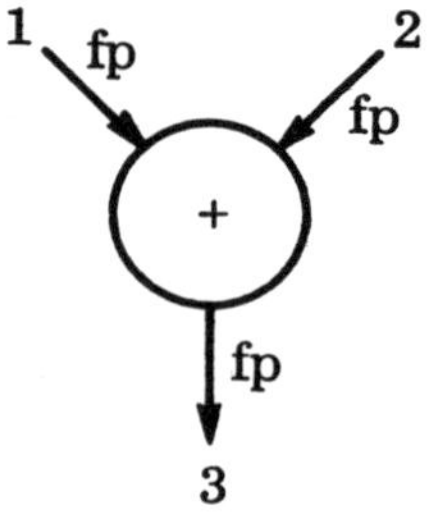

f: $v_3 \leftarrow V_1 + V_2$;

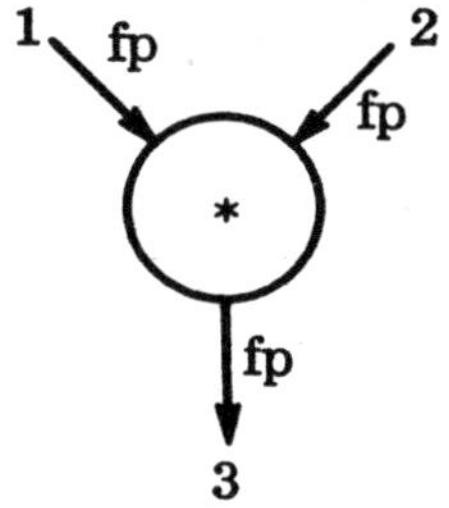

f: $v_3 \leftarrow V_1 \cdot V_2$;

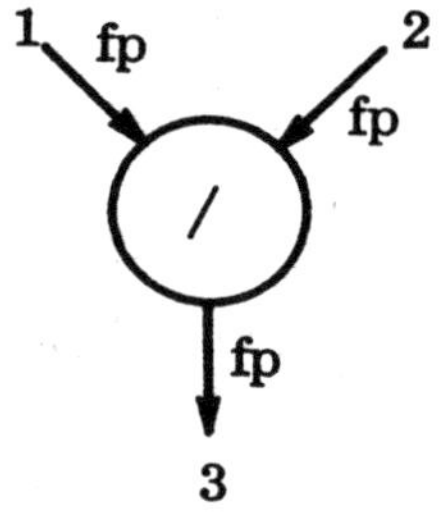

f: $v_3 \leftarrow$ <u>if</u> $V_2 \neq 0$ <u>then</u> V_1/V_2 <u>else</u> φ;

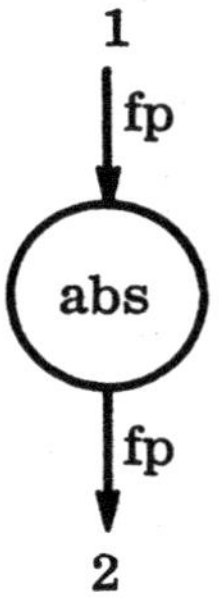

f: $v_2 \leftarrow$ <u>if</u> $V_1 < 0$ <u>then</u> $-V_1$ <u>else</u> V_1;

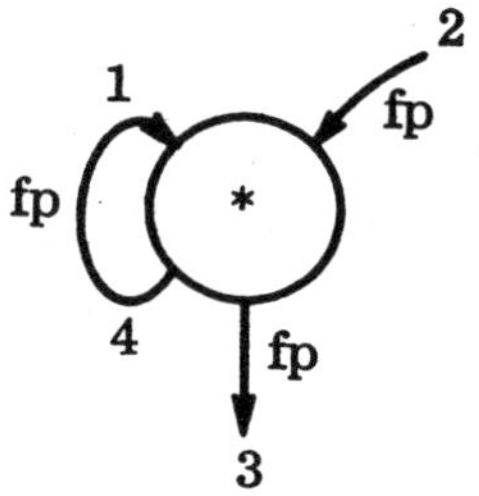

f: $v_3 \leftarrow V_1 \cdot V_2$;
$v_4 \leftarrow V_1$;

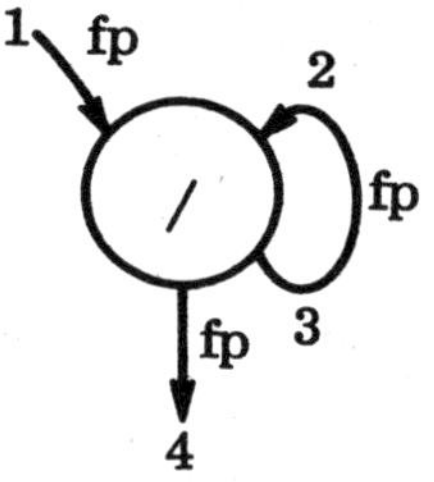

f: $v_4 \leftarrow$ <u>if</u> $V_2 \neq 0$ <u>then</u> V_1/V_2 <u>else</u> φ;
$v_3 \leftarrow V_2$;

The last two nodes are given so that a constant may be used in the multiplication and division, respectively.

Graph Program for Numerical Integration The procedure *Trapezoid* (Fig. 1) accepts as its inputs an element of type *p* containing the left-hand endpoint and the function evaluated there, an element of type *fp* containing an estimate of integral between the 2 given endpoints, and another element of type *p* containing the right-hand endpoint and the function evaluated there. *New Approx* (Fig. 2) subdivides the interval, and using the trapezoid rule it obtains an approximation for

Trapezoid:

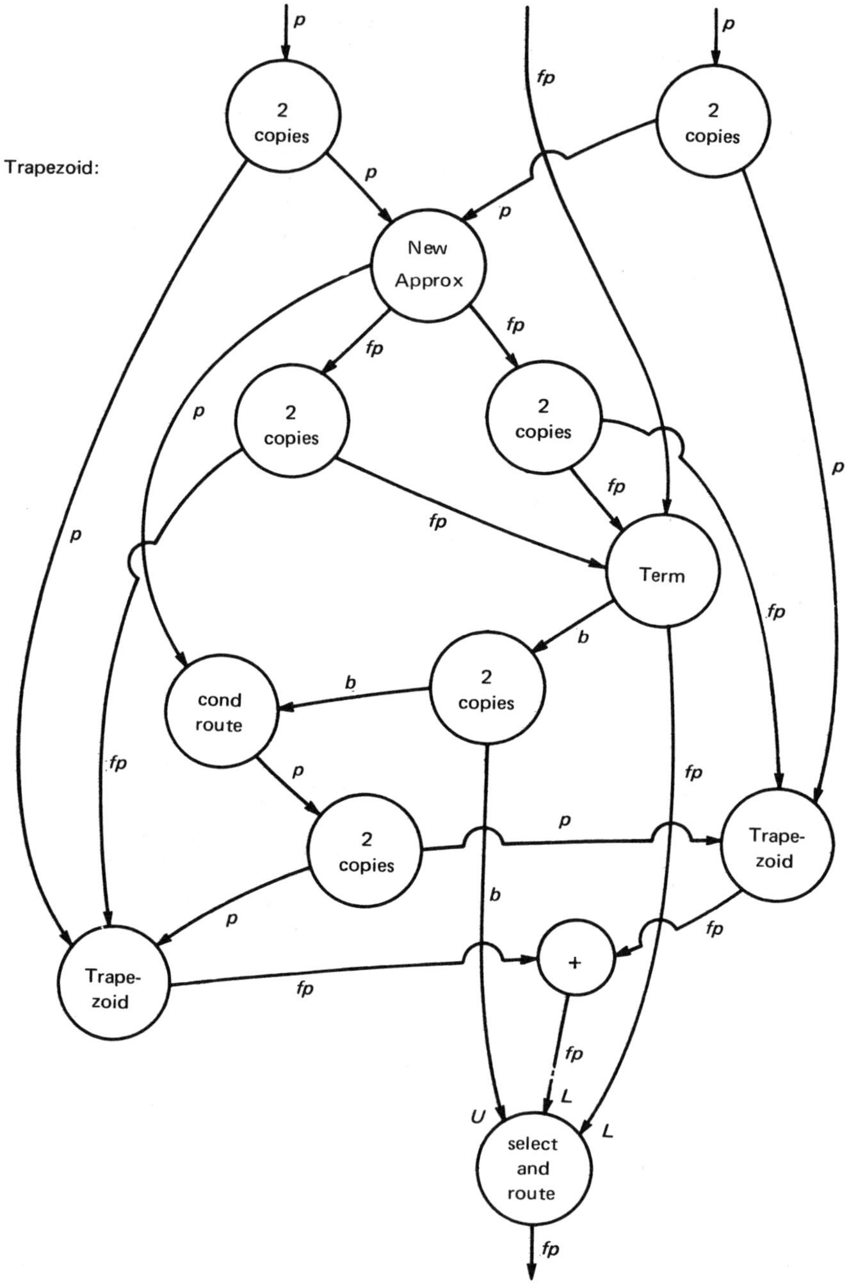

Fig. 1.

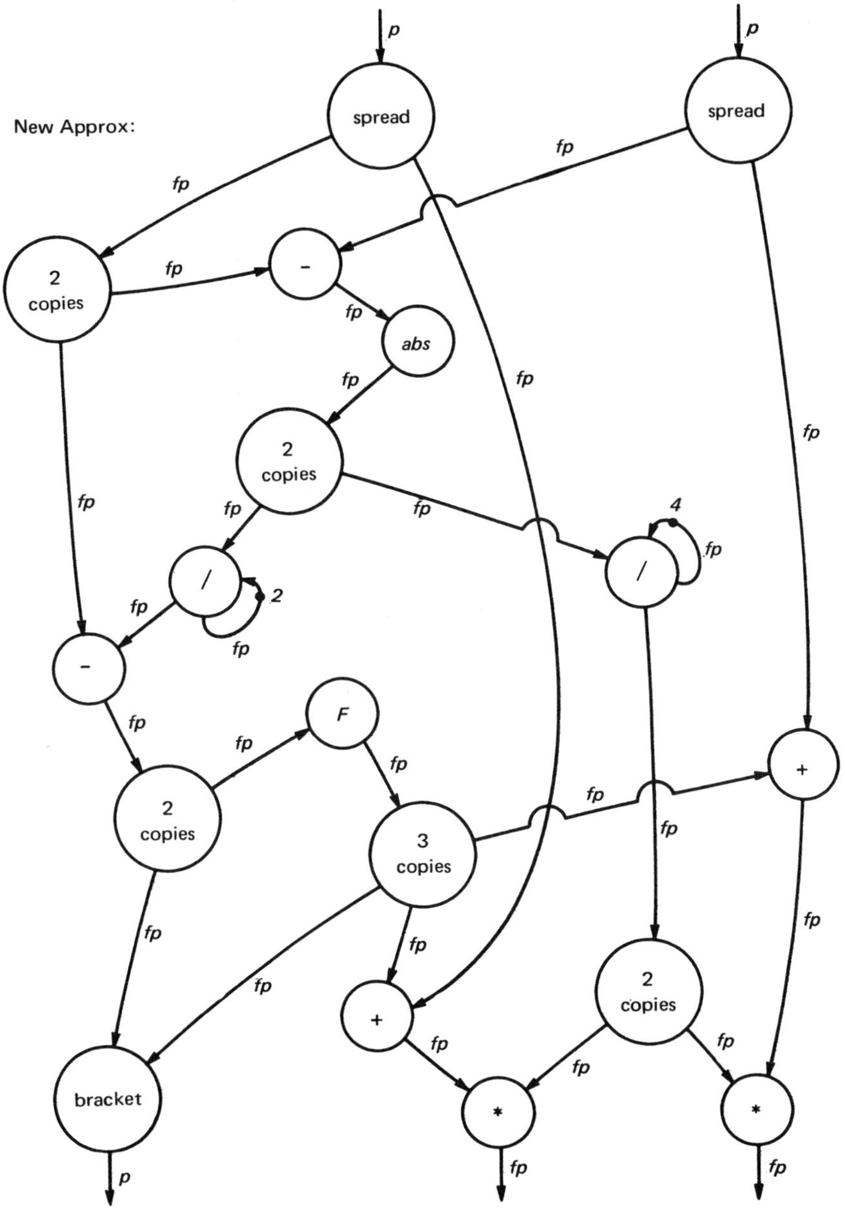

New Approx:
p
spread
p
spread
fp
2
copies
-
abs
2
copies
/
2
4
/
-
F
2
copies
3
copies
+
+
2
copies
bracket
*
*
p

Fig. 2.

Term:

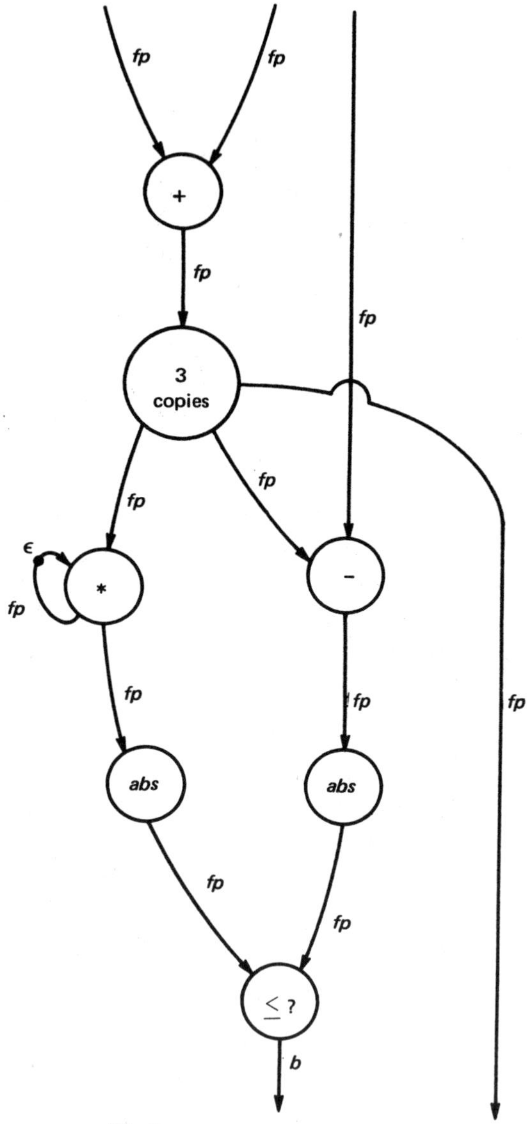

Fig. 3.

each subinterval. It outputs an element of type *p* containing the midpoint of the interval and the function evaluated there, an estimate of the integral over the left-hand half of the interval, and a similar estimate for the right-hand half. The node labelled *F* is a call on the routine to evaluate the function which is being integrated. *Term* (Fig. 3) is used to determine whether the integral has been approximated sufficiently close, or whether *Trapezoid* should be recursively called for each of the subintervals.

The major source of parallelism in this program comes from being able to execute several copies of the procedure *Trapezoid* simultaneously. An additional source of parallelism comes from being able to simultaneously execute more than one primitive node within each of the procedures. The use of recursion to express parallel computations has also been used in an algorithm for sorting and an algorithm in symbolic data processing.

THE MODEL AS A BASIS FOR MACHINE DESIGN

The model described in this paper is *determinate;* that is, the results of the computation do not depend on the number of processors used on the computation, the relative speeds of any of the processors, or the sequence in which the computation steps are performed from among those ready to be executed. Furthermore, all computable functions can be represented in the model. The above properties are stated as theorems and proved in the author's thesis.[5] A parallel processor system designed to execute graph programs would insure the user that the results of his computation would be the same regardless of system configuration. Such a system would allow processing resources to be dynamically allocated to the computation. This could be governed either by the demand for resources by the computation or by the availability of resources within the system.

The model given in this paper allows the computation to be described in as much detail as is desired, including going down to the bit level. There is no distinction between what is done in hardware and what is done in software. This is true for the way data is described as well as for the primitive nodes. Presumably the primitive nodes would correspond to functions performed by the hardware. With LSI semiconductor technology, it is likely that more complex functions will be included as basic operations performed by the hardware. This model not only allows such primitive nodes to be added, but may also indicate the kind of functions which such a system should have.

In the model there was no mention of input or output. It was assumed that all initial data was placed on appropriate input edges, and that the results of the computation were left on appropriate output edges.

Since each queue is unbounded in length and since there are no referenced variables, this is an adequate treatment of input/output. Both a card reader and a printer operate on a first-in first-out queue of data, and could be treated as primitive nodes within the model.

The primitive nodes are defined to accept on each edge any element of data of the correct type, and to produce a well defined output. In the model there are no undefined conditions allowed, even though operations such as dividing by 0 or taking the *car* of an atom may be mathematically undefined. In each case, the primitive node is defined to output some result, even though it may be the null element. Such conditions would usually be considered as runtime errors and would result in an interrupt. They are usually treated external to a programming language and would be functions handled by the compiler. It would be possible for the primitive nodes to detect an undefined condition and to communicate this outside the model in a manner similar to the way program interrupts are presently handled.

In many cases an error condition can be detected within a graph program. However, in the model there is no equivalent of a go to statement, and hence the computation cannot be halted by simply transferring to an exit. Any communication which is to take place within a program must be done through the flow of data. Thus an execution can be terminated only when there are no further computations to be performed. It is assumed that in a system in which graph programs could be executed that there would be a supervisor for the system through which communications regarding run-time conditions could take place. It would not be part of the model, but would be in addition to the model.

Within the model we have made reference to copying of procedures and also copying of data. In an actual system this copying need not be performed. Reentrant code may be used to eliminate the copying of procedures and at the same time allowing more than one processor to access the same code. Separate data areas must be established, however. The data will probably be primarily treated as read-only data. When more than one node is to have access to a piece of data, a scheme similar to that proposed by Van Horn may be used to insure determinacy.[6] A count associated with each value (or possible queue of values) would indicate the number of nodes having access to that value. After a node has read a value the count would be decremented. If the count is greater than 1, then the value may not be written over. A count of 0 would indicate the storage area could be released. There will be cases where a certain amount of copying may be necessary, but in general unnecessary copying can be reduced considerably.

A computational model such as the one described in this paper raises many questions concerning the design of a parallel processor system, the efficient allocation of resources in such a system, and the design of higher-level programming languages for describing parallel computations. Such questions should be considered in the design of future systems.

References

1. M. E. Conway, "A Multi-Processor System Design," *Proc. AFIPS 1963 FJCC,* 24: (1963), 139–46.
2. N. Wirth, "Program Structures for Parallel Processing," *Comm ACM 9:* (May, 1966), 320-21.
3. H. W. Bingham, D. A. Fisher, and E. W. Riegel, *Automatic Detection of Parallelism in Computer Programs,* AD 662274, November, 1967.
4. R. M. Karp and R. E. Miller, "Properties of a Model for Parallel Computations: Determinacy, Termination, Queueing," *SIAM J. Appl. Math, XIV* (November, 1966), 1390–411.
5. Duane A. Adams, "A Computation Model with Data Flow Sequencing" (Ph. D. dissertation, Stanford University, 1968), *Technical Report No. CS 117* (Stanford University [December, 1968]).
6. E. C. Van Horn, Jr., "Computing Design for Asynchronously Reproducible Multiprocessing." *MAC-TR-33,* November, 1966.

CHAPTER 17

RECOGNITION AND REPRESENTATION OF PARALLEL PROCESSABLE STREAMS IN COMPUTER PROGRAMS

M. J. Gonzalez and C. V. Ramamoorthy

University of Texas
Austin, Texas

Abstract

Since there are many real-time applications which demand computational rates exceeding that afforded by present day hardware technology, there exists a need to detect and multiprocess independent segments of a single program. In this paper some of the available techniques for the detection, recognition, and execution of parallel processable streams of a computer program are surveyed. Two techniques are then developed, one for implicit recognition and representation of parallel processable segments, and the other for parallel processing of substructures within a task. In the former, a conventional program is represented by its program graph, and by analysis of the graph and the input-output relationships of segments in it, an identification of all parallel processable tasks in the program and the sequence in which these tasks must be executed is derived. In the second technique, the maximum parallel processable form of a compound arithmetic expression is derived, and the execution sequence which minimizes the number of intermediate storages of the results is given. It is shown that this technique is particularly suitable for computers having multiple arithmetic elements with large register arrays.

This work was supported by NASA Grant NGR 44-012-144.

INTRODUCTION

Motivation

Computational processes have traditionally been approached in what is essentially a serial or sequential manner. The entire process—beginning with the writing of a program and concluding with the actual execution of the program—has been based on the assumption that an instruction stream is processed a single step at a time.

Initial efforts in the field of information processing with regard to computers were limited mainly to technological improvements of existing facilities. Some of the goals in this area included an increase in computational speed and a decrease in physical size and power consumption. After these efforts were successfully launched, however, it was realized that the serial approach to computational processes did not result in efficient use of available resources. This realization culminated in techniques such as multiprogramming and time-sharing.

Although these techniques did yield higher resource utilization, the speed of computational processes was now a function of technological limitations. Further progress in this area resulted in the concept of parallel processing, a concept whereby several segments of an instruction stream are processed concurrently.

Common to any discussion on parallel processing is an assumption of the existence of a large number of processing units and memories controlled by an operating system. In the past, the prohibitive expense implied by this assumption sealed the fate of any progress in this area. State-of-the-art advances, however, and, in particular, anticipated advances generated by large scale integration have given fresh impetus to research in this area. Given a program with a large degree of inherent parallelism, the theoretical time required for problem solution can be reduced " . . . by a factor at least as great as the number of processors[1]."

The motives for parallel processing include the following:

1) Real-time urgency. Parallel processing can increase the speed of computation beyond the limit imposed by technological limitations.
2) Reduction of turnaround time of high priority jobs.
3) Reduction of memory and time requirements for "housekeeping" chores. The simultaneous but properly interlocked operations of reading inputs into memory and error checking and editing can reduce the need for large intermediate storage or costly transfers between members in a storage hierarchy.

4) An increase in simultaneous service to many users. In the field of the computer utility, for example, periods of peak demand are difficult to predict. The availability of spare processors enables an installation to minimize the effects of these peak periods. In addition, in the event of a system failure, faster computational speeds permit service to be provided to more users before the failure occurs.

The subject of parallel processing can be divided into three separate problem areas. The first problem occurs at the time a program is written, and involves an explicit indication of parallelism within a program. In some cases this indication can be accomplished by means of the programming language itself. The second problem involves detection at compilation time of the segments of a program which can be processed concurrently. The final problem involves communication and utilization of this information by the computing system.

This first section of the paper introduces the general subject of parallel processing by defining the concepts of explicit and implicit parallelism and discussing problems associated with these areas.

The second section "Algorithms for Detection of Subtask Parallelism," is primarily a survey of existing techniques for detection of subtask parallelism as exemplified by arithmetic expressions. This section concludes with the introduction of a new technique for detecting subtask parallelism in arithmetic expressions.

The major portion of this paper is contained in "Parallel Task Detection at a Global Level." Here a new technique is introduced for detecting parallel processable segments within a single program. The technique has been incorporated in the form of a FORTRAN recognizer program. This recognizer inputs a source program also written in FORTRAN, and outputs the parallel processable segments of the source program. Details, limitations, and overall flow charts of the recognizer are also contained in this section.

Throughout the remainder of this paper two thoughts should be kept in mind. In order for parallel processing to be truly effective, the overhead which accompanies this mode of operation must be minimized. Thus, any approach to this subject must include machine organization which can take advantage of parallel processable streams. In addition, it is felt that successful recognition of these streams at compilation time will minimize overhead during execution time.

Hierarchies of Parallelism

Within an individual program, parallelism can exist at several levels—from the level of statements or groups of statements of procedural languages to the level of microoperations. Discussion in this paper will be confined to *intraprogram parallelism* as opposed to *program parallelism*. The latter term refers to the type of processing in which independent programs are processed concurrently; it is intended to be synonymous with the conventional meaning of multiprocessing in the multiprogrammed environment. Intraprogram parallelism, on the other hand, refers to the type of processing in which a single program can be partitioned into tasks that can be performed in parallel. Two types of intraprogram parallelism will be considered: *global* and *local*.

In the global type of parallel processing, a program is partitioned into tasks that can be performed in parallel. Throughout this discussion the terms *task* and *process* are intended to mean a selfcontained portion of a computation that once initiated can be carried out to its completion without the need for additional inputs. The completion of a task is significant in that its occurrence can initiate the execution of another set of tasks. An example of such a task would be a single FORTRAN statement or a set of such statements. Consider, for example, the

```
                        •
                        •
    Read 100, A, B, C, D
                        •
                        •
    X = A**2 - 2.0*A*B + B**2
    Y = C**2 + 2.0*C*D + D**2
                        •
                        •
                        •
  10 Z = (A*B) + (C*D)
                        •
                        •
                        •
    Call SUB1 (A, B, E)
    Call SUB2 (C, D, F)
                        •
                        •
    End
```

Fig. 1. Sample FORTRAN Program Illustrating Global and Local Parallelism

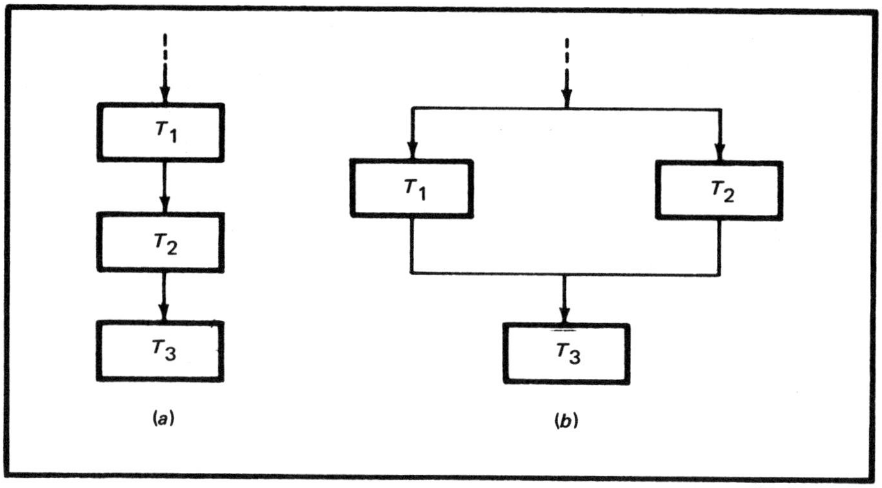

Fig. 2. Sequential and Parallel Execution of Two Tasks

hypothetical FORTRAN program shown in Fig. 1 which illustrates both global and local parallelism.

Because the two arithmetic expressions have independent input sets, they can theoretically be executed in parallel. This is an example of independent tasks consisting of single statements. The two subroutine CALL statements, on the other hand, are examples of independent tasks that can consist of many statements. Again because of the disjoint input sets, both of these statements could theoretically be executed in parallel.

The type of reasoning suggested here has been investigated by several sources; a reference prominent in this area is that due to Bernstein[2]. Consider several tasks, T_1, of a sequentially organized program illustrated by a flow chart as shown in Fig. 2(*a*). If the execution of task T_3 is independent of whether tasks T_1 and T_2 are executed sequentially or in parallel as shown in Fig. 2(*b*), then parallelism is said to exist between T_1 and T_2.

With reference to Fig. 1, parallelism on a local level can be illustrated by statement 10. As shown in Fig. 3, "subtasks" 10*a* and 10*b* within the task depicted by statement 10 can be executed sequentially, (*a*) or in parallel (*b*) without altering the outcome at the completion of the task.

A second illustration of local parallelism is given by the individual subtasks within subroutine SUB1. Conceivably, the subroutine could itself consist of statements having the same form as the example arithmetic expressions within the calling program. These subtasks within the subroutine could themselves be executed in parallel.

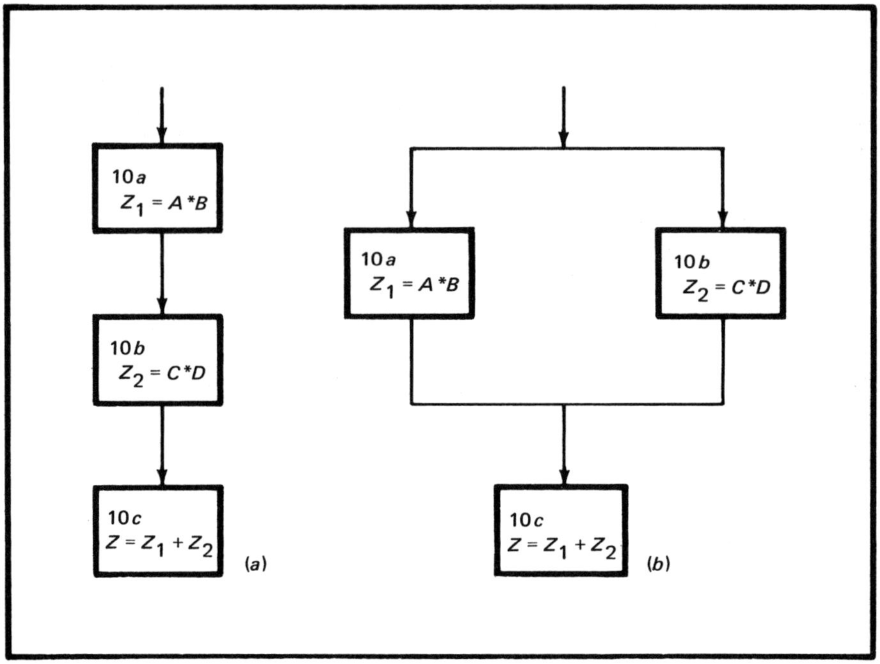

Fig. 3. Illustration of Parallelism on a Local Level

Explicit and Implicit Parallelism

The problem of recognizing parallel processable tasks within a program is difficult primarily because programs are usually written in what is essentially a serial manner. The solution to this problem can be approached from two directions. The first approach places the burden upon the programmer and requires an *explicit* indication of parallelism. This information is generated by means of additional instructions within the programming language itself. The second approach relieves the programmer of any additional duties, and relies totally upon indicators existing in the program itself, and is known as *implicit* parallelism.

Needless to say, each approach has certain advantages. If parallel processing is ever to be truly successful, it can only do so by avoiding the overhead that a complex compiler and a supervisory program would create. By allowing the programmer to indicate the parts of a program that can be executed in parallel, much of this undesirable overhead can be prevented. This approach, however, is of little value when one considers the many thousands of programs already in existence, since it is safe to assume that most would not be recoded in order to take advantage of inherent parallelism. This problem would be minimized, however, if parallelism within these programs could be detected using implicit recognition techniques. This cannot be accomplished, however, without a corresponding increase in system complexity.

A problem common to both approaches is that concurrently executable processes may compete for the same resources (memory, I/O, etc.). Thus, the methods of indicating parallelism to the computing system and the benefits derived by parallel processing are intimately related to the overall architecture of the computing system, the task management algorithms, etc.[3].

When parallel tasks use the same hardware resources like memory and I/O, access to which is serial, they have to be queued up for service. The solution is to overlap the parallel processes such that the critical resource will not create a bottleneck. Dijkstra, Knuth, and Coffman have developed efficient scheduling procedures for using common resources.[4,5,6]

When parallel tasks manipulate common data (e.g., one task inputs data whereas another edits the data), the sequence and the ordering of the operations with respect to each other may be important. Thus when sequentially coded tasks are prepared for parallel processing, it is important to determine whether or not the outputs of parallel paths are the same regardless of the order and sequence of execution. If the results of this determination are dissimilar, then task ordering information must be incorporated into the execution of these tasks.

Explicit Parallelism

In the explicit approach to parallelism, the programmer himself indicates the portions of a program that can be executed in parallel. This is normally done by providing the programmer with additional instructions that identify the parallel processable streams. One of the most frequently mentioned explicit approaches to parallelism is the FORK and JOIN.[7] FORK is an instruction that indicates the initiation of parallel tasks; JOIN is an instruction that initiates a new task only after the processes initiated by the FORK are completed and merge into the JOIN. The use of these instructions is illustrated in Fig. 4.

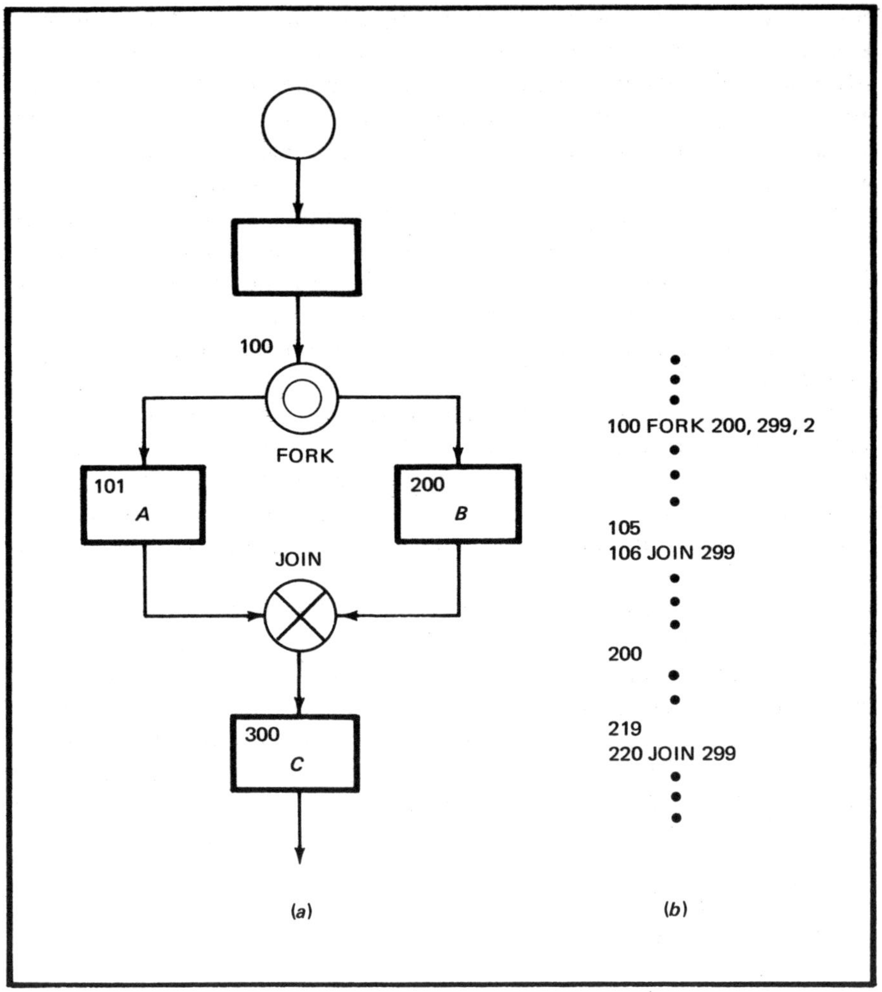

Fig. 4. FORK and JOIN (a) Flow Chart, (b) Instructions

The FORK at location 100 means: "Set the contents of location 299 to 2; then FORK to 101 and 200." The "2" in the instruction indicates the number of processors that the FORK at location 100 will activate (if they are available). As a processor completes its instructions, it branches to location 299 where it decrements the counter by 1. If the result is not zero, the processor is released. The processor which sets the counter to zero simply transfers to location 300 and continues. A distinct advantage of this approach is that it is simple and natural.

The basic premise of the PARALLEL FOR technique is that FOR statements in ALGOL programs and similar constructs in other languages permit exploitation of parallel activity.[8] As an example, consider the addition of two $I \times J$ matrices $M1$ and $M2$. Mathematically this would be represented as

$$M = M1 + M2$$

In ALGOL, the instructions to do this would read as shown in Fig. 5. Since the pair-wise additions are independent, they can be performed in

```
FOR K1 : = 1 STEP 1 UNTIL I DO

FOR K2 : = 1 STEP 1 UNTIL J DO

M [K1, K2] : = M1 [K1, K2] + M2 [K1, K2]
```

Fig. 5. ALGOL Instructions for Addition of Two Matrices

parallel. The implementation of this process as a PARALLEL FOR is shown in Fig. 6. $B1$ initializes the generation of the values of the parameters of the FOR statement, and establishes the new parallel path counter (PPC). PREP is a function performed before a new set of parallel paths begins. If other PPC's exist for this process, they are pushed down, which allows sets of parallel paths to be nested. PREP sets the new PPC to 1. $B2$ is the stepping or incrementing function. $B3$ is the end test and transfers to JOIN when all paths have been started. $B4$ starts a new parallel path. $B5$ is a parallel path. $B6$ is the JOIN. The function AND (L) is a simple two-way fork; it requests the controller to start a parallel sequence at L and add 1 to the current PPC for this process. JOIN is a function used to terminate a set of parallel paths. When executed, it reduces the current PPC for the process by 1. If PPC is then zero, it is popped up and processing continues. If PPC is not zero, meaning that more paths remain to be completed, it releases the processor executing the JOIN. For the example given, the implementation of this process would result in J simultaneous pair-wise additions as specified by the inner FOR statement.

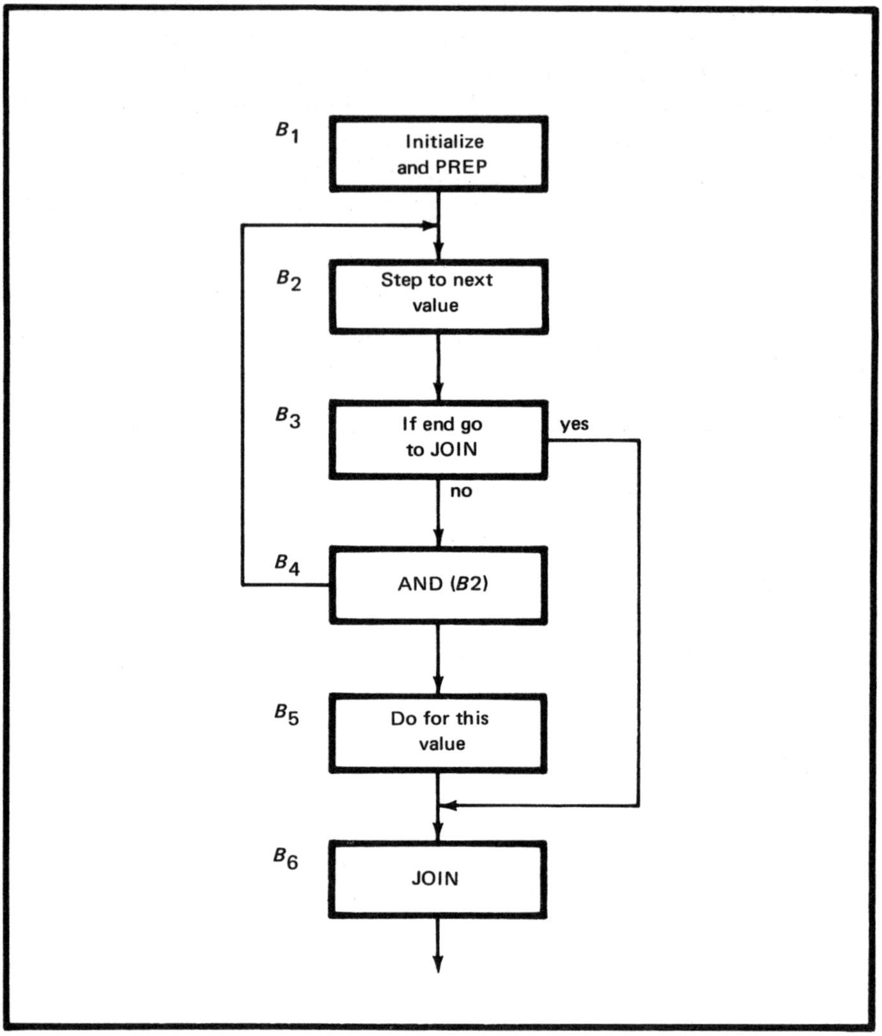

Fig. 6. Implementation of a PARALLEL FOR Process

An attractive feature of this scheme is that it is independent of the number of processors available. At any given time the maximum number of active paths equals the number of available processors. When a processor terminates its duties, it is either released or it initiates a new parallel path.

The FORK and JOIN and the PARALLEL FOR are examples of explicit indication of parallelism. Because of the relative simplicity of indicating inherent parallelism explicitly, there has been an increasing amount of effort in this direction. PL/1, for example, provides the TASK option with the CALL statement that indicates concurrent execution of parallel tasks. Languages like TRANQUIL,[9] an ALGOL-like language, have been designed to exploit parallelism in algorithms to be implemented in array processors such as the ILLIAC IV.

Sutherland has proposed a graphical language (intended primarily for display) that exhibits parallelism in a computation.[10] The computation is modeled as a directed graph in which a node represents a primitive function or a function of previously defined functions and primitives. The directed branches represent direction of control and data flow. The operational parallelism is indicated by multiple branches going to different functional nodes. The method is simple and direct, but the representation can be unwieldy in large computations.

Another approach to this problem is suggested in Single Assignment Languages.[11] Here all the statements are assignments. Only a single statement is allowed to assign a value to a variable. Since the assignments indicate the functional relationships between the variables and the generated functions, the order of execution of the various quantities is also defined. The parallel operations are thus implicitly specified by this representation.

Since explicit indicators of parallelism are generally formulated at the time of coding, it is highly desirable that these indicators and related symbols be simple and natural. Some considerations associated with these indicators are given here.

Because programs are usually represented by flow charts, any discussion on explicit indicators can be simplified by referring to flow charts.[12,13,14] Necessary to such a representation are nodes and branches which define computational procedures. A *task node* represents a FORK operation. A *decision node* represents a computation whose outcome determines the next sequence of tasks to be undertaken. The JOIN node is the point of convergence of two or more parallel tasks; that is, control will move on to the next sequence of operations only after all the converging tasks are completed. The branches of a graph indicate the permissible transitions to the next set of computations as indicated by their orientation (direction).

In addition to these indicators, additional primitives should be included to protect the use of common data. The statement LOCK ($X1$, $X2$, ...), for example, makes the data variables $X1$, $X2$, ..., the exclusive property of the branch issuing the LOCK statement; the dual statement UNLOCK ($X1$, $X2$, ...) releases the previously locked statements.

Some of the problems associated with the FORK and JOIN and similar type statements[15] include the fact that not all paths from a FORK statement have to converge to a single JOIN statement.[16] Thus the conventional FORK statement must be modified to permit nonconverging paths to release their processors after their completion. This can be accomplished by means of the IDLE statement, which releases a processor and initiates no further action.

Another problem arises when two or more concurrent tasks seek access to the same critical resource. Simultaneous access can destroy useful data if it is not properly controlled. One way to prevent this is to *lock* out common data from other processes and thus give its use to a single process exclusively. If an interrupt occurs when LOCK is in effect, however, the common data may never permit its use even to high priority processes. The solution is either to inhibit the interrupt until an UNLOCK is issued by the process and/or warn the process of the occurrence of an interrupt and allow it some delay to unlock the data. Furthermore, a process can *lock* the data and go into an endless loop due to errors. An obvious solution would be to allow the system to limit the time during which a lock can be effective.[17]

Implicit Parallelism

From the outset it must be stated that this approach to parallel processing is associated with sophisticated compiling and supervisory programs and the related overhead. Its attractiveness lies in that it is independent of the programmer, and existing programs would not have to be modified to take advantage of inherent parallelism. An algorithm written by Fisher utilizes the input and output sets of each process within a program to determine essential ordering, and thus inherent parallelism.[18] Consider the flow diagram of the process shown in Fig. 7. The algorithm requires as input:

1) The number of processes to be analyzed. For this example each box in the flow diagram represents a process.
2) The input and output set for each process. The input set for process G, for example, consists of the variables v and w. The output set consists of the single variable z.
3) The given permissible ordering ${}_{0}S$ among the processes. This relation, illustrated in Fig. 7(b), is simply a representation of the possible paths which exist between the processes.

4) Any initially known essential order ${}_{o}T$ among the processes. This relation includes arcs connecting conditionals to their various conditional successors. The flow diagram of this example consists of only one conditional, process D. Thus process D and its possible successors E and F comprise the relation ${}_{o}T$ [Fig. 7(*c*) .].

Given this information the algorithm generates:

1) T, the essential serial ordering relation, and
2) S, the covering for the essential serial ordering relation.

This relation represents T without those arcs that directly join processes which are joined indirectly by a sequence of two or more arcs in T.

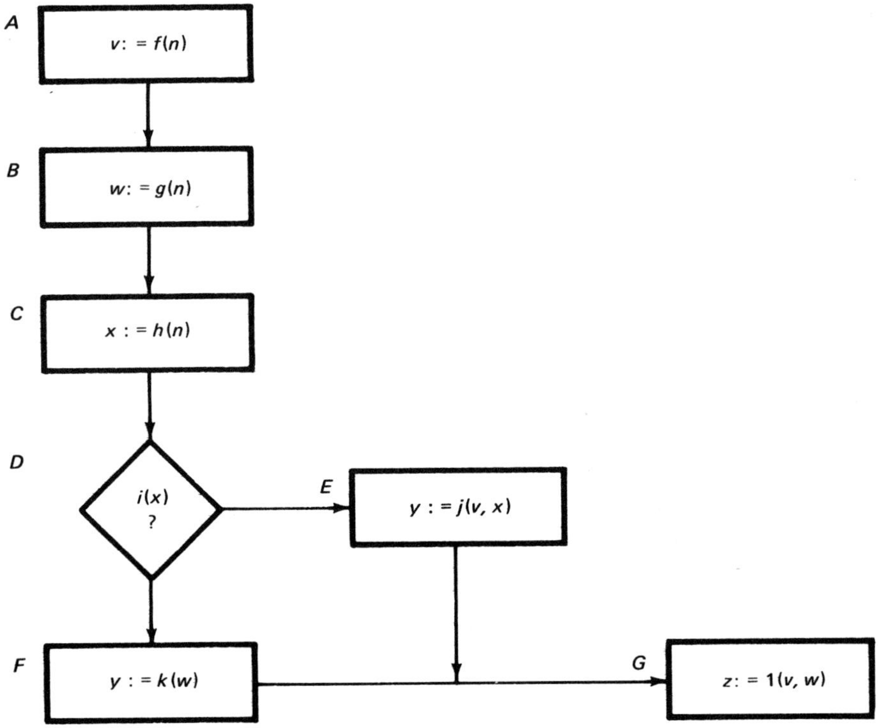

(*a*) Flow Diagram

Fig. 7. Graphical Representation of a Program

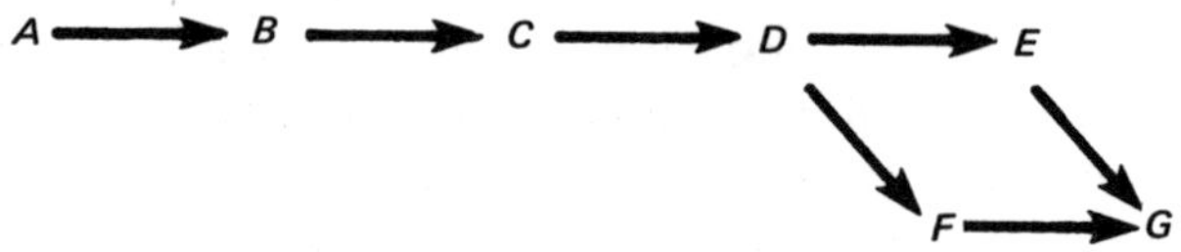

(*b*) ${}_oS$: Given Permissible Ordering Relation

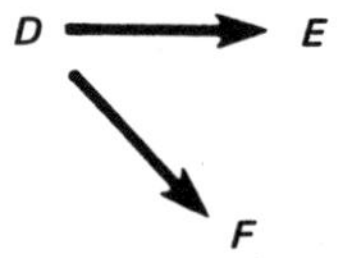

(*c*) ${}_oT$: Known Essential Serial Ordering

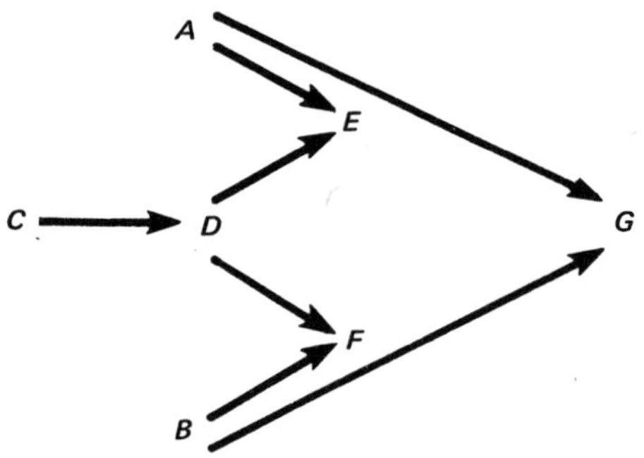

(*d*) S: Cover for the Essential Serial Order

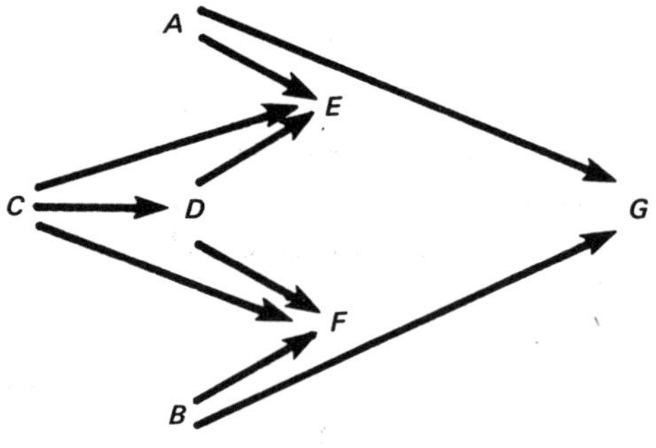

(*e*) T: Essential Serial Ordering Relation

Fig. 7. Graphical Representation of a Program (cont'd)

Once the essential serial ordering among the processes is known, parallelism inherent in the overall process can be determined.

Reference to Fig. 7(*d*) shows that processes *A, B,* and *C* can be processed simultaneously. Similarly, processes *D* and *G* can be processed after the results of processes *A, B*, and *C* are available.

The implication from the relation S is that the overall process can be accomplished in 3 units of time as shown below.

Unit of Time	Processes Computed
1	*A, B, C*
2	*D, G*
3	*E* or *F*

A possible shortcoming of the algorithm is that it fails to expressly provide any indication of flexibility in the ordering of processing among the processes. For example, the 2 variations which follow can also be accomplished in 3 units of time with no violation of the essential serial order.

Unit of Time	Processes Computed
1	*A, C*
2	*B, D*
3	*G, E* or *F*
1	*B, C*
2	*A, D*
3	*G, E* or *F*

Thus, if only 2 processors were available for the 3 units of time, either of the 2 orderings shown above could get the job done. These alternatives are not apparent from the covering relation S.

Further discussion on implicit parallelism will be continued in the following two sections and will take two distinct forms. "Algorithms for Detection of Sub-Task Parallelism" will be confined to parallelism on a local level as exemplified by subtasks within expressions. Following this discussion a technique will be introduced for detecting parallelism on a global level.

ALGORITHMS FOR DETECTION OF SUBTASK PARALLELISM

This section will be devoted to implicit recognition of subtask parallelism within arithmetic expressions. The first half of the section will present a brief survey of existing algorithms which are designed to accomplish this purpose.[19] The second half will introduce a technique which approaches the problem in a somewhat different manner. The arithmetic expression which will be used as an example for each algorithm is given below.

$$A + B + C + D*E*F + G + H$$

Throughout this discussion, the usual precedence between operators will apply. In order of increasing precedence, the precedence between operators will be + and –, * and /, and ↑, where ↑ stands for exponentiation.

Hellerman's Algorithm[20]

This algorithm assumes that the input string is written in reverse Polish notation and contains only binary operators. The string is scanned from left to right replacing by temporary results each occurrence of adjacent operands immediately followed by an operator. These temporary results will be considered as operands during the next passes. Temporary results generated during a given pass are said to be at the same level and therefore can be executed in parallel. There will be as many passes as there are levels in the syntactic tree. The compilation of the expression listed above is shown in Fig. 8.

Although this algorithm is simple and fast, it has two pronounced shortcomings. The first is that it requires the input string to be in Polish notation; the second is its inability to handle operators that are not commutative.

Stone's Algorithm[21]

The basic function of this algorithm is to combine 2 subtrees of the same level into a level that is 1 higher. For example, A and B, initially of level 0, are combined to form a subtree of level 1. The algorithm then searches for another subtree of level 1 by attempting to combine C and D. Since precedence relationships between operators prohibit this combination, the level of subtree $(A + B)$ is incremented by 1. The algorithm now searches for a subtree of level 2 by attempting to combine C, D, and E. Since this combination is also prohibited, subtree $(A + B)$ is incremented to level 3. The next search is successful, and a subtree of level 3 is obtained by combining C, D, E, and F. Both of these subtrees are then combined to form a single subtree of level 4.

i	Input String after the *i*th Pass	Temporary Results Generated During *i*th Pass
0	*AB* + *C* + *DE***F** + *G* + *H* +	
1	*R*1 *C* + *R*2 *F** + *G* + *H* +	*R*1 = *A* + *B* *R*2 = *D***E*
2	*R*3 *R*4 + *G* + *H* +	*R*3 = *R*1 + *C* *R*4 = *R*2**F*
3	*R*5 *G* + *H* +	*R*5 = *R*3 + *R*4
4	*R*6 *H* +	*R*6 = *R*5 + *G*
5	*R*7	*R*7 = *R*6 + *H*

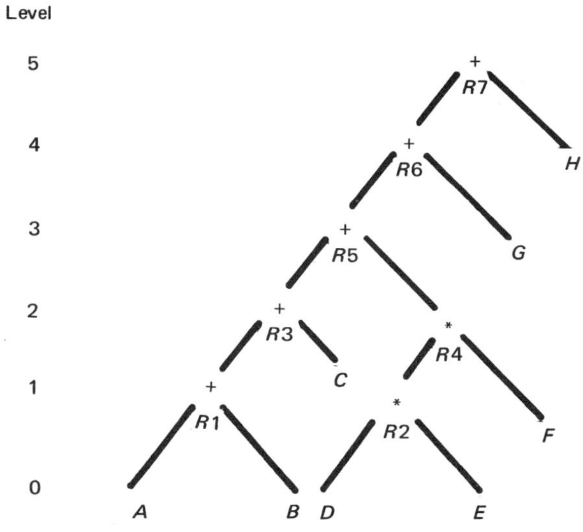

Fig. 8. Parallel Computation of $A + B + C + D*E*F + G + H$ Using Hellerman's Algorithm

In a similar manner the subtree $(G + H)$, originally of level 1, is successively incremented until it achieves a level of 4; at that time it is combined with the other subtree of the same level to form a final tree of level 5.

The algorithm yields an output string in reverse Polish which does not expressly show which operations can be performed in parallel. Even though the output string is generated in a single pass, the recursiveness of the algorithm causes it to be slow, and at least an additional pass would be required to specify parallel computations.

Squire's Algorithm[22]

The goal of this algorithm is to form quintuples of temporary results of the form: Ri (operand 1, operator, operand 2, start level = max [end level op. 1; end level op. 2], end level = start level + 1).

All temporary results which have the same start level can be computed in parallel. Initially, all variables have a start and end level equal to zero.

Scanning begins with the rightmost operator of the input string and proceeds from right to left until an operator is found whose priority is lower than that of the previously scanned operator. In the example, the scan would yield the substring

$$D*E*F + G + H .$$

Now a left to right scan proceeds until an operator is found whose priority is lower than that of the left-most operator of the substring. This yields D^*E^*F. At this point a temporary result $R1$ is available of the form

$$R1(D,*, E, O, 1) .$$

The temporary result, $R1$ replaces one of the operands, and the other is deleted together with its left operator. The new substring is then

$$R1^*F + G + H.$$

The left to right scans are repeated until no further quintuple can be produced, and at that time the right to left scan is reinitiated. The results of the process are shown in Fig. 9.

Initial String: $A + B + C + D*E*F + G + H$

Right to Left Scan	Left to Right Scan
$D*E*F + G + H$	$R1*F + G + H$
	$R2 + G + H$
$A + B + C + R2 + G + H$	$R3 + C + R2 + G + H$
	$R4 + R3 + R2 + H$
	$R4 + R5 + R2$
	$R6 + R2$
	$R7$

Quintuples	Op. 1	Operator	Op. 2	Start	End
R1	D	*	E	0	1
R2	F	*	R1	1	2
R3	A	+	B	0	1
R4	C	+	G	0	1
R5	H	+	R3	1	2
R6	R4	+	R5	2	3
R7	R2	+	R6	3	4

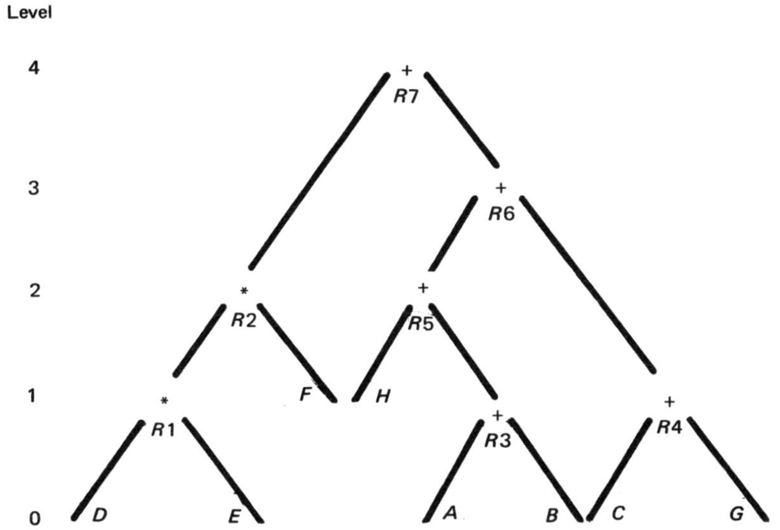

Fig. 9. Parallel Computation of A + B + C + D*E*F + G + H Using Squire's Algorithm

Although the example shows the algorithm applied to an expression containing only binary operators, the algorithm can also handle subtraction and division with a corresponding increase in complexity.

A significant feature of this algorithm is that Polish notation plays no part in either the input string or the output quintuples. Because of the many scans and comparisons the algorithm requires, it becomes more complex as the length of the expression and the diversity of operators within the expression increase.

Baer and Bovet's Algorithm[2 3]

The algorithm uses multiple passes. To each pass corresponds a level. All temporary results which can be generated at that level are constructed and inserted appropriately in the output string produced by the corresponding pass. Then, this output string becomes the input string for the next level until the whole expression has been compiled. Thus the number of passes will be equal to the number of levels in the syntactic tree. During a pass the scanning proceeds from left to right and each operator and operand is scanned only once.

The simple intermediate language which this algorithm produces is the most appropriate for multiprocessor compilation in that it shows directly all operations which can be performed in parallel, namely those having the same level number. The syntactic tree generated by this algorithm is shown in Fig. 10.

Observations

Of the four algorithms mentioned here, Squire's, and Baer's and Bovet's algorithms are the two most powerful as well as the most complex, although the former algorithm requires more passes than the latter. Although Hellerman's algorithm is the simplest and most rapid, it is handicapped by its requirement for a Polish input string and its limitation to binary operators. Stone's algorithm requires the fewest passes, but its recursiveness causes it to be slow. In addition, the reverse Polish string that it generates is not in a form suitable for multiprocessor utilization.

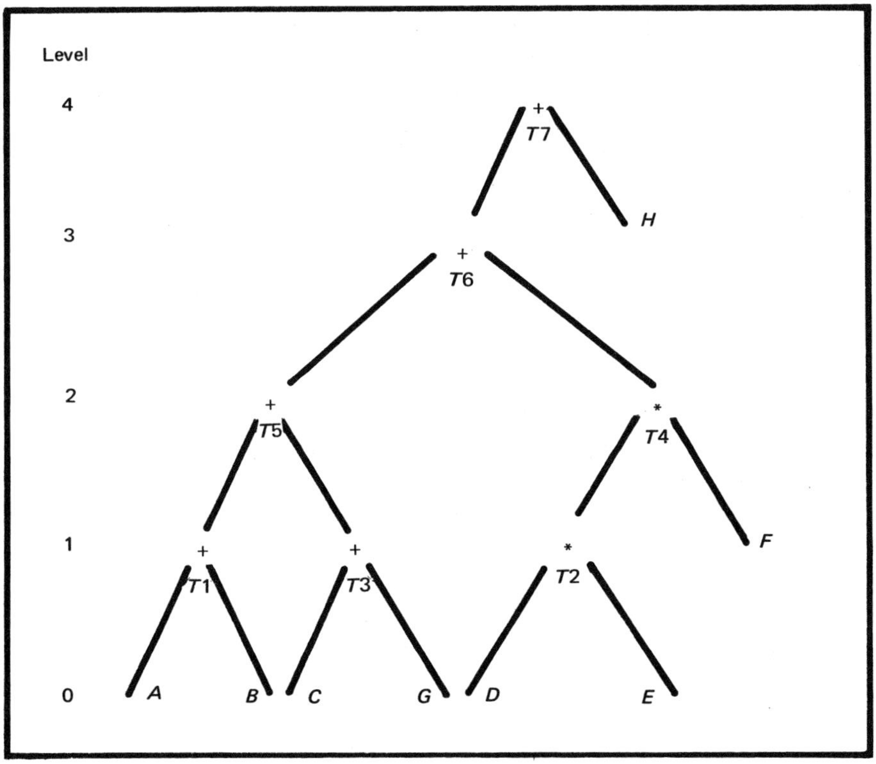

Fig. 10. Parallel Computation of $A + B + C + D*E*F + G + H$ Using Baer's and Bovet's Algorithm

A New Algorithm for Detection of Subtask Parallelism

The paragraphs which follow will introduce a technique whose goals are (1) to produce a binary tree that illustrates the parallelism inherent in an arithmetic expression, and (2) to determine the number of registers needed to evaluate large arithmetic or Boolean expressions without intermediate transfers to main memory.

This technique is prompted by the fact that existing computing systems possess multiple arithmetic units that can contain a large number of active storages (registers). In addition, the superior memory bandwidths of the next generation of computers will simplify some of the requirements of this technique.

In the material presented below, a complex arithmetic expression is examined to determine its maximum computational parallelism. This is accomplished by repeated rearrangement of the given expression. During this process, the given expression in reverse Polish form is also tested for "well formation," i.e., errors and oversights in the syntax, etc.

The arithmetic expression that was used as a model earlier will also be used here, namely $A + B + C + D * E * F + G + H$. The details of the algorithm follow.

1) The first step is to rewrite the expression in reverse Polish form and to reverse its order.

$$+ H + G + * F * E D + C + B A .$$

2) Starting with the rightmost symbol of the string, assign a weight to each member of the string based on the following procedure: Assign to symbol S_i the value $V_i = (V_i - 1) + R_i \quad i = 1,2, \ldots, n$
where $R_i = 1 - \delta(S_i)$ given that
$\delta(S_i) = 0$ if S_i is a variable
$\delta(S_i) = 1$ if S_i is a unary operator
$\delta(S_i) = 2$ if S_i is a binary operator
and $V_{i-1} = V_{i-2} + R_{i-1}$, $V_{i-2} = V_{i-3} + R_{i-2}$, etc.,
such that $V_{i-(i-1)} = V_1 = R_1$, and $V_0 = 0$.
Using this procedure, the following expression results:

	Root Node								V_m						
i	15	14	13	12	11	10	9	8	7	6	5	4	3	2	1
S_i	+	H	+	G	+	*	F	*	E	D	+	C	+	B	A
V_i	1	2	1	2	1	2	3	2	3	2	1	2	1	2	1

Note that for a "well formed expression" of n symbols $V_n = 1$.

3) At this point the root node of the proposed binary tree can be determined. Thus the given string can be divided into 2 independent substrings. To determine the root node, draw a line to the left of the first symbol with a weight of 1 ($i = 11$, $S_i = +$, $V_i = 1$) to the left of the symbol with the highest weight, V_m($i = 7$, $S_i = E$, $V_i = V_m = 3$). The 2 independent substrings consist of the strings to the left and to the right of this line. The root node will be the leftmost member of the string to the left of the line ($i = 15$, $S_i = +$, $V_i = 1$). Note that V_i also equals 3 for $i = 9$;

however, V_m is chosen from the earliest occurrence of a symbol with the highest weight.

4) The next step is to look for parallelism within each of the new substrings. Consider the rightmost substring. Form a new substring consisting of the symbols within the values of $V_i = 1$ to the right and to the left of V_m. Transpose this substring with the substring to the right of it whose leftmost member has a weight of $V_i = 1$.

Initial Rightmost	S_i	+	*	F	*	E	D	+	C	+	B	A
Substring	V_i	1	2	3	2	3	2	1	2	1	2	1

Final Rightmost	i	11	10	9	8	7	6	5	4	3	2	1
Substring	S_i	+	+	C	+	B	A	*	F	*	E	D
	V_i	1	2	3	2	3	2	1	2	1	2	1

This procedure is repeated until the initial V_m occupies the position $i = 2$ in the substring. For this example this is already the case. Thus the rightmost substring is in the proper form.

5) The transposition procedure of step 4 is applied next to the leftmost substring. However, since the leftmost substring of this example consists of only 2 operands and 1 operator, no further operations are necessary.

6) The resultant binary tree is shown in Fig. 11. The numbers assigned to each node represent the final weight V_i of the symbol as determined in steps 1–5 above.

Some observations and comments on this algorithm are given below.

1) The two branches on either side of the root node can be executed in parallel. Within each main branch, the transposition procedure of step 4 yields supplementary root nodes. The subbranches on each side of the supplementary nodes can be executed in parallel.

2) The number of levels in the binary tree can be predicted from the Polish form of the original string.

   No. Of Levels = MAX (Number of 1's; V_m) in the substring (rightmost or leftmost) containing V_m.

3) The tree is traversed in a modified postorder form.[24] The resulting expression is

$$D * E * F + A + B + C + G + H$$

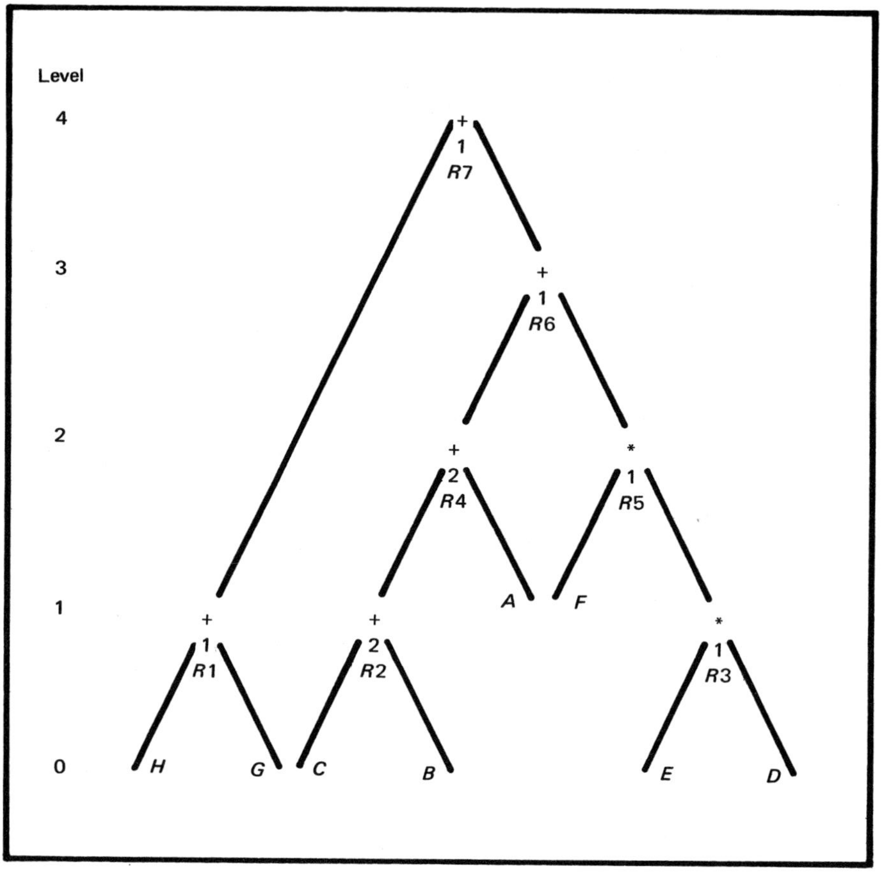

Fig. 11. Binary Tree for Parallel Computation of $A + B + C + D*E*F + G + H$

4) An added feature of this technique is that the number of registers required to evaluate this expression without intermediate *store* and *fetch* operations is obtained directly from the binary tree. This information is provided by the highest weight assigned to any node within the tree. Thus, for this example the expression could be evaluated using at most two registers without resorting to intermediate stores and fetches.

5) This technique of recognizing parallelism on a local level has been applied to a single instruction, in particular, an arithmetic expression. It is worthwhile mentioning that each variable within the expression can itself be the result of a processable task. Thus

this technique can be extended to a higher level of parallel stream recognition, i.e., global parallelism.

6) No formal proof has been obtained for this algorithm. It has, however, been applied to many examples of varying complexity with satisfactory results.

PARALLEL TASK DETECTION AT A GLOBAL LEVEL

Background Information

The problem presented here is that of compiling a conventionally written program for multiprocessing. This involves the detection of the parallel processable segments and an indication of what tasks can be initiated in parallel and what tasks must be completed before the start of the next sequence of tasks. A graph theoretic model is presented that seems to provide enough generality that it is independent of the language, the application, the mode of compilation, and the number of processors in the system. Some graph terminology will be explained next to facilitate further discussion.

Computational processes are modeled by oriented graphs in which the vertices (nodes) represent single tasks, and the oriented edges (directed branches) represent the permissible transition to the next task in sequence. Again, as before, a single task may represent one or more statements in a computational procedure and the associated data.

A directed graph, which consists of n vertices and a set of oriented edges, can be represented in a computer by its connectivity matrix C of dimension $n \times n$ such that its generic term C_{ij} is a "1" if and only if there is a directed edge from node i to node j, and it is "0" otherwise.[25] The properties of the directed graph and hence of the computational process it represents can be studied by simple manipulations of the connectivity matrix.

A graph consisting of a set of vertices and edges is said to be *strongly connected* if and only if any node in it is reachable from any other. A *subgraph* of any graph is defined as consisting of a subset of vertices with all the edges between them retained. A *maximal strongly connected* (MSC) *subgraph* is a strongly connected subgraph that includes all possible nodes which are strongly connected with each other. Given a connectivity matrix of a graph, all its MSC subgraphs can be determined simply by well known methods.[26] A given program graph can be *reduced* by replacing each of its MSC subgraphs by a single vertex, and retaining the edges connected between these vertices and others. After the reduction, the *reduced graphs* will not contain any strongly connected components in them.

The paragraphs that follow will describe the sequence of operations needed to prepare, for parallel processing in a multiprocessor computer, a program written for a uniprocessor machine. It should be pointed out that some of the indicated operations can be performed simultaneously during compilation or assembly.

1) The first step is to derive during compilation the program graph which identifies the sequence in which the computational tasks are executed. The program graph is represented in the computer by its connectivity matrix. An example of such a graph is shown in Fig. 12(*a*).
2) By an analysis of the connection matrix, the maximal strongly connected components are derived by simple operations.[27] This type of component is illustrated by tasks *c* and *e* in Fig. 12(*a*). Each maximal strongly connected component (subgraph) is next considered as a single task, and the new graph, called the reduced or condensed graph, is constructed. This reduced graph is illustrated in Fig. 12(*b*). The condensed version of the program graph does not contain any strongly connected components, i.e., it has no loops.
3) From an input-output analysis of each task, it can be determined if some tasks can be performed in parallel. If two tasks exist in which the input parameters of one are not a function of the output parameters of the other, and vice versa, then by definition the two tasks can be executed in parallel. Thus from the input-output set analysis of the tasks, the previously obtained reduced graph is transformed into a *parallel task* graph, which indicates parallel tasks. Whereas the reduced graph shows only permissible transitions between tasks, the parallel task graph gives consideration to the input-output relationships between tasks. In this latter graph, two or more branches emanating from a single vertex represents a FORK operation, and the conjunction of two or more branches into a vertex is equivalent to a JOIN operation. Figure 13 illustrates the parallel task graph for the program graph of Fig. 12.

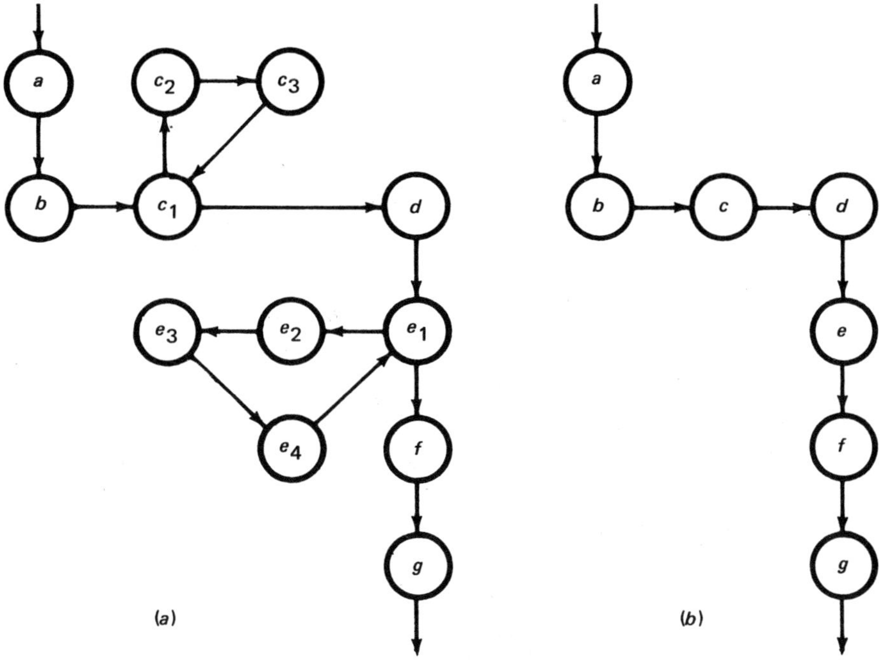

Fig. 12. Graphical Representation of a Computational Process

4) From the connectivity matrix of the parallel task graph, the sequences in which the tasks must be performed are determined. The method used to determine which tasks can be performed in parallel uses the connectivity matrix of the final form of the program graph and the process of "precedence partitions".[28] The method uses a series of eliminations (partitions) of columns (and corresponding rows) in the matrix which contain only zeros. Whenever more than one column can be removed from the matrix during a given elimination, the tasks represented by the node names of the eliminated columns can be executed in parallel. The series of eliminations used to determine the parallel processable tasks in the example are illustrated in Fig. 14.

Reference to Fig. 14 indicates that during the first step only node *a* is eliminated; the second step eliminates node *b*; during the third iteration nodes *c* and *d* are eliminated; the fourth iteration eliminates nodes *e* and *f*; the fifth iteration eliminates node *g*. Thus the precedence partitions are (*a*), (*b*), (*c*,*d*), (*e*, *f*), and (*g*). This indicates that task *a* must

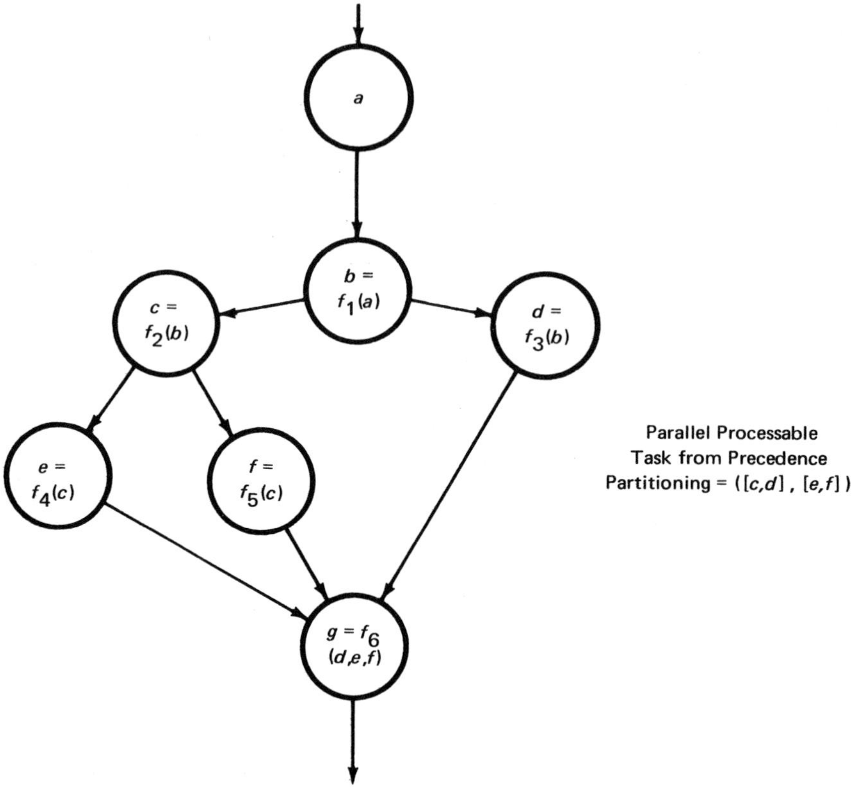

Fig. 13. Representation of Parallel Processable Tasks of Fig. 12

precede b, which in turn precedes the tasks (c,d) which can be done in parallel, etc.

A FORTRAN Parallel Task Recognizer

As mentioned in the introductory chapter, programs written in FORTRAN readily lend themselves to investigation of parallel processing techniques. In order to determine the degree of applicability of the method described in the previous section, it was decided to apply the method to a sample FORTRAN program. This was accomplished by writing a program whose input consists of a FORTRAN source program; its output consists of a listing of the tasks within the source program which can be executed in parallel. The program written to accomplish this parallel task detection is known in its final form as a FORTRAN Parallel Task Recognizer.

The recognizer, also written in FORTRAN, does not investigate the actual structure of the source program language. Instead it relies on indicators generated by the way in which the program is actually written, as suggested by Bernstein.[29] For example, consider

$$X1 = A + B$$
$$X2 = C + D\,.$$

Because the right-hand side of the second expression does not contain a parameter generated by the computation which immediately precedes it, the two computations can be accomplished in parallel. If, on the other hand, the expressions were rewritten, the termination of the first computation would have to precede the initiation of the second as in

$$X1 = A + B$$
$$X2 = X1 + C\,.$$

The key to the determination of the essential ordering, of course, is the presence of the equality sign in each of the expressions. In effect,

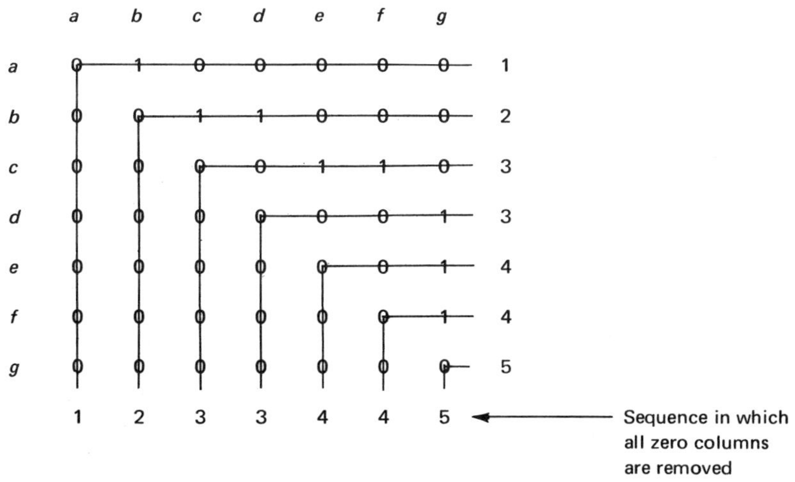

Precedence Partitions = ([*a*] , [*b*] , [*c*,*d*] , [*e*,*f*] , [*g*])

Fig. 14. Precedence Partitions of the Program Graph of Fig. 12

the "right-hand side" of the second equality must be compared to the "left-hand side" of the first equality.

Although the implicit indicator here—the equality sign—is readily apparent, other (though perhaps less obvious) indicators exist in other statements. Consider the arithmetic *If* statement

$$IF(X - Y)3,4,5 \; .$$

Although no equality sign is present here, the right-hand and left-hand suggestion is still present. The parameters within the parentheses in the statement above must be compared to the outputs of preceding statements in order to determine essential order. Thus X and Y in this statement are "right-hand" parameters. An identical argument holds for the *Print* statement

$$\text{PRINT } 100, X, Y .$$

The *Read* statement, on the other hand, consists entirely of "left-hand" parameters. This means that unless the Read statement is the successor of a conditional, or the explicit successor of a previous statement, it can be executed concurrently with any statement that does not require its parameters with the exception of another Read statement.

A subroutine *Call* statement may consist entirely of "right-hand" variables, or it may contain "right-hand" and "left-hand" variables. Consider the sequence of

$$X1 = A + B$$
$$X2 = C + D$$
$$\text{CALL ALPHA } (X1, X2, X3) \; .$$

The implication here is that $X3 = f(X1, X2)$. Thus in a manner analogous to an arithmetic expression, $X1$ and $X2$ are "right-hand" variables, and $X3$ is a "left-hand" variable.

The *Do* statement is unique in that it does not contain any parameters, nor does it of itself generate any parameters. The repeated series of statements executed as a result of the Do statement form a "loop," however, and due to the graph considerations given earlier, all of the statements within a loop are considered as a single vertex in the reduced flow chart of a computational process. This condition is also made necessary because the rules of FORTRAN require that a Do loop be entered only through the Do statement. Statements within a Do loop, however, can be examined for inherent parallelism in a manner similar to that applied to statements outside of a Do loop.

Because it was necessary to place an initial upper bound on the scope of this project, the recognizer in its present form possesses certain limitations. Thus only a subset of the entire FORTRAN set is considered for recognition. As an example, the *Go To* instruction is not part of the subset of instructions that are recognized. This is not to say that the recognizer does not permit transfer operations altogether, since the three-branch arithmetic If statement is an acceptable form. If statements, however, may themselves generate loops within the program. To eliminate these loops it would be necessary to analyze the connectivity matrix in the manner mentioned earlier before beginning the process of precedence partitions. The recognizer does not presently perform this analysis.

If it is determined that two subroutines can be executed concurrently, no consideration is given to the fact that both may access memory locations established by *Common* or *Equivalence* statements.

The recognizer assumes that statements within the source program are written correctly according to the rules of FORTRAN. Thus the absence of a comma immediately after the format statement number in a *Print* statement, for example, would cause the recognizer to eliminate that statement from the recognition process.

Some limitations were placed on the recognizer merely to simplify programming and to make it more manageable in terms of memory requirements. Thus the input program can have no more than 99 executable statements, and no statement can have more than 4 "right-hand" or "left-hand" parameters. The recognizer cannot accept imbedded blanks within a statement, nor will it accept statements that require more than one standard punched-card form. Source program statement numbers may be 1 to 5 digits in length, but the statement number must terminate in column 5 on the punched card representing the source program statement.

The assumptions and limitations listed above are not all inclusive. An attempt will be made to list as many of these limitations as possible in subsequent paragraphs as the recognizer is discussed in detail.

Theory of Operation

The recognition process consists of 3 phases. In Phase I, the program statements are accepted by the recognizer one at a time. The first consideration made after this Read operation is the determination of whether or not the statement is executable. This is done by comparing the statement type to the standard types accepted by the recognizer. A summary of the statement types accepted by the recognizer is given below.

Nonexecutable	*Executable*
Format	Do
Real	Read
Common	Three-Branch Arithmetic If
Integer	Print
Dimension	Continue
	Call
	End
	Arithmetic Expressions

In addition, the recognizer will accept comment cards and all-blank cards.

As each executable statment is read, it is analyzed for "right-hand" and "left-hand" variables as suggested earlier. As the variables within the statement are obtained, they are placed in tables containing variables only of the same kind, i.e., "right hand" or "left hand." An important distinction is made depending on whether or not the statement is within a Do loop. A "left-hand" variable within a Do loop, for example, is not placed in the same table as a "left-hand" variable outside of a Do loop. Thus, depending on its location within a program, a variable can be placed in one of four tables.

The location of a variable within a table is determined by the Executable Statement Number (ESN) assigned to the statement containing the variable at the time the statement is read. The tables mentioned here are dimensioned (50 × 7). All variables of the same type within a statement will occupy the same line but different columns of a particular table.

An example of what has been discussed here is given below. Assume that the statement shown is the fourth executable statement which has been read. Assume further that the statement is not located within a Do loop. The entries in the appropriate tables are shown below the statement.

The first column in each table corresponds to the number of entries which have been made in that table. The second column contains the Executable Statement Number. The third column contains the number of variables contained in that entry. Columns 4–7 contain the actual variables; they correspond in number to the entry in column 3.

The Do statement generates an entry in a table which records the limit of the Do loop, i.e., the statement number. The If statement also generates entries in a table which records the statement numbers of the three possible If statement successors.

$$X1 = A + B * (C - D)$$

Left-Hand Table Outside Do

2	4	1	*X*1			

Right-Hand Table Outside Do

1	4	4	*A*	*B*	*C*	*D*

In addition, the address field of each executable statement is scanned to determine if it is a possible successor of an If statement or the limit of a Do loop. When a match is found, the ESN of the statement under study is entered in the appropriate table.

Throughout this Read process, nonexecutable statements are ignored. The Read process is continued until an End statement is reached. At that point the recognizer is ready to begin Phase II of the recognition process–generation of the connectivity matrix. A flow chart of Phase I of the recognition process is shown in Fig. 15.

The goal of the recognizer during the second phase is the generation of the connectivity matrix of the source program. This is accomplished by analyzing the variables in each of the 4 tables generated during Phase I. In general, variables in the "right-hand" tables are compared to entries in the "left-hand" tables. When a match is found, a "1" is entered at the appropriate point in the connectivity matrix.

The key to this matching process is the executable statement number. An entry in a "right-hand" table will not be compared to an entry in a "left-hand" table when the ESN of the latter exceeds the ESN of the former. The reasoning here is that a "right-hand" entry cannot affect the statements which succeed it in the sequential process.

(For the sake of convenience and brevity, the terms "right-hand" and "left-hand" will be written as RH and LH, respectively, throughout the remainder of this discussion.)

The unique structure of Do loops creates an additional important consideration during Phase II. Recall that a RH variable may exist within a Do loop or outside of a Do loop. For this reason it is necessary to compare RH variables to variables in both LH tables.

If a RH variable within a Do depends on the outcome of a statement that precedes the Do statement, the entry in the connectivity matrix is from the statement containing the LH variable to the Do statement, and not to the statement containing the RH variable. Similarly, if a RH

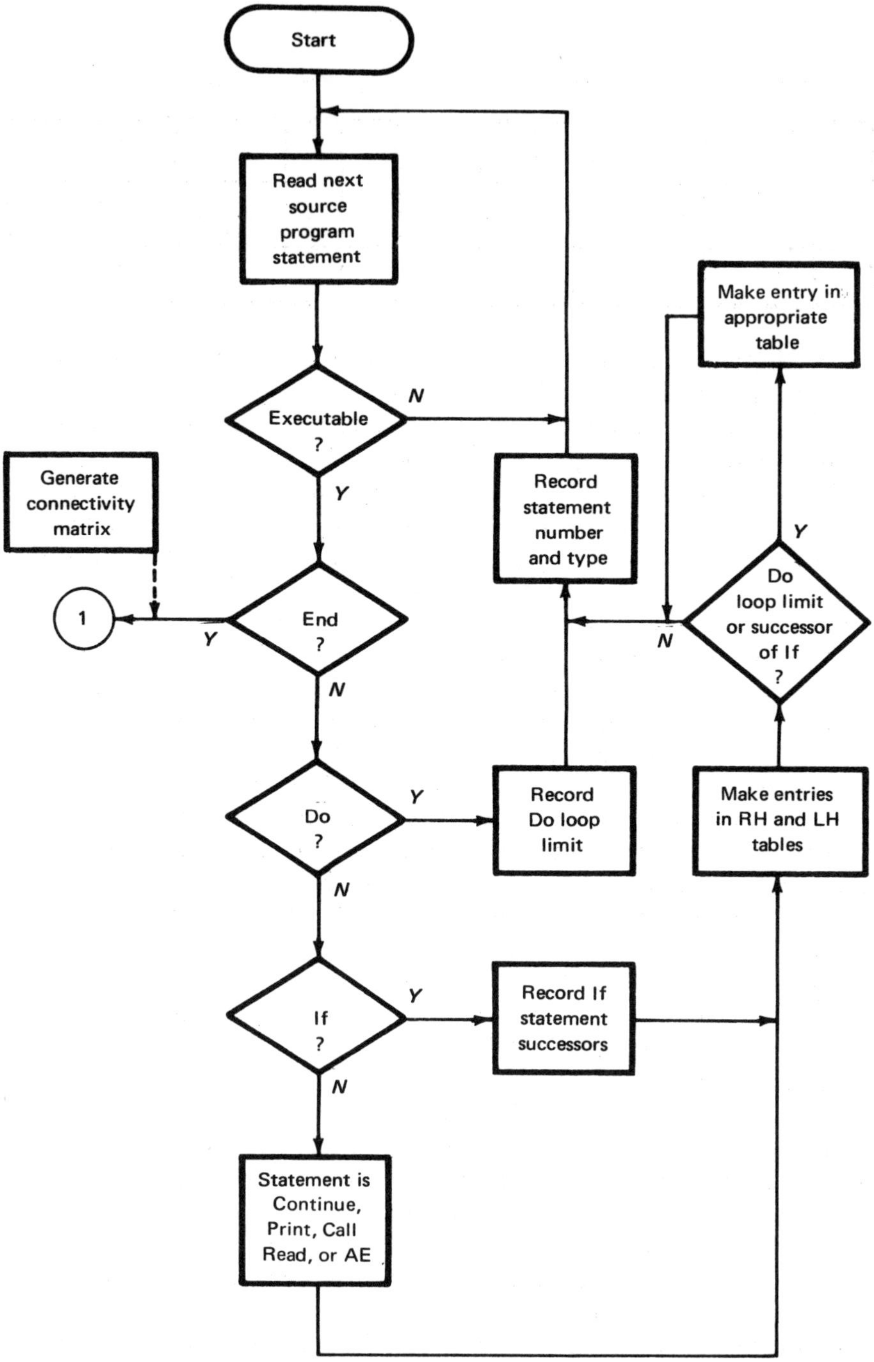

Fig. 15. Phase I of the Recognition Process

variable outside of a Do loop depends on the outcome of any state-within a Do block, the entry in C is from the Do statement to the statement containing the RH variable, and not from the statement generating the required outcome. In effect, then, an entire Do loop is considered as either a predecessor or successor of a particular statement. This is in line with the graph policies presented earlier.

This consideration may produce a disparity, however, when the time comes to implement the parallel processable blocks generated by the recognizer. For example, if a source program contains 20 executable statements, 15 of which are contained within a Do loop, a high degree of potential parallelism is lost if the statements within the Do are not themselves considered for possible concurrent execution. For this reason, in addition to the overall connectivity matrix, the recognizer will generate a matrix for each Do block and output the components of the block which may be executed in parallel. In its present form, the recognizer will not accept more than 5 Do blocks, nor will it recognize imbedded Do's, i.e., a Do within a Do.

A final consideration important to the matching process is created by the If statement. As long as the successors of an If statement located within a Do loop are also located within the loop, no complications arise. The relationship between the If statement and its successors will be reflected in the connectivity matrix for that Do loop. If, on the other hand, a possible successor of this If statement is located outside of the Do, then their relationship will show up in the matrix for the overall program.

If it is determined that for a given program N tasks can be initiated simultaneously, and one of these tasks is a successor of an If statement, then the concurrent execution of all N tasks may not be permissible. Only those tasks within the block which would normally follow the successor of the If statement can be executed in parallel. The flow chart for Phase II of the recognition process is shown in Fig. 16.

Phase III of the recognition process is devoted to the generation of the parallel processable tasks within the source program. This is accomplished by performing precedence partitions on the connectivity matrices generated in Phase II. Precedence partitions were discussed in the first part of this section.

The output is in the form of a list of blocks. Each block contains the statement numbers of the statements within the program which can be executed concurrently.

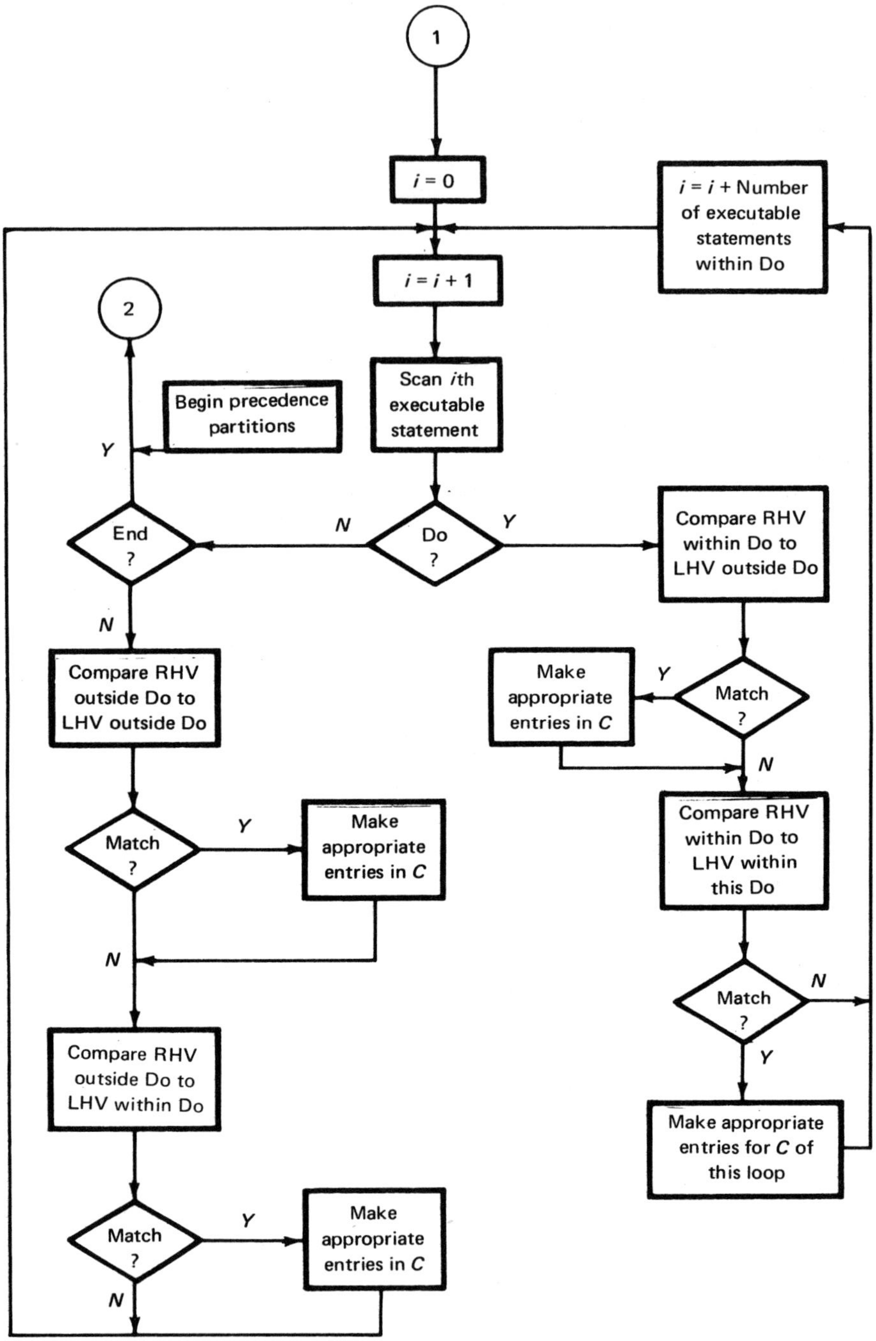

Fig. 16. Phase II of the Recognition Process

Observations

This section has introduced a method for recognizing parallelism on a global level. The method has been implemented in the form of a FORTRAN recognizer. On a limited basis, at least, it has been shown that the method of precedence partitions can be successfully employed to determine parallel processable tasks within a computational process. The recognizer itself possesses both major and minor limitations which must be eliminated before the recognition techniques used here can be applied to programs more complex in nature than the example program.

Some of the limitations of the recognizer can be eliminated quite easily. Recognition of a more complete set of FORTRAN instructions, for example, can be accomplished during Phase I by simple expansion of the recognition process. The generation of the connectivity matrix during Phase II, however, would be somewhat more complicated because of the many types of loops and branches that can be created by several instructions in the FORTRAN set.

In order to accommodate a source program which has more than 99 executable statements, it would be necessary to expand the size of the appropriate tables. Similar action would also be necessary to accommodate executable statements with more than 4 "right-hand" or "left-hand" parameters.

Because FORTRAN subroutines are compiled separately, it may be desirable to write portions of the main program in the form of subroutines if memory limitations were to become a problem. In its present form, the main program is the largest component of the recognizer program and requires $51{,}556_8$ words in memory.

As was stated earlier, the method of precedence partitions is independent of the source program language. The recognition process, however, is based on the language used by the source program. Thus the recognizer cannot accept a source program written in a language other than FORTRAN. Furthermore, it is not possible to implement a recognition process that is independent of the source program language.

It is quite likely, however, that FORTRAN is not the best language to use in implementing a recognizer program. Further work in this area may determine that a list processing language, for example, would be considerably more efficient for work of this nature.

More important, however, is a need to determine the benefits accrued, if any, by applying these techniques to the fundamental problem—the reduction of total processing time. Before this type of information can be obtained, it will be necessary to make exhaustive measurements in a manner similar to that proposed by Russell and Estrin.[30] It is not enough to merely determine that two tasks can be executed concurrently; it must be determined whether or not this concurrent execution will actually result in a more efficient utilization of system resources.

The concept of parallel processing can be easily defended on an intuitive basis. Proof of its actual worth, however, is much harder to obtain. That parallel processing will play a vital role in the future of information processing cannot be denied. Much work remains to be done before it can live up to its promise.

Some of the subjects covered here have been dealt with in an expanded form by the authors;[31] as have the conditions developed by Bernstein for parallel execution of two tasks.[32]

References

1. R. K. Richards, *Electronic Digital Systems* (New York: John Wiley & Sons, Inc., 1966), p. 267.
2. A. J. Bernstein, "Analysis of Programs for Parallel Processing," *IEEE Transactions on Electronic Computers,* EC-15 No. 5 (October, 1966), 757-63.
3. J. B. Dennis, "Programming Generality, Parallelism and Computer Architecture," *Proc. IFIPS Congress '68,* C1-C7, 1968.
4. E. W. Djkstra, "Solution of a Problem in Concurrent Programming Control," *Communications of the ACM,* VIII, No. 9 (September, 1965), 569.
5. Donald Knuth, "Additional Comments on a Problem in Concurrent Programming Control," *Communications of the ACM,* IX, No. 5 (May, 1966), 321-22.
6. E. G. Coffman and R. R. Muntz, "Models of Pure Time Sharing Disciplines for Research Allocation" (Proc. ACM National Conference, 1969).
7. Melvin E. Conway, "A Multiprocessor System Design," *Proc. Fall Joint Computer Conference,* 23: (1963), 139-46.
8. J. A. Gosden, "Explicit Parallel Processing Description and Control in Programs for Multi- and Uni-Processor Computers," *Proc. Fall Joint Computer Conference,* XXIX (1966), 651-60.
9. Norma E. Abel *et al.,* "TRANQUIL: A Language for an Array Processing Computer," *Proc. Spring Joint Computer Conference,* 1969, pp. 57-68.
10. W. R. Sutherland, "On-Line Graphical Specification of Computer Procedures" (Technical Report No. 405 [Lincoln Laboratory, M. I. T., 1966]).
11. L. G. Tesler and H. J. Enea, "A Language Design for Concurrent Processes," *Proc. Spring Joint Computer Conference,* 32: (1968), 403-8.
12. Dennis, *op. cit.*

13. C. V. Ramamoorthy, "A Dynamic Look Ahead & Program Segmentation System for Multi-Programmed Computers" (Proc. ACM National Conference, 1966).
14. D. E. Martin and G. Estrin, "Models of Computational Systems," *IEEE Transactions on Electronic Computers*, Vol. EC-XVI, No. 1 (February, 1967).
15. Ascher Opler, "Procedure-Oriented Statements to Facilitate Parallel Processing," *Communications of the ACM,* VIII, 5 (May, 1965), 306-7.
16. Gosden, *op. cit.*
17. Jack B. Dennis and Earl C. Van Horn, "Programming Semantics for Multi-Programmed Computations," *Communications of the ACM,* IX, No. 3 (March, 1966), 143-55.
18. D. A. Fisher, "Program Analysis for Multiprocessing" (Burroughs Corporation, May, 1967).
19. J. L. Baer and D. P. Bovet, "Compilation of Arithmetic Expressions for Parallel Computation," *Proc. IFIPS, '68,* B4-B10, 1968.
20. H. Hellerman, "Parallel Processing of Algebraic Expressions," *IEEE Transactions on Electronic Computers,* Vol. EC 15: No. 1 (February, 1966).
21. Harold S. Stone, "One-Pass Compilation of Arithmetic Expressions for a Parallel Processor," *Communications of the ACM,* 10: No. 4 (April, 1967), 220-23.
22. J. S. Squire, "A Translation Algorithm for a Multiprocessor Computer" (Proc. 18th ACM National Conference, 1963).
23. Baer and Bovet, *op. cit.*
24. D. Knuth, "The Art of Computer Programming," *Fundamental Algorithms* (Reading, Mass.: 1968), Addison-Wesley, 1: 316.
25. C. V. Ramamoorthy, "Analysis of Graphs by Connectivity Considerations," *Journal of the ACM,* XIII, No. 2 (April, 1966), 211-22.
26. Ramamoorthy, "Analysis of Graphs by Connectivity Considerations," *op. cit.*
27. *Ibid.*
28. C. V. Ramamoorthy, "A Structural Theory of Machine Diagnosis," *Proc. Spring Joint Computer Conference,* (1967), pp. 743-56.
29. Bernstein, *op. cit.*
30. E. C. Russell and G. Estrin, "Measurement Based Automatic Analysis of FORTRAN Programs," *Proc. Spring Joint Computer Conference,* 1969.
31. C. V. Ramamoorthy and M. J. Gonzalez, "Recognition and Representation of Parallel Processable Streams in Computer Programs–II (Task/Process Parallelism)." (Proc. ACM National Conference, 1969).
32. Ramamoorthy and Gonzalez, "A Survey of Techniques for Recognizing Parallel Processable Streams in Computer Programs," *Proc. Fall Joint Computer Conference,* 1969.

CHAPTER 18

PREPARATION AND EVALUATION OF COMPUTER PROGRAMS FOR PARALLEL PROCESSING SYSTEMS

J. L. Baer and E. C. Russell

University of California
Los Angeles, California

INTRODUCTION

The modeling of computations for parallel processing can be divided into four general areas (1) those general models concerned mostly with formal aspects of parallel processing without regard to actual programing considerations, (2) models that incorporate new programing languages designed to enhance parallel processing, (3) models that incorporate extensions of existing sequential programing languages, and (4) models that attempt to detect and represent parallelism in existing sequential languages.

Each approach has considerable merit. The general models (1) allow us to gain deeper insight into the nature of parallel processes and to prove the validity of such processes. At this point in time, new languages for parallel processing are embryonic. We need additional information and experience with parallel processing to be able to state

---

This research was supported by the Atomic Energy Commission (AT [11-1] Gen. 10 Project 14), and Office of Naval Research Information Systems Branch (N00014-67-A-0111-0016). Since its beginning, this project has been directed by Professor G. Estrin. The authors wish to thank him for his support and his guidance.

just which aspects of parallelism need to be explicitly stated and which should be left to machine (compiler) recognition. Experience with such languages will also teach us the most "natural" means of expression of parallelism.

At present, extensions to existing languages are useful in reformulating existing programs, however such extensions usually require complete recoding of existing programs. Models which attempt to detect parallelism in existing sequential languages can play an important role in this transition period. Large existing programs can be made operational on existing (or simulated) parallel processing systems, and observations of system behavior can be made while reformulation of the program is in process. Nor have we learned all there is to know about programs for sequential processors. Some of the modeling for parallel processing has given insight into optimization techniques for single processors.

What follows is a review of some of the research in each of these areas, concluding with a statement of the approach followed in this study.

MODELS OF PARALLEL COMPUTATIONS

Formal Models of Parallel Processes

Karp and Miller[1] present a graph-theoretic model of computations in which properties of parallel execution are studied. In their model, each node of the graph corresponds to an operation in the computation, and each branch represents a queue of data directed from one node to another. The results of one node are placed in first-in-first-out queues associated with its out-branchings. The eligibility for initiation of a computation step is determined by the lengths of the queues on branches directed into its associated node. Severe restrictions on the type of sequence control which can be represented within the model arise from the fixed queue discipline, the fact that each data queue has a unique *source* and a unique *sink*, and the fact that data dependent conditional transfers cannot be represented.

The properties of the model with which Karp and Miller are particularly concerned are (1) to prove that the model is determinate, i.e., for a given input, the program will yield a unique output independent of relative processor speeds; (2) a test to determine whether a given computation will indeed terminate; (3) a procedure for finding the number of performances of each computation step; and (4) finally, the amount of temporary storage required for the data queues associated with branches of the graph together with conditions for the queue lengths to remain bounded.

In his dissertation, Rodriguez[2] continues along the same lines as Karp and Miller in investigating determinacy and what he terms "hang-up states" of the computation. A hang-up state is a condition in which the computation has progressed to a situation in which no node is active and some arc is still active (thus the computation cannot proceed).

Rodriguez bases his model directly on that of R. Turn (See "Presentation of the Model"), introducing additional types of nodes for pure control information so that decision elements do not transmit any data but only enable or disable computations.

Languages Designed for Parallel Processing

The basic concept of parallel coding was described by Conway.[3] He suggests the notion of a pair of machine language instructions, FORK and JOIN. The FORK instruction has the form—*L* FORK*A,J,N,*—which is interpreted as "Invoke a second procedure which begins at *A*, to run concurrently with the procedure which begins at $L + 1$, set the total number of concurrent processes (N) to be reunited at J." Then either $L + 1$ or A may contain additional FORK instructions of the form FORK *B*. Each FORK instruction initiates two parallel processes and the process may be continued indefinitely. The JOIN instruction has two forms "JOIN *J*" or "JOIN *J,B*," which are interpreted as, "wait for everything which is to be concluded at *J* before continuing" or, "if anything to be finished at *J* is still in process, branch to *B*." This is a machine language level extension.

The design of array computers has led to specification of languages which include facilities for topological description of problem solution spaces. The SOLOMON Computer[4] and its successor, ILLIAC IV,[5] both have languages which allow the programmer to make explicit data linkages between individual processing elements. Thus one can link the elements in the two dimensional array to represent planar, linear, toroidal, cylindrical, or spherical geometries. This type of language would seem to depend a great deal on the actual configuration of the hardware and thus be of limited general use.

Models Using Extensions

Opler,[6] Gosden,[7] Tesler,[8] and Anderson[9] have proposed various extensions to basic ALGOL 60 to allow description of parallel processes. Besides the customary FORK and JOIN statements, the PARALLEL FOR statement would seem to provide the most significant key to their statement of parallel processing. These three adjuncts account for most of the literature on the subject of language extensions for parallelism.

Transformation Techniques

The analysis of existing programs in order to discover parallelism in the algorithm has received considerable attention in the literature. The transformation can be made at several levels. At the extremes are the individual instructions and entire programs. Subexpression analysis has been performed to discover parallelism within arithmetic expressions. On the other extreme, independent programs are scheduled on multiprocessors. In this case, the only conflict is in competition for resources within the system, as there is no interdependence of processes. Most of the investigations have settled on the statement as the unit for parallel computation. Interstatement dependence is well defined in languages such as ALGOL, FORTRAN, and PL/I.

Fortran Statement Analysis

Bernstein[10] presents a model for FORTRAN language programs and a set of conditions which determine whether or not two successive portions of a given program can be executed in parallel and still produce the same results. He considers two machine models, one in which each processor communicates directly with a common main memory and one in which each processor has a slave memory through which information coming from or going to main memory is buffered.

The conditions which he derives are developed in terms of input/output data sets. Variables are classified into four categories:

1) locations which are only fetched during computation
2) locations which are only stored during computation
3) locations which are first fetched and then stored

4) locations which are first stored and then fetched.

These are denoted as sets W_i, X_i, Y_i, Z_i respectively.

If three processes are to be performed as in Fig. 1, the conditions for independence are as follows:

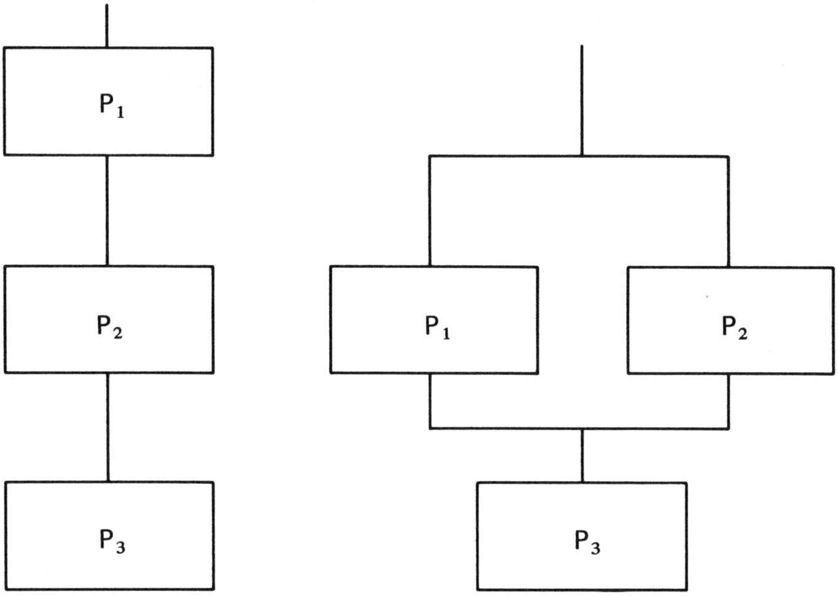

Fig. 1. Sequential Program vs. Parallel Program

1) $(W_1 \cup Y_1 \cup Z_1) \cap (X_2 \cup Y_2 \cup Z_2) = \Phi$

2) $(X_1 \cup Y_1 \cup Z_1) \cap (W_2 \cup Y_2 \cup Z_2) = \Phi$

3) $X_1 \cap X_2 \cap (W_3 \cup Y_3) = \Phi$.

Loops are discussed in terms of these same independence relations. If the data on successive iterations can be shown to satisfy relations (1–3), then the iterations are shown to be independent and amenable to parallel treatment.

The UCLA Modeling Procedure

The work reported herein is an outgrowth of the variable structure computer project at UCLA.[11] The research project which started as the design of a fixed plus variable structure computer system has evolved into a many-faceted basic digital computer research project. With the advent of microminiaturized components, many of the system designs proposed have become realizable with commercially available subsystems. Special purpose computer system design has reached a high degree of automation. Research in this area by Turn,[12] Mandell,[13] and Potash,[14] has produced tools for transforming well formed design-level descriptions of programs into sufficient documentation for verification of design and implementation.

The area of problem modeling has been investigated by many persons. Within the scope of this project several persons have investigated designs for special purpose hardware configurations to solve particular problems. Bussell[15] looked at pipeline computer techniques with restructurable subsystems for application to the solution of partial differential equations. Martin[16] investigated n^{th} root computing methods. Mandell[17] investigated special purpose designs for bivariate analysis. Wood[18] modeled the MINTZ numerical weather prediction program to explore how different system configurations and particularly how parallel systems would improve overall computation time.

In each of these cases, application was made to a particular design (or set of designs) for one problem. Turn[19] proposed a general model for computations which has served as the basis for much further work in this area. He provided a scheme for representing a computation and the system on which the computation was to be performed. With both the program and system described in his nomenclature, an assignment of computations to subsystems was performed in such a way as to attempt to optimize solution time. Martin[20] made many refinements and improvements in the computational model and thoroughly tested the model with regard to sensitivity to variations and imprecisely known parameters. But still he could only test as many programs as he could model manually. Baer[21] has made more improvements in both the model and the processes for analysis using the model. Bovet[22] has extended and refined Turn's memory allocation schemes to optimally allocate several levels of memory hierarchy.

The entire modeling procedure consists of the following:

1) preparation of problem description[23]
   a) structural description

   b) vertex attributes
   c) measurement of data-dependent attributes
2) preparation of processor model[24]
   a) memory hierarchy
   b) process times
   c) number of processors
3) prediction of performance and assignment of computations to processors[25,26]
   a) bounds for maximum parallelism
   b) "A priori" assignment and sequencing on parallel processor systems
4) allocation of memory[27]
   a) memory profile for each individual processor (or common memory if such is the configuration).

The UCLA modeling procedure is block-diagrammed in Fig. 2.

PRESENTATION OF THE MODEL

The model is a directed graph, $G(W,U)$ where W is the set of vertices $\{w_1, \ldots, w_n\}$, and U is the set of ordered pairs (or arcs) $u_k = (w_i, w_j)$. We assume that the reader is familiar with the concepts of elementary graph theory such as path, cycle, loop, (immediate) predecessor and (immediate) successor. A vertex represents a computational task, either a single operation, or a sequence of elementary operations, or even a subroutine depending on the degree of detailed representation that one wants to be pictured by the graph. One can assume in general that a computational task requires some input data S_i which it will transform into output data S_o. Therefore, a vertex is the representation of the transformation

$$\{S_{01}, S_{02}, \ldots, S_{0k}\} = F\{S_{i1}, S_{i2}, \ldots, S_{ih}\} .$$

An arc u_k linking 2 vertices w_i and w_j implies that the operations represented by w_j will obtain some or all of its input data from w_i; u_k represents the transfer of control from w_i to w_j.

To be able to start, the computational task represented by w_i may need either all the sets of data generated by its immediate predecessors, or the data generated by only one of these immediate predecessors exclusively of all of the others, or by some but not all of the immediate

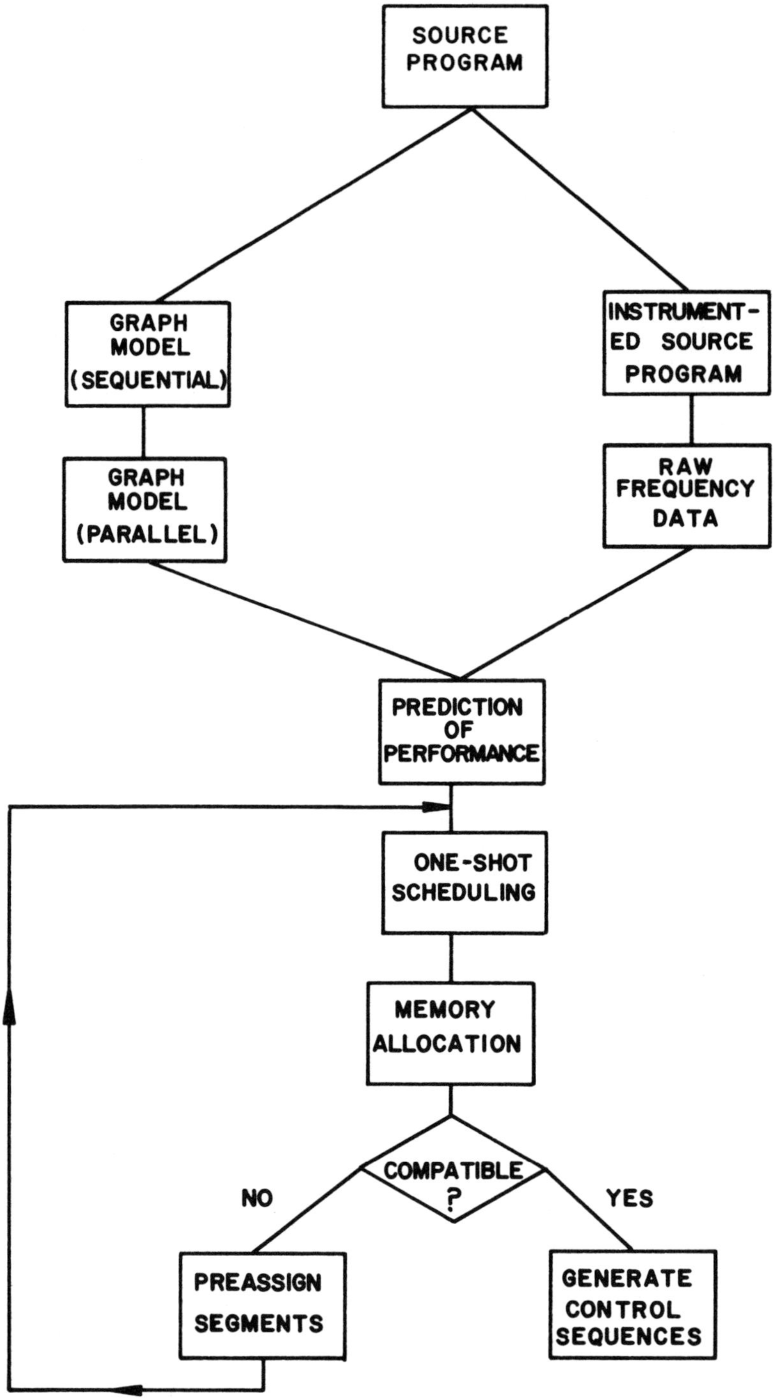

Fig. 2. The U.C.L.A. Modeling Procedure

predecessors. In the first case, the vertex w_i will be labeled as having *AND-input* logic and graphically represented as in Fig. 3(*a*); in the second case, it will be labeled as having *EOR-input* logic (exclusive-or) with the representation of Fig. 3(*b*).

In order to restrict ourselves to these two types of logic only, dummy vertices, i.e., vertices representing no-operations, have to be inserted when compound conditions arise as in the third case above. For example, Fig. 3(*c*) shows the insertion of dummy vertices $\omega_i{'}$ and $\omega_i{''}$.

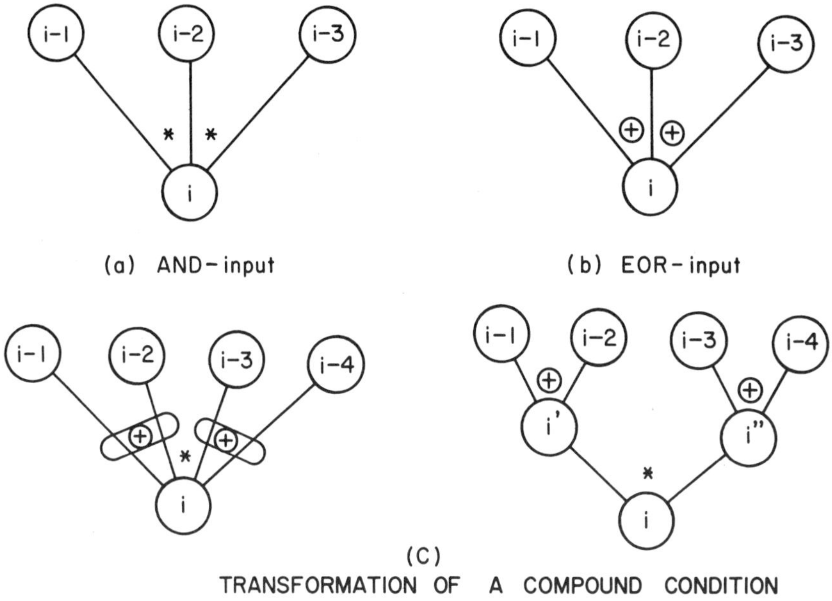

Fig. 3. Input Logic

In nontrivial computer programs, decisions by means of tests have to be taken. The vertices representative of these operations will be called *branching vertices.* That is more than one arc (or bundle of arcs) will be incident out from these vertices in mutually exclusive fashion. These vertices will also be labeled as having *EOR-output* logic (cf., Fig. 4[*a*]). Since one of the main goals of this model is to show the inherent parallelism present in the computer program it represents, there will also be vertices which will generate data required unconditionally by their immediate successors. We will label these vertices as having *AND-output* logic (cf., Fig. 4[*b*]). Compound conditions, by

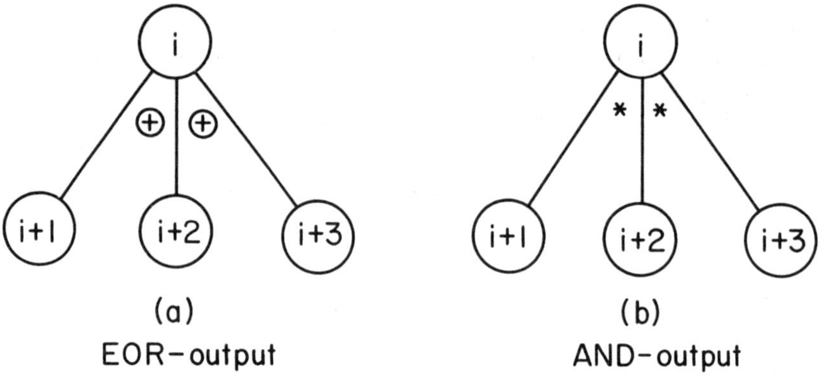

Fig. 4. Output Logic

mixing the two above cases, can appear, and they are eliminated by introduction of dummy vertices as for the input logics.

Along with the structure of the program as pictured by its directed graph and the logics associated with the vertices, certain parameters must be given in order to define more completely the computational task.

Vertex Attributes

Each vertex w_i is given a number i which can be considered as its name (cf., "Transformation of Graphs in Sequential Form" for its choice); an input and an output logic in order to define the "syntactic" structure of the program and a series of attributes:

1) – t_i time required for one execution of w_i on one processor
2) – e_i frequency of execution of w_i
3) – p_i probability of reaching w_i from the initial vertex w_1 (supposed unique)
4) – $a_i = e_i/p_i$ activity of w_i
5) – $r_i = a_i p_i t_i$ weight of w_i
6) – l_i the temporary data storage requisite for the computation
7) – s_i the block of storage required to contain the program instructions.

Arc Attributes

Each arc is defined by its origin and terminus and its probability q_{ij} of occurrence once its origin has been reached (this probability is 1 if the origin vertex is of AND-output logic and less than 1 if it is a branching vertex).

Boolean Matrix Representations of the Model

A pictorial model is not presently suitable for processing by digital computer. Many authors have proposed techniques for representing directed graphs by Boolean matrices.[28, 29, 30] The ease with which such matrices can be manipulated and the compactness of their storage requirements make them very convenient for computer representation of graph models.

Connection Matrix

The connection matrix is used to represent all intervertex connections. For an n vertex graph, C is an $n \times n$ matrix such that $C_{ij} = 1$ if and only if there is an arc directed from vertex w_j to w_i. All feedback paths of cycles are separately represented in a list of the form of a two column array $FB_{n1} = w_i$ and $FB_{n2} = w_j$ for the representation of a cycle which has vertex w_i as head and w_j as tail.

Precedence Matrix

The precedence matrix, derived from the connection matrix, is an $n \times n$ matrix P such that $P_{ij} = 1$ if and only if there is a path from vertex w_j to vertex w_i.

Representation of Data Dependency

In the model of programs under consideration, data is not associated (stored) with the vertex. In order to describe the correspondence between data variables and the vertices of the graph, two matrices are generated. The *input data set, I,* is a Boolean matrix of size $n \times m$ for an n vertex graph with m named variables. $I_{ij} = 1$ if and only if variable j is *used* by vertex w_i. The *output data set, O,* is a Boolean matrix of the same dimensionality as I in which $0_{ij} = 1$ if and only if variable j is *set* by the execution of vertex w_i.

TRANSFORMATION OF GRAPHS IN SEQUENTIAL FORM

As a specific example of the application of this process, we shall now consider the analysis of programs written in the FORTRAN IV language. FORTRAN was selected as the first target for analysis because there exists a considerable number of large (in the sense of computation time) programs written entirely in FORTRAN. Several of these programs have been analyzed by previous investigators (Bussell,[31] Turn,[32] Martin,[33] Wood,[34] and Russell[35]), but in each case the transformation from program to graph was made manually.

NOMENCLATURE

Vertex Definition

The application of the model to FORTRAN will be made at the statement level. That is, each executable FORTRAN statement will be assigned to a unique vertex. Other choices could be made: for example, if more than one vertex per statement were permitted, then each arithmetic operator could be assigned to a unique vertex and analysis of parallel arithmetic expressions could be performed. This would result in graphs with large numbers of vertices. On the other hand, a single vertex could be made to represent several statements or even an entire subprogram. This would provide for smaller graphs and would generally obscure potential parallelism. We choose to make an initial analysis at the statement level. A refined analysis can then be performed on these portions of the program which consume significant portions of execution time, whereas less significant portions can be collapsed into fewer vertices.

Each program will contain one origin vertex (w_1) which is predecessor to all other vertices in the graph. (A program with multiple entries will be assigned one pseudoorigin with immediate outbranchings to the various entry points.)

Each program will contain one *terminal vertex* (w_n) which is successor to all other vertices in the graph. (Multiple exits from a program will be tied to this one terminus.)

For each vertex, the *input data set*, I_i, consists of the names of those variables which are referenced by the statement represented by the vertex. For example, for an arithmetic assignment statement this includes all variables in the expression and any subscripts.

For each vertex, the *output data set*, 0_i, consists of the names of those variables which are modified by the execution of the statement represented by the vertex.

Arc Definitions

Three categories of connections between vertices are established; sequential connections, logical connections, and loops.

Each executable FORTRAN statement which does not explicitly specify a successor statement implicitly "falls through" to the next executable statement in the program. These connections, of the form (w_i, w_{i+1}), are designated *sequential connections.* There can be only one sequential successor vertex for each vertex.

Each executable FORTRAN statement which does specify explicitly one or more successor statements produces connections which are designated *logical connections.* The connections are of the form (w_i, w_j) where $i \neq j$.

Some of the arcs of a program represent the return path of a programmed loop. Of course, any one arc in a cycle can be selected as the return path, but often one is recognizable from the syntax of the language (e.g., the FORTRAN *Do* statement). Each arc which closes a programmed loop is designated as a *feedback connection.*

When a choice is possible for the feedback arc, a triangularization algorithm is applied to the connection matrix such that the remaining nonzero elements on or above the diagonal represent feedback arcs. This procedure amounts to a renumbering of the vertices for ease of identification of cycles.

Experimental Measurements

Two attributes of the vertices of a graph are determined from experimental measurement. These are (1) vertex frequency number e_i and (2) vertex single-execution-time, t_i.

Vertex Activity in a Sequential Program

Vertex frequency can be determined by making some minor modifications to the source program before execution. These modifications are similar to those outlined in G. Estrin.[36] For these experiments, however, only vertex activity and loop activity are desired and not all arc activity as in the reference. Because of this, the artifact introduced is considerably reduced. The required measurements are as follows:

1) the number of times the program is entered
2) the number of times each labeled statement is executed
3) the number of times the auxiliary statement of "logical if" is executed.

From these measurements, the entire set of vertex frequencies and loop factors may be calculated.

The detailed application of this process to FORTRAN has been treated elsewhere.[37] In that paper all the procedures are presented for representing FORTRAN programs by directed graphs and for introducing measurements artifact into the source program.

Implementation

The original implementation of the entire evolution process for FORTRAN programs consisted of five steps: three in phase ϕ plus two for analysis.[38] What will be described herein is an improved version of this process which includes a single pass over the source FORTRAN program.[39]

Preprocessor Phase

The preprocessor phase accomplishes the following:

1) Generate an instrumented version of the source FORTRAN program for execution to collect measurements.
2) Assign vertex numbers to FORTRAN statements.

3) Generate connection matrix, dictionary, input and output matrices, and vertex attributes.
4) Generate logical connection list and feedback arc list.

The connection matrix can be generated as the source program is processed. Recalling that only sequential connections are represented in C, both coordinates of C_{ij} are known when the arc is generated. The logical connection matrix is not as simple and generation of it must be accomplished in two parts. First, a list of arcs of the form (w_i, w_k) where k is a FORTRAN external formula number is generated during the source scan. Then the w_j corresponding to w_k is substituted for it, and the element L_{ij} set to one $(L_{ij} = 1)$.

The dictionary of variables is generated, and entries are added as the variables are encountered in the source program. The entries in the dictionary consist of:

1) variable name (1–6 alphameric characters, first one alphabetic)
2) type (single digit)
   a) integer
   b) real
   c) logical
   d) complex
3) precision (in bytes)
4) size-product of dimensions

The variables are added to the dictionary as they are encountered in the program. A variable may appear in several declaration statements without producing a conflict. For example, a variable may be dimensioned in a DIMENSION statement and then appear in a "type" statement, or it may appear in a type statement with dimension. When a variable appears for the first time in an executable statement, all unspecified attributes are given default values.

The correspondence of the variables to the input/output sets is determined from the dictionary. Each column of I or O represents one variable. The column number corresponds to the entry number on the dictionary. The input and output sets are generated statement by statement. Any variables not previously defined are entered into the dictionary as they are encountered.

The vertex attributes which can be generated in the preprocessor are:

1) single-execution time

   This is in effect a statement-by-statement compilation of the program. The sum of the average execution times of the instructions which would be required to execute each statement is used to represent the average execution time of the statement. When a statement references another subprogram, the execution time for the subprogram is added to the vertex execution time. Thus there is an implied ordering of the analysis of subprograms. The lowest level programs must be analyzed before programs which use them.

2) input/output logic
The input/output control logic for a vertex is stored in a single signed digit according to the table below.

Table 1
Input/Output Logic

Input	Output	Logic
EOR	EOR	-1
EOR	AND	0
AND	EOR	+1
AND	AND	+2

In the analysis of a source FORTRAN program control logic is assigned as follows:

a) All numbered statements and the first statement after a *Do* statement are given EOR input logic.
b) All other statements get AND input logic.
c) Statements with multiple outbranchings receive EOR output logic.
d) All others receive AND output logic.

3) program storage requirement
The number of words of storage required for the instructions representing the computation at each vertex is computed with the execution time.

4) temporary storage requirement
The number of words of storage for intermediate results is computed in the compilation process.

5) external formula number
Only the original statement labels are preserved for ease of correlating output with the original program. For uniform data-handling, every statement is given an external statement label. If none was originally present, the negative of the vertex

number is used. No conflict with user-assigned numbers can occur since FORTRAN does not allow negative statement numbers. (These negative statement numbers are for internal use only and are not made a part of the instrumented FORTRAN program.)

Execution Phase

The execution of an instrumented program differs from a noninstrumented in only the following ways:

1) Some or all of the FORTRAN routines have been preprocessed to insert emitter calls.
2) An emitter routine must be included. This routine contains the table of counters for all the emits which have been inserted into the source program. For efficiency this routine can be made to modify the calling sequence the first time it is called so as to access and modify the counter from "in line" code thus reducing the number of instructions executed as measurement artifact.
3) Emitter counts must be output at the end of the run. On Sigma 7 this was accomplished by modifying the system so as to dump the emit table upon encountering the next job card. On /360 the same thing was accomplished by writing a small interface routine which passed control to the main routine of the user's program and to which control returned from the user's program upon completion.

Post Processor Phase

The post processor must correlate the emit table with the vertex table. For a single program this correlation is direct. Emitter counters were selected to correspond directly to the vertex numbers. When more than one routine has been instrumented, an offset is computed for each program giving the emitter counter to vertex correspondence. Once this correlation is completed, the evaluation of program execution on a single processor is performed, as described in the section "Evaluation of Computer Programs for Parallel Processing." After sequential analysis the transformation to parallel form is performed as described in the following section.

TRANSFORMATION OF GRAPHS FOR PARALLEL PROCESSING

Algorithm for Eliminating Nonessential Sequential Connections

Given the model of programs presented in section "Presentation of the Model," and applied to FORTRAN in the section "Transformation of Graphs in Sequential Form," the task now is to develop a procedure for generating models of programs for execution on multiprocessors. One of the prime constraints imposed upon this process is the preservation of the *determinacy* of the program. Preserving the determinacy of a program means guaranteeing that a unique result will be produced for each execution of the program given the same initial conditions and input data, regardless of the number of individual processors or their relative processing speeds.

The constraints upon the introduction of parallelism in a program are as follows:

1) All control logic explicitly stated in the program is preserved.
2) For each vertex, it is assumed that all input variables are obtained before any output variables are generated.
3) Data dependency is preserved.

Separation of Control Flow Dependency from Data Dependency

The logical control matrix L is used in order to implement the first condition, (that all explicit control logic be preserved). Recall that $L_{ij} = 1$ if and only if there is an explicit transfer of program control from vertex w_j to vertex w_i. Thus L becomes a subset of the previously defined connection matrix, C. Thus we modify the definition of C to include only those sequential connections which are implicit in the language, such as the FORTRAN rule, "After completion of this statement, the 'next' executable statement is performed." These are the connections which are candidates for elimination in an attempt to gain parallelism.

The basic statement of the algorithm is as follows:

1) Assume the following matrices are available describing the program for a single processor:

a) the precedence matrix, P
b) the logical connection matrix, L
c) the input variable matrix, I
d) the output variable matrix, O.

Also one new matrix, L_p, the parallel logical control matrix is to be defined. Initially set $L_p = L$.

2) For each $P_{ij} = 1$, examine the set

$$(I_j \cap 0_j) \cup (I_j \cap 0_i) \cup (0_i \cap 0_j) .$$

If this set is empty, there is no data dependency between vertices w_i and w_j, and thus precedence relations between them need not be preserved. In order to record this, perform the following:

a) set $P_{ij} = 0$
b) *set* $L_{\bar{p}_i} = L_{\bar{p}_i} \cup L_{\bar{j}}$, in order to preserve the logical control conditions on w_i (where $A_{\bar{i}}$ is the ith row of the boolean matrix A).

3) When all reductions of the type specified in step (2) are completed, we are left with an incomplete precedence matrix. The goal now is to generate the connection matrix representing the parallel form of the graph from this precedence matrix.

Minimization of the Number of Arcs

1) generate the second-order precedence matrix,

$$P_2 = \bigcup_{i=2}^{n} P^i$$

2) generate the parallel connection matrix, eliminating redundant connections,

$$C_2 = (P \cap \widetilde{P_2}) \cup L_p .$$

Now, C_2 will contain only those connections that are based on data dependency or the original logical control conditions as preserved in L_p. This procedure is the same as that of Simoes-Peroira[70] with the addition of the preservation of logical control connections. These connections often violate the rule of a minimal number of connections. As an example of such a situation, consider the following representation of a simple conditional statement.

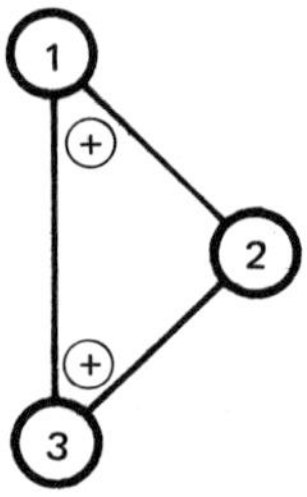

Vertex w_1 represents a conditional test statement. If the test is true, vertex w_2 is executed. If the test is false, vertex w_2 is not executed. In either case, vertex w_3 is to be executed. By the reduction procedure of Simoes-Peroira, the connection from w_1 to w_3 directly would be eliminated because of the existing path from w_1 to w_2 to w_3. This would not be a complete representation of the program, since it is possible to execute w_3 without executing w_2. Hence, the need for the preservation of the logical connection matrix, L_p .

Examples

The following small example is shown in complete detail to make explicit the transformation process. First, an incomplete graph is produced to illustrate the need for the preservation of control logic.

Connection Matrix, C

	1	2	3	4	5	6	7	8
1	.							
2	1	.						
3	.	1	.					
4	.	.	1	.				
5	.	1	.	.	.			
6	.	.	.	.	1	.		
7	.	.	.	1	.	1	.	
8	.	.	.	.	.	.	1	.

Logical Connection Matrix, L

	1	2	3	4	5	6	7	8
1	.							
2	.	.						
3	.	1	.					
4	.	.	.	.				
5	.	1	.	.	.			
6	.	.	.	.	.	.		
7	.	.	.	1	.	.	.	
8	.	.	.	.	.	.	.	.

Input Matrix, I

	1	2	3	4	5
1	.	.	.	.	.
2	.	1	.	.	.
3	.	.	.	.	.
4	.	.	.	.	.
5	.	.	.	.	.
6	.	.	.	.	.
7	.	.	1	1	.
8	.	.	.	.	1

Output Matrix, O

	1	2	3	4	5
1	.	1	.	.	.
2	.	.	.	.	.
3	.	.	1	.	.
4	.	.	.	1	.
5	.	.	1	.	.
6	.	.	.	1	.
7	.	.	.	.	1
8	.	.	.	.	.

Precedence Matrix, P

	1	2	3	4	5	6	7	8
1	.							
2	1	.						
3	1	1	.					
4	1	1	1	.				
5	1	1	.	.	.			
6	1	1	.	.	1	.		
7	1	1	1	1	1	1	.	
8	1	1	1	1	1	1	1	.

Parallel Precedence Matrix, P_p

	1	2	3	4	5	6	7	8
1	.							
2	1	.						
3	0	0	.					
4	0	0	0	.				
5	0	0	.	.	.			
6	0	0	.	.	0	.		
7	0	0	1	1	1	1	.	
8	0	0	0	0	0	0	1	.

The second order precedence matrix P_2 generated from P_p is null for this example. Those precedence relations which were broken by application of step (2) are shown as explicit zeros in P_p.

The parallel control logic matrix is generated at the same time as the parallel precedence matrix (step 2*b*).

Parallel Logic Matrix, L_p

	1	2	3	4	5	6	7	8
1	.							
2	.	.						
3	.	1	.					
4	.	1	.	.				
5	.	1	.	.	.			
6	.	1	.	.	.	.		
7	.	.	.	1	.	.	.	
8	.	.	.	.	.	.	.	.

Parallel Connection Matrix, C_p

	1	2	3	4	5	6	7	8
1	.							
2	1	.						
3	.	1	.					
4	.	1	.	.				
5	.	1	.	.	.			
6	.	1	.	.	.	.		
7	.	.	1	1	1	1	.	
8	.	.	.	.	.	.	1	.

The final parallel connection matrix, C_p, is generated in step (3*b*). The initial and final graphs are shown in Fig. 5.

The logic of interconnection of the parallel form is complex and can be determined from the sequential graph. As will be seen in the following, the parallel graph shown (Fig. 5) cannot be represented with simple logic, and transformation to allow such representation is discussed.

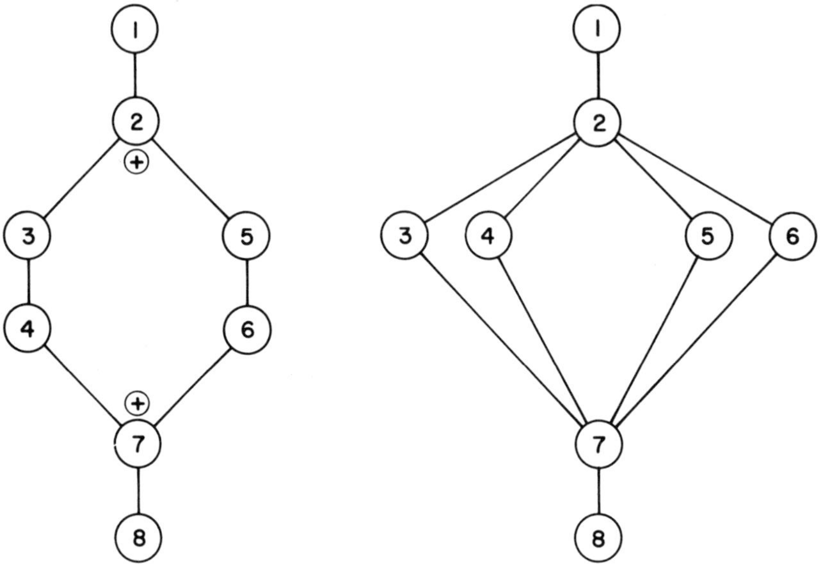

a) Sequential Graph b) Equivalent Parallel Graph

Fig. 5. Transformation from Sequential to Parallel Graph

Algorithm for the Simplification of Control Logic/ Statement of the Problem

In the original graph model proposed by Turn, a multilevel complex control logic was specified for each vertex.[40] In this logic, input and output connections could be combined in any arbitrary combination of AND and EOR logic. D. F. Martin constrained the logic functions to simple AND and EOR for the purposes of simplifying a priori assignment computations such as probability and path length.[41] Martin's constraint required introduction of pseudovertices. In the process of generating a parallel description from a sequential one, the logic may become complex. The example from the previous section as transformed is shown in Fig. 6.

Treatment of Loops/Detection of Potential Parallelism

Several investigators, A. J. Bernstein and H. W. Bingham *et al.*, have presented methods of detecting parallelism in the execution of a programmed loop.[42,43] In particular, the question raised is, "Can successive iterations of the loop be executed simultaneously or is there some inherent dependency upon the sequential execution?"

The first approach to answering this question is very conservative. If we form aggregate input data sets and output data sets for the vertices contained in a loop, then the test for independence of successive loops is to ascertain that the aggregate input and output data sets are disjoint.

Let vertices $w_{i_1}, \ldots, w_{i_2}$ be contained in a loop, L, then

$$I_L = \bigcup_{w_i \epsilon L} I_i$$

$$O_L = \bigcup_{w_i \epsilon L} O_i$$

If $I_L \cap O_L = \phi$, then successive loops of L are independent.

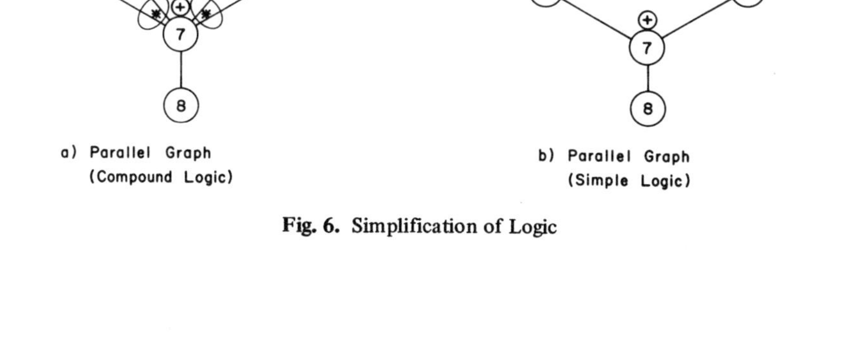

Fig. 6. Simplification of Logic

While the above constraint is sufficient for the guarantee of equivalent-parallel processing, it is more restrictive than necessary. From the definitions given for I and O, an entire array is represented by a single element in either of these sets. In loop structures, it may be (and in fact is usually the case) that only specific elements of an array are referenced on a specific iteration.

Thus, additional parallelism is possible if additional information can be gained about the subscript usage in arrays. The conditions necessary for a parallel representation of a loop are as follows:

1) When a subscripted element is in the intersection of I and O, subscript analysis is required to determine which specific elements are referenced. If it can be shown that they are unique for each iteration, then parallelism is possible.
2) When step (1) still does not eliminate conflicts, it is necessary to examine the order in which variables are referenced.
   a) If a conflicting variable is always set before it is used, the conflict is artificial, and the storage for the variable may be replicated in each processor.
   b) If subscript arithmetic occurs, no statement about potential parallelism can be made prior to the execution of the program.

When a loop is converted to a parallel form, it is necessary to modify the activity number and total computation time. Whenever a loop is transformed into parallel form, the activity number of each vertex in the loop must be divided by the replication factor for the loop.

The replication factor for each vertex is the product of the replication factors for all parallel loops in which the vertex is contained.

The activity number may become noninteger in this process. This is due to decision branching within the loop. Some of the vertices are executed fewer times than the loop factor. Thus when the individual cycles are to be assigned to separate processors, some of the vertices will (probably) never be executed. It is not possible to know a priori which branch will be taken in each replication.

The final result of the algorithm is a model showing all the recognized parallelism in the program. The main purposes in preparing this model are to study automatic assignment and sequencing of programs, and prediction of performance for parallel processing systems.

EVALUATION OF COMPUTER PROGRAMS FOR PARALLEL PROCESSING

In the two preceding sections, we have seen how a source program was analyzed automatically in order to be in the form of the model discussed in the section, "Presentation of the Model." We now are going to review how this model can be used to evaluate the performance of a program if run on a homogeneous multiprocessor system.

Cyclic-Acyclic Transformation

Since we are dealing with a model, we may have to part from a real representation of the phenomenon we describe. The first step in this direction is to transform the cyclic graph in an acyclic graph by removing the feedback connections. The motivation behind this transformation is that it will be possible to derive some analytic criteria of performance from the acyclic graph that would not be possible from the cyclic graph. This cyclic-acyclic transformation must obey to two constraints:

1. obtain a legal graph (see the definition of legality later in this section)
2. transform the graph in such a way that the mean path length of the acyclic structures obtained by removing the feedback arcs are equal to the mean path length of the original structures when the latter are considered by themselves and not embedded in more complex structures.[44]

This second requirement implies that prediction of performance derived from the acyclic graph will remain true (on the average) for a cyclic graph, that is the closer representation of the program.

This mean path length cyclic-acyclic transformation is performed by considering that the time of execution of a vertex w_i is not t_i but $e_i\ t_i$, i:e its weight r_i and by removing the feedback connections. In doing so, we obtain a directed acyclic bilogic graph (from now on DAB graph). The process is summarized in boxes 1, 2, 3 of the flow chart of Fig. 7.

However, we will see in the following that the arc probabilities q_{ij} of traversing the arc (w_i, w_j) when w_i is a branching vertex (EOR logic)

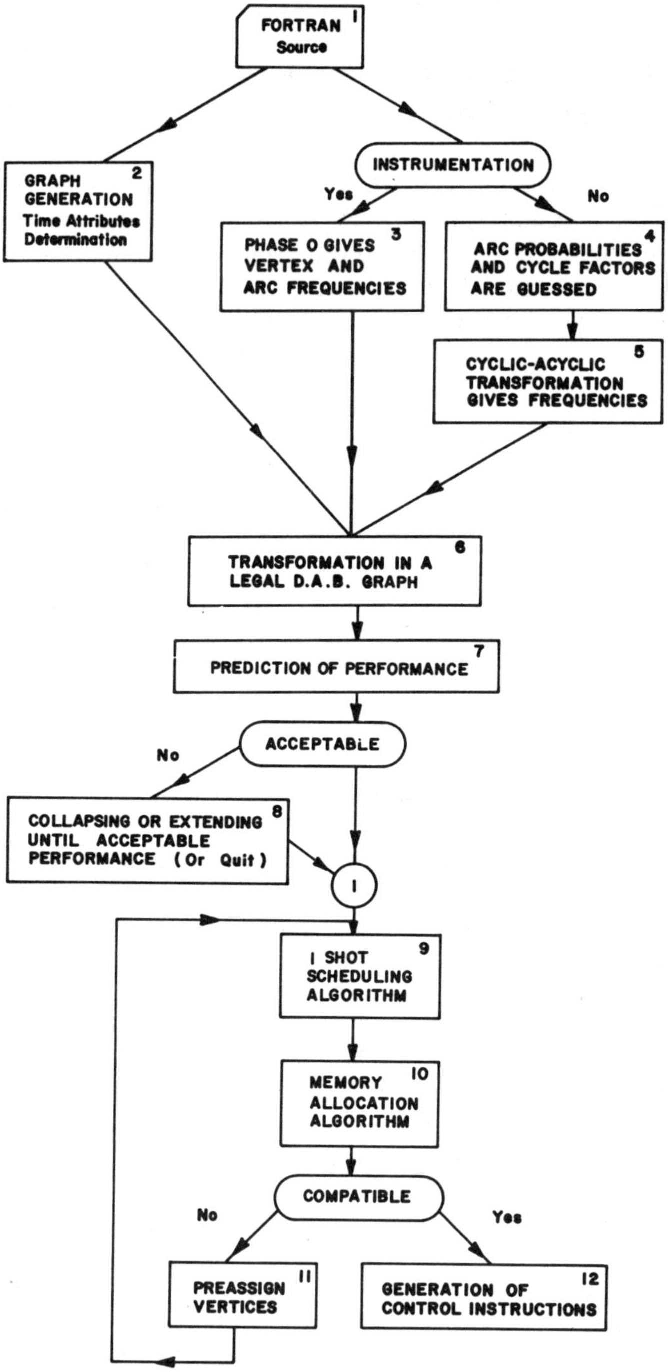

Fig. 7. Flow Chart of the UCLA Modeling Project

are necessary in some algorithms. They can be determined by instrumenting the arc frequencies say f_{ij} and

$$q_{ij} = \frac{f_{ij}}{\sum\limits_{w_j \epsilon C_i} f_{ij}}$$

where C is the connection matrix with the feedback arcs removed and C_i^+ its ith column.[45]

Prior to the implementation of the instrumentation process, this cyclic-acyclic transformation was performed analytically.[46,47]

Boxes 4 and 5 of Fig. 7 represent this alternate transformation. We can note also that the knowledge of the q_{ij} allows us to determine the probabilities p_i which are defined as the probabilities of reaching the w_i from w_1 in the acyclic graph.[48] These p_i can be computed as[49]

$$p_1 = 1$$

$$p_i = \sum_{j=1}^{m} p_{ij} q_{i-j,i} \quad w_i \text{ has EOR-input logic}$$

$$p_i = \min_{j=1,2,m} \quad (p_{i-j}\ q_{i-j},_i) \quad w_i \text{ has AND-input logic}$$

$2 \leq i \leq n$; w_{i-j} $(j = 1,2, \ldots, m)$ are the immediate predecessors of w_i.

Legality of a DAB Graph

Since we are now dealing with acyclic graphs, the logic associated with the vertices of $G(W, U)$ allows us to construct a set of subgraphs $A(W', U')$ of $G(W,U)$ as follows:[50]

*I*1) The set W' always contains the initial vertex w_1 and the terminal vertex w_n of $G(W, U)$.

*I*2) If w_i belongs to W' and has AND output logic, U' contains all arcs of U which are incident out of w_i.

*I*3) If w_i belongs to W' and is a branching vertex of W, one and only one arc of U incident out of w_i belongs to U'.

*I*4) A vertex w_i with AND input logic belongs to W' if and only if all arcs of U incident into it in $G(W,U)$ belong to U'.

*I*5) A vertex w_i with exclusive-OR input logic belongs to W' if and only if one and only one arc of U incident into it in $G(W,U)$ belongs to U'.

*I*6) Any arc belonging to U' such that only one of its extremities belongs to W' is suppressed from U'.

If the above six conditions are satisfied, we define $A(W';U')$ as an *AND-type subgraph* of $G(W,U)$.[51]

We will say that a DAB graph is *legal* if

1) the graph has a single initial vertex and a single terminal vertex
2) the constructive rule *I*2 to *I*6 define an algorithm, i.e. always terminate in a subgraph $A(W';U')$ with W' and U' satisfying *I*1 to *I*6.

Since the number of AND-type subgraphs can be extremely large, rules have been defined to test the legality of a given graph.[52] We will not report them here but in order to give a flavor consider the example graph of Fig. 8(*a*).

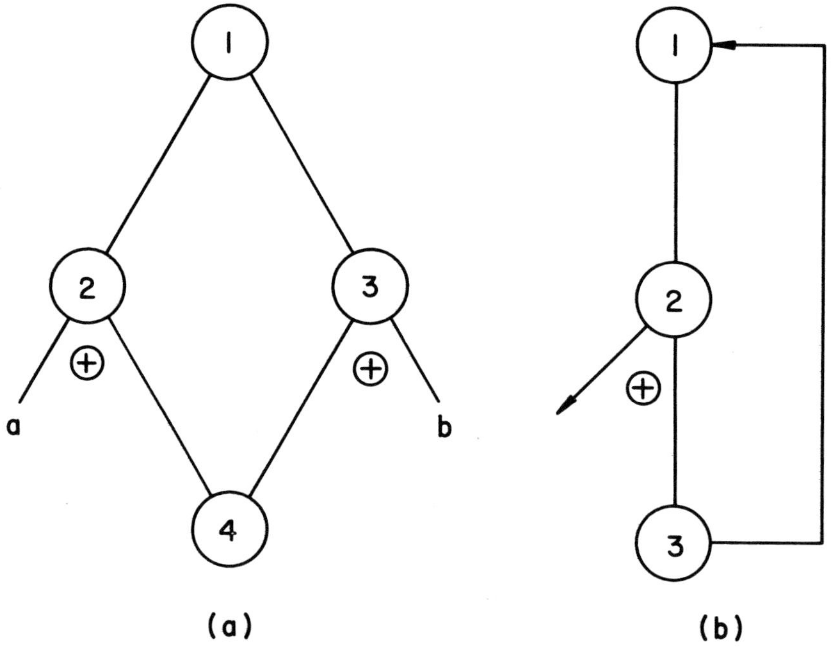

Fig. 8. Example of Illegal Graph

This graph is illegal since w_n being unique there must exist a common successor say w_k to w_2 and w_3. This successor can be reached if arc a and not arc b is traversed, or if arc b and not arc a is traversed, or if both arcs a and b are traversed. Therefore the input logic of w_k is of inclusive OR logic, and such a structure is not allowed. Practically this prevents some parallelization such as, "Test and see if at least one element of an array is 0, and if so go to vertex w_k." However such a statement translated into FORTRAN yields

```
        DO 10 I = 1, N
        IF (A[I]) 10, 20, 10
10      CONTINUE
           .
           .
           .
           .
20       . . . .
```

which cannot be parallelized (cf., "Transformation of Graphs for Parallel Processing"). This is because our original statement has been altered in the FORTRAN version in, "Find the first element of the array which is null," that which implies sequential processing and the graph of Fig. 8(*b*).

The first part of the legality requirement, i.e. unicity of w_1 and w_n, may imply some changes in the structure of $G(W, U)$ after removal of the feedback connections, since some pseudoheads or pseudotails may appear. This problem has been treated by E. C. Russell and J. L. Baer.[53,54]

The whole process of checking the legality of the DAB graph is represented by box 6 of Fig. 7.

First Prediction of Performance—Bounds for Maximum Parallelism

A rapid overview of the amount of parallelism shown in a DAB graph can be obtained by drawing a vertex histogram where the ordinate represents computation time, and the abscissa represents vertex relative importance (e.g. sorted in order of descending value).[55,56] It is then easy to recognize which, if any, are dominant vertices and to check their precedence relationship, thus giving a first insight into the amount of parallelism one can expect to achieve with a given configuration.

Moreover, one can give lower and upper bounds on the number of processors required for maximum parallelism. In Martin and Estrin,[57] algorithms are given for computing the mean path length of the graph where the vertices are scheduled on one processor (P_{uni}) and on as many processors as required for leaving the graph undistorted (P_{und}). That is to say, a graph is scheduled in an undistorted fashion if there is always a processor available when a computation can be initiated.

We have

$$P_{uni} = \sum_{i=1}^{n} p_i a_i t_i = \sum_{i=1}^{n} e_i t_i = \sum_{i=1}^{n} r_i$$

while P_{und} has to be computed through the inductive procedure

$$g_1 = t_1$$

$$g_i = p_i \left[a_i t_i + \max \left(\frac{(g_{i-j})}{p_{i-j}} \right) \right] \qquad \text{(AND-input)}$$

$$g_i = p_i a_i t_i + \sum_j g_{i-j} q_{i-j,i} \qquad \text{(EOR-input)}$$

where the w_{i-j} are the immediate predecessors of w_i or equivalently

$$g_1 = r_1$$

$$g_i = r_i + \max (g_{i-j}\, q_{i-j,i}) \qquad \text{(AND-input)}$$

$$g_i = r_i + \sum_j g_{i-j}\, q_{i-j,i} \qquad \text{(EOR-input)}$$

and $P_{und} = g_n$.

The lower bound is then

$$\alpha = \left\lceil \frac{P_{uni}}{P_{und}} \right\rceil .$$

In order to get the upper bound β, the basic idea is to find the cut in the graph with the maximum number of vertices such that all vertices determined by that cut can be performed in parallel. As we readily see, this procedure is complicated by the presence of mutually exclusive vertices. The detailed algorithm can be found in J. L. Baer and G. Estrin.[58]

Suppose $\alpha = 2$ (and in fact $P_{uni}/P_{und} = 1 + \epsilon$) and β rather large. A look at the histogram could show that there exists a small number of dominant vertices and recognition of these vertices would certainly reveal that they cannot be executed concurrently. If these vertices are representations of subroutines, they in turn can be represented by DAB graphs, therefore extending the original graph. If the histogram does not present dominant vertices there must be a long sequence of vertices which have to be executed serially, and this could be checked by computing the average number of immediate successors of vertices with AND-output logic. Then, either another approach to the algorithm must be taken or the situation is hopeless for parallel processing because of the nature of the computations involved.

At the other extreme, if α is not too small and β very large, there must be a large number of vertices with a very small average weight. (This can be checked very easily on the histogram.) Regrouping of these vertices into a smaller number of vertices, whenever possible, will allow less overhead time spent during both execution and generation of the program as well as minimization of possible transfers depending on the memory organization.

As a last example, suppose $\alpha = n + 1$, and that there are n dominant vertices, with no precedence relationship, in the histogram. Thus, these n vertices should have a large probability of having to be performed in the same AND-type subgraphs. As a consequence, one should see a decrease in mean path length inversely proportional to the number of processors until $n/2$ processors and then practically no decrease between $n/2$ and n processors.

This process is represented by boxes 7 and 8 of Fig. 7.

"A priori" Scheduling

One can view the compilation of a source program for execution on a homogeneous multiprocessor as the sequence of three basic procedures: recognition of the inherent parallelism in the program yielding for example a bilogic graph; scheduling of the computational tasks, as represented by a DAB graph drawn from the previous phase by a suitable acyclic-cyclic transformation, in a suboptimal fashion; and generation of machine language executable by the multiprocessor system. Here we are interested in the second phase.

In a series of works culminated by Martin's dissertation, the UCLA group implemented an "a priori" scheduling with a DAB graph as a starting point.[59,60,61,62] The function to minimize was the mean path length of the scheduled graph. An iterative scheduling was performed in the following fashion:

1) Perform an initial assignment by equalizing the number of vertices assigned to each processor.
2) Perform an initial sequencing on each processor, grouping the vertices by their number of predecessors and within these groups applying an urgency rule (the vertex with greater urgency is sequenced first).
3) Reassign single vertices and resequence with the same policy. If the mean path length is reduced, keep the new schedule or else restore the previous one.

Step (3) above was performed successively for all vertices for a given iteration. The urgency rule chosen was, "Urgency number proportional to the length of the longest path from the vertex to the terminal vertex." The number of iterations had to be kept low because of the time consuming procedure.

Some comments are appropriate here to give the reasons why this scheduling algorithm was not retained for our studies of parallelism.

With an "a priori" scheduling, one would like to perform, at compilation time, some of the overhead functions usually devoted to a dynamic scheduling scheme at execution time. In order to gain execution speed, this phase must be rapid, and thus a one-shot algorithm is mandatory. Moreover, some weaknesses appeared in the iterative algorithm such as:

1) The criterion function, i.e., the mean path length, is not very reliable when a large distortion is present (small number of processors).
2) The computing time of an iteration is proportional not only to the number of vertices but also to the number of processors.
3) The vertex exchange as described above is limited to the rescheduling of one vertex at a time (or possibly a cluster) from one processor to another while it could be more advantageous to swap several vertices.

On the other hand, the scheduling rule, and consequently the urgency rule, is of primary importance when performing a one-shot algorithm. In fact, some anomalies have been reported where some urgency rules can have the effect of increasing the running time of a particular job when the number of processors is augmented. Therefore, our goals were:

1) Perform a one-shot scheduling which can yield at least as good and generally better results than a purely dynamic schedule.
2) This algorithm must be fast and its rapidity independent of the number of processors.
3) Numerous urgencies should be tested such as:
   a) F.I.F.O.
   b) Local, that is dependent only on the vertex at hand and some local information on the graph.
   c) Global, that is dependent on the whole graph structure.
4) The algorithm must be such that it allows some tasks to be preassigned either to a given processor (e.g., in a variable structure environment), or to the same processor as a given predecessor (in order to meet the requirements of a feasible memory allocation).[63]
5) As a by-product, the algorithm will also give an estimation of the mean path length.

We can summarize the scheduling process as follows.[64] We will say that a vertex is *candidate* for scheduling if all of its immediate predecessors have been scheduled. A machine P_i is *available* if its *associated clock* C_i has value 0. All candidates are in a *queue* ranked by their urgency numbers. All available machines form a *pool.* When a machine is not in the pool, it means that there is at least one vertex *"running"* on it. The set of all vertices currently running on a machine P_i forms the set of *temporarily assigned* vertices on P_i.

The algorithm can be compared to a discrete event simulation. When a machine becomes available, the following scheduling rule is applied. Select the first candidate in the queue, say w_j, and temporarily assign it to P_i, incrementing P_i's clock by the number of cycles required by w_j, i.e. ($C_i = C_i + r_j$). Then find the first candidate in queue, if any, which is mutually exclusive with all temporarily assigned vertices on P_i. If such a candidate exists, say w_k, it becomes temporarily assigned to P_i ($C_i = C_i + r_k$), and the process is repeated.

In case of preassigned vertices, the above rule is biased to consider first all vertices preassigned on P_i, then all those which have no preassigned commitments.

When no machine is available, the one with the smallest running time left on it, τ, becomes available. The global clock is incremented (TIME = TIME + τ), all associated clocks decremented by τ, and the above scheduling rule is activated unless the queue is empty. In this latter case, or if all preassignments prevent any action, then another machine is freed by the same process. *Time* is a measure of the mean path length.

Boxes 9 and 11 of Fig. 7 represent this process. It was found that in order to get an improvement with regard to the dynamic scheduling, urgencies linked both to the structure of the graph and the weight of the vertices had to be taken into account.

Memory Allocation Algorithm

Bovet[65] has studied the problem of memory allocation with the aid of DAB graphs for two types of memory organization; central organizations where the processors share the pool of memory units which constitute the main memory, and local organization where each processor possesses its own memory. Starting with the scheduled graph and the name and dimension of each space taking entities associated with each vertex (input, local, output), the memory allocation algorithm (cf. box 10) performs an efficient package and detects possible main memory overflows. This latter detection may imply a new scheduling until compatibility is reached.

Once the scheduling has been performed, the control instructions must be generated. Of course now, the cyclic bilogic graph must be considered. We give the arcs three possible states namely IDLE, ENABLED and DISABLED, and in addition give two possible states to the vertices namely BLOCKED and AVAILABLE. We assume that upon execution of a vertex on a processor, control is given to an executive routine which has access to all information pertaining to the structure of the graph, the states of all arcs and vertices and the "a priori" scheduling. The executive routine has a sequence counter SEQCOUNT as well as an IDLE/BUSY switch associated with each processor. At initialization time, all processors are free, all arcs are in the IDLE state, all vertices in the AVAILABLE state. Then the initial vertex w_1, assigned on P_1, is executed with SWITCH (1) turned on

BUSY and from then on the OUTEX procedure described below is called as after the execution of any vertex.

OUTEX Procedure

Let w_i be the vertex which has been executed and j the index of the processor on which it is assigned.

1) The state of w_i is changed from AVAILABLE to BLOCKED. All arcs incident into w_i take an IDLE state.
2) If w_i has:
   a) AND-output logic.
   All arcs incident out from w_i take the state ENABLED.
   b) EOR-output logic.
   The arc chosen during execution (this may alter slightly the code generated for the branching vertices in order for the executive to be able to recognize arcs instead of transfer instructions) becomes ENABLED, and all other arcs take DISABLED states.
3) If w_i is head of a cycle, say C_i, then: For each machine $k \neq j$, let SEQ(i+) be the sequence attribute of the vertex w_m, if any, such that w_m is a successor of w_i, belongs to C_i, and such that there exists no vertex belonging to C_i, successor of w_i, assigned on machine k, and with a sequence attribute less than SEQ (m). Then, if w_m exists, the sequence counter associated with machine k takes the value:

   SEQCOUNT (k) = min (SEQCOUNT [k], SEQ [i+]).

   Also the states of all vertices belonging to C_i become AVAILABLE.
4) If w_i is tail of a cycle C_i of which w_h is the head, and if the feedback arc (w_h, w_i) is chosen then:
   a) PROC(i)= PROC(h) then SEQCOUNT(j)= SEQ(h) – 1.
   b) PROC(i) = $k \neq j$, then SEQCOUNT(k) = min (SEQCOUNT[k], SEQ[h]). In both cases, the state of w_h becomes AVAILABLE.
5) SEQCOUNT(j) = SEQCOUNT(j)+ 1.
   SWITCH(j) = IDLE.

At the end of an OUTEX procedure—as well as the end of an INTEX procedure—the executive scans in turn the states of the SWITCH. For any processor j which is free the INTEX procedure described below is called. If all processors have been scanned without bringing any vertex into execution, the executive is left until it is triggered again by a call to an OUTEX procedure.

INTEX Procedure

Let w_i be the vertex such that PROC(i) = j, and SEQCOUNT(j) = SEQ(i).

1) If the state of w_i is BLOCKED then:
   All arcs incident into w_i take IDLE states.
   All arcs incident out from w_i take DISABLED states.
   SEQCOUNT(j) is increased by 1 and the procedure INTEX is called again.
2) If the state of w_i is AVAILABLE, and w_i is of AND-input logic:
   a) If any arc incident into w_i is of state DISABLED, then w_i takes a BLOCKED state.
   All arcs incident into w_i take IDLE states unless these arcs are delayed entries, in which case they keep their former state. All arcs incident out from w_i take a DISABLED state. SEQCOUNT(j)is increased by 1, and the procedure INTEX is called again.
   b) If all arcs incident into w_i have an ENABLED status then: execution of w_i is initiated (the term execution is voluntarily vague and may mean for example loading and execution, or setting an instruction counter and passing from a "wait" state to a "run" state).
   SWITCH(j)= BUSY
   c) If no incident arc is DISABLED, and at least one is IDLE, the procedure INTEX is called again.
3) If the state of w_i is AVAILABLE, and w_i is of EOR-input logic:
   a) If all arcs incident into w_i are DISABLED, cf. (2*a*) above.
   b) If one arc incident into w_i is ENABLED, cf. (2*b*) above.
   c) If at least one arc incident into w_i has an IDLE state, and no arc incident into w_i has an ENABLED state, then cf. (2*c*) above.
4) If the situation described in (3*a*) above arises, and w_i is head of a cycle C_i, then in addition all vertices of the cycle C_i which are not termini of delayed entries take a BLOCKED state.

The two procedures described above form the core of a possible executive routine. Naturally more instructions may be necessary depending on memory structure and I/O procedures. However, these two routines also describe how a simulation program for the cyclic graph could be written. They present some similarity with the dynamic schedules of Lampson.[66] The process is summarized in box 12 of Fig. 7.

(Note that only boxes 1–9 and box 11 have been at present implemented.)

CONCLUSION

In this paper, we have reviewed the U.C.L.A. approach in modeling computer programs for parallel processing systems. The main originality and the strongest feature of the procedure is that, starting with already existing programs written in FORTRAN, one can automatically derive the model and predict the performance of the program if run on a system of N homogeneous parallel processors.

The type of parallel processing recognition demonstrated is by its very nature limited. The technique of starting from a program written for a single processor has natural deficiencies. The difficulty with subscript analysis in the recognition of potentially replicative loops prevents much parallelism which could perhaps be more naturally expressed in a language designed for parallel processing. Another difficulty is the "linearizing" of independent sections of the program via explicit transfers of control.

The positive value of the procedure is the fact that an automatic analysis of existing programs can be performed without any reprogramming. Also the graph model and the transformation performed provide a great deal of insight into the program structure. The procedure, applied to single-processor programs, could be embedded into compilers thereby providing some optimization with regard to elimination of unreachable code, minimization of control transfers etc.

As for any model, a certain number of simplifying assumptions were made for both the prediction of performance and the "a priori" scheduling (e.g. mean time values for the computational task, cyclic-acyclic transformation). However, since the starting point is a real program, they are much less stringent, for example, than those used in queueing theory.

Finally, the question of the usefulness of such a modeling project may arise. The principal argument in its favor is the fact that it is feasible to start from a source program written in FORTRAN and go through all steps described in Figs. 1 and 7. Thus, it is an efficient means to test if computer programs written in sequential languages are amenable to general parallel processing. If the answer is negative, the next step should be to test sequential languages augmented by a few instructions of the FORK and JOIN type. If again the answer is negative, and there is a good chance that it will be so if the basic algorithm has not been changed, then the programming philosophy must be rethought. Some studies have already been started in this direction for general parallel processor systems and the Illinois Group for the ILLIAC IV.[67,68,69] However, our approach is certainly valuable for an interim period due to the magnitude of the library of FORTRAN programs. It will also give a clear insight into the problems encountered when designing software for the multiprocessor system.

References

1. R. M. Karp and R. E. Miller, "Properties of a Model for Parallel Computations: Determinacy, Termination, Queuing," *JSIAM,* 14: (November, 1966), 1390–411.
2. J. Rodriguez, "A Graph Model for Parallel Computations" (Sc. D. thesis, Massachusetts Institute of Technology, 1967).
3. M. R. Conway, "A Multiprocessor System Design," *AFIPS FJCC,* 24: (1963), 139–46.
4. J. C. Murtha, "Highly Parallel Information Processing Systems," *Advances in Computers,* 7: (1966), 1-116.
5. D. J. Kuck, "ILLIAC IV Software and Applications Programming," *IEEE Transactions on Electronic Computers,* C-17: (August, 1968), 758-69.
6. A. Opler, "A Procedure-Oriented Language Statement to Facilitate Parallel Processing," *CACM,* 8: (May, 1965), 306-7.
7. J. A. Gosden, "Explicit Parallel Processing Description and Control in Programs for Multi- and Uni-Processor Computers," *AFIPS FJCC,* 29: (1966), 651-660.
8. L. G. Tesler and H. J. Enea, "A Language Design for Concurrent Processes," *AFIPS SJCC,* 1968, pp. 403-8.
9. J. P. Anderson, "Program Structures for Parallel Processing," *CACM,* 8: (December, 1965), 786-88.
10. A. J. Bernstein, "Analysis of Programs for Parallel Processing," *IEEE Transactions on Electronic Computers,* EC-XV, (October, 1966), 757-62.

11. G. Estrin, "Organization of Computer Systems–The Fixed Plus Variable Structure Computer," *Proc. WJCC,* 17: (1960), 33–40.
12. R. Turn, "Assignment of Inventory in a Variable Structure Computer" (Ph.D. dissertation, University of California, Los Angeles, 1963).
13. R. L. Mandell, "Tools for the Construction of Design Automation Systems" (Ph.D. dissertation, University of California, Los Angeles, 1968).
14. H. Potash, "A Digital Control Design System" (Ph.D. dissertation, University of California, Los Angeles, 1969.
15. B. Bussell, "Properties of a Variable Computer System in the Solution of Parabolic Partial Differential Equations" (Ph.D. dissertation, University of California, Los Angeles, 1962).
16. D. F. Martin, "N-th Root Computing Methods" (M.S. thesis, University of California, Los Angeles, 1962).
17. R. L. Mandell, "A Study of Some Methods of Performing Aitken-Neville Bivariate Interpolation with a Fixed-Plus-Variable Structure Computer" (M.S. thesis, University of California, Los Angeles, 1963).
18. G. R. Wood, "Application of a Restructurable Computer System to Numerical Weather Prediction Computations" (M.S. thesis, University of California, Los Angeles, 1966).
19. Turn, *op cit.*
20. D. F. Martin, "The Automatic Assignment and Sequencing of Computations on Parallel Processor Systems" (Ph.D. dissertation, University of California, Los Angeles, 1966).
21. J. -L. Baer, "Graph Models of Computations in Computer Systems"(Ph.D. dissertation, University of California, Los Angeles, 1968).
22. D. P. Bovet, "Memory Allocation in Computer Systems" (Ph.D. dissertation, University of California, Los Angeles, 1968).
23. E. C. Russell, "Automatic Program Analysis" (Ph.D. dissertation, University of California, Los Angeles, 1969).
24. Turn, *op. cit.*
25. Martin, "The Automatic Assignment . . . ," *op. cit.*
26. Baer, *op. cit.*
27. Bovet, *op. cit.*
28. R. T. Prosser, "Applications of Boolean Matrices to the Analysis of Flow Diagrams," *AFIPS EJCC,* 1959.
29. Turn, *op. cit.*
30. Rodriguez, *op. cit.*
31. Bussell, *op. cit.*
32. Turn, *op. cit.*
33. Martin, "The Automatic Assignment . . . ," *op. cit.*
34. Wood, *op. cit.*
35. E. C. Russell, "Automatic Assignment of Computational Tasks in a Variable Structure Computer" (M.S. thesis, University of California, Los Angeles, 1963).

36. G. Estrin *et al.*, "SNUPER COMPUTER–A Computer in Instrumentation Automaton," *AFIPS SJCC*, 30: (Spring, 1967).
37. E. C. Russell and G. Estrin, "Measurement Based Automatic Analysis of FORTRAN Programs," *AFIPS, SJCC*, 1969.
38. Estrin *et. al.*, "SNUPER COMPUTER . . . ," *op. cit.*
39. Russell, "Automatic Program . . . ," *op. cit.*
40. Turn, *op. cit.*
41. Martin, "The Automatic Assignment . . . ," *op. cit.*
42. Bernstein, *op. cit.*
43. H. W. Bingham, D. A. Fisher, and W. L. Semon, "Detection of Essential Ordering Implicit in Compiler Language Programs," TR-67-1 (Burroughs Corporation, February, 1967).
44. D. F. Martin and G. Estrin, "Models of Computational Systems, Cyclic to Acyclic Graph Transformations," *IEEE Transactions on Electronic Computers*, EC-16: (February, 1967), 70–9.
45. Estrin *et. al.*, "SNUPER COMPUTER . . . ," *op. cit.*
46. Martin and Estrin, "Models of Computational Systems . . . ," *op. cit.*
47. J. L. Baer and G. Estrin, "Frequency Numbers Associated with Directed Graph Representations of Computer Programs" *2nd Hawaii International Conference on Systems Sciences*, 1969, pp. 139–42.
48. D. F. Martin and G. Estrin, "Models of Computations and Systems-Evaluation of Vertex Probabilities in Graph Models of Computations," *JACM*, 14: (April, 1967), 281-99.
49. J. L. Baer, D. P. Bovet, and G. Estrin, "Legality and Other Properties of Graph Models of Computations," JACM, 1970.
50. Baer, Bovet, and Estrin, *op. cit.*
51. Martin, "The Automatic Assignment . . . ," *op. cit.*
52. Baer, Bovet, and Estrin, *op. cit.*
53. Russell, "Automatic Program Analysis," *op. cit.*
54. Baer, *op. cit.*
55. Russell, "Automatic Program Analysis," *op. cit.*
56. Russell and Estrin, *op. cit.*
57. D. F. Martin and G. Estrin, "Path Length Computations on Graph Models of Computations," *IEEE Trans on Computers*, C-18 (June 1969) pp. 530–536.
58. J. L. Baer and G. Estrin, "Bounds for Maximum Parallelism in a Bilogic Graph Model of Computations," *IEEE* Trans on Computer C-18 (November 1969) pp. 1012–1014.
59. Turn, *op. cit.*
60. Russell, "Automatic Assignment of Computational . . . ," *op. cit.*
61. Martin, "The Automatic Assignment . . . ," *op. cit.*
62. D. F. Martin and G. Estrin, "Experiments on Models of Computations and Systems," *IEEE Transactions on Electronic Computers*, EC-16: (February, 1967), 59–69.
63. Bovet, *op. cit.*

64. J. L. Baer and G. Estrin, " 'A Priori' Scheduling of Bilogic Graph Models of Computations," *INTERNET 1969,* pp. 119-127.
65. Bovet, *op. cit.*
66. B. W. Lampson, "A Scheduling Philosophy for Multiprocessing Systems," *Comm. Ass. Comp. Mach.* 11: (May, 1968), 347-60.
67. G. S. Shedler, "Parallel Numerical Methods for the Solution of Equations," *Comm. Ass. Comp. Mach.,* 10: (May, 1967), 286-91.
68. J. Nievergelt, "Parallel Methods for Integrating Ordinary Differential Equations," *Comm. Ass. Comp. Mach.* 7: (December, 1964), 731-33.
69. Kuck, *op. cit.*
70. J. M. S. Simoes-Peroira, "On the Boolean Matrix Equation $m = Vi^d = {}_1M^1$," *JACM* 12: (July 1965) 376-382.

CHAPTER 19

PARALLELISM EXPOSURE AND EXPLOITATION

Earl W. Reigel

Burroughs Corporation
Paoli, Pennsylvania

Abstract

An algorithm is presented for detecting parallelism between compiler language statements in a serially written program. A hardware mechanism is described that controls the sequencing of the partially ordered instructions using the order information generated by the detection algorithm. This mechanism permits initiation of concurrently executable instructions and is asynchronous in that there is no inessential fixed order of instruction initiation and the sequencing control is independent of variable instruction duration.

INTRODUCTION

At any point in time, the speed of digital circuits is limited by economic or feasibility considerations. As a result, the study of parallelism has received a large amount of attention as an alternative solution to increase computer power. The potential advantages gained from parallelism include:

1) increased throughput
2) expandability
3) reliability

The research reported in this paper was supported in part under contract DA-28-043-AMC-02463(*E*) by the United States Army Electronics Command.

The help of H. W. Bingham and D. A. Fisher, co-workers at Burroughs Corporation, is gratefully acknowledged.

4) reduced waiting time
5) reduced memory capacity
6) reduced memory-time product.

Another motivation for the study of parallelism is the advances in component technology which emphasize the use of regular structures. The technology advances also result in reduced logic costs. The relative costs of logic and memory dictate the use of more logic if it reduces memory requirements.

There are several levels of parallelism to be studied:

1) nanosequencing
2) microprogramming
3) instruction parallelism
   a) within expressions
   b) between statements
4) task parallelism
   a) routines (co-, sub-, parallel-)
   b) repeat statements
5) multiprogramming.

The study of parallelism may be separated into the following areas:

1) existence
2) exposure
   a) explicitly specified by programmer
   b) detected by program
3) exploitation
   a) hardware techniques
   b) software methods.

This paper is concerned with the exposure and exploitation of parallelism at the instruction level; in particular, the parallelism existing between statements.

EXPOSURE

At the instruction level automatic detection methods should be used for the following reasons:

1) Too great a burden would be placed on the programmer to determine and explicitly express all the parallelism existing in the problem.

2) The concurrent instruction execution relation is not transitive. This fact plus the added labor for choosing between alternatives may lead to many errors.
3) The overhead costs (development and execution) of incorporating effective detection algorithms into compilers is not significant and therefore would provide a more efficient solution.
4) Programs already written need not be rewritten but simply recompiled.

The algorithm described here for automatic detection of implicit parallelism at the between-statement level may be extended and applied at higher levels. In the following, a process is defined as being composed of an address space, a set of transformations, and control information describing the order in which the transformations are to be performed. A process may be interpreted at several levels (e.g., statement, routine, program). The input and output sets in the address space of a process are used to determine the partial order between processes. An implementation of the algorithm has been used to detect parallelism at the between-statement level. The implementation program is written in B5500 Extended ALGOL and has been executed on the B5500 system at Burroughs Corporation, Paoli, Pennsylvania.

The block diagram of the program (Fig. 1) shows the major sections of the program and the outputs of these sections. Figure 2 contains a list of the symbols used in the description of these sections.

Language Recognizer

The function of the language recognizer is to scan the character string of an input program, recognize declared symbols in the input program, and generate a corresponding program in an intermediate language. Prior to the input of the program to be analyzed, however, a description of the source language in which the program is written must be entered. The source language for the programs analyzed is a subset of Burroughs B5500 Extended ALGOL.

The operation of the language recognizer is initiated by a call made by the process generator. Each call causes the generation and return of one instruction in the intermediate language. Each intermediate language instruction is composed of (1) an operand that is the internal representation of some symbol in the input program, and (2) an operator that describes the semantic category of the operand.

Process Generator

The function of the process generator is to transform the intermediate language components received from the language recognizer into processes. A specification of the level of parallelism analysis to be

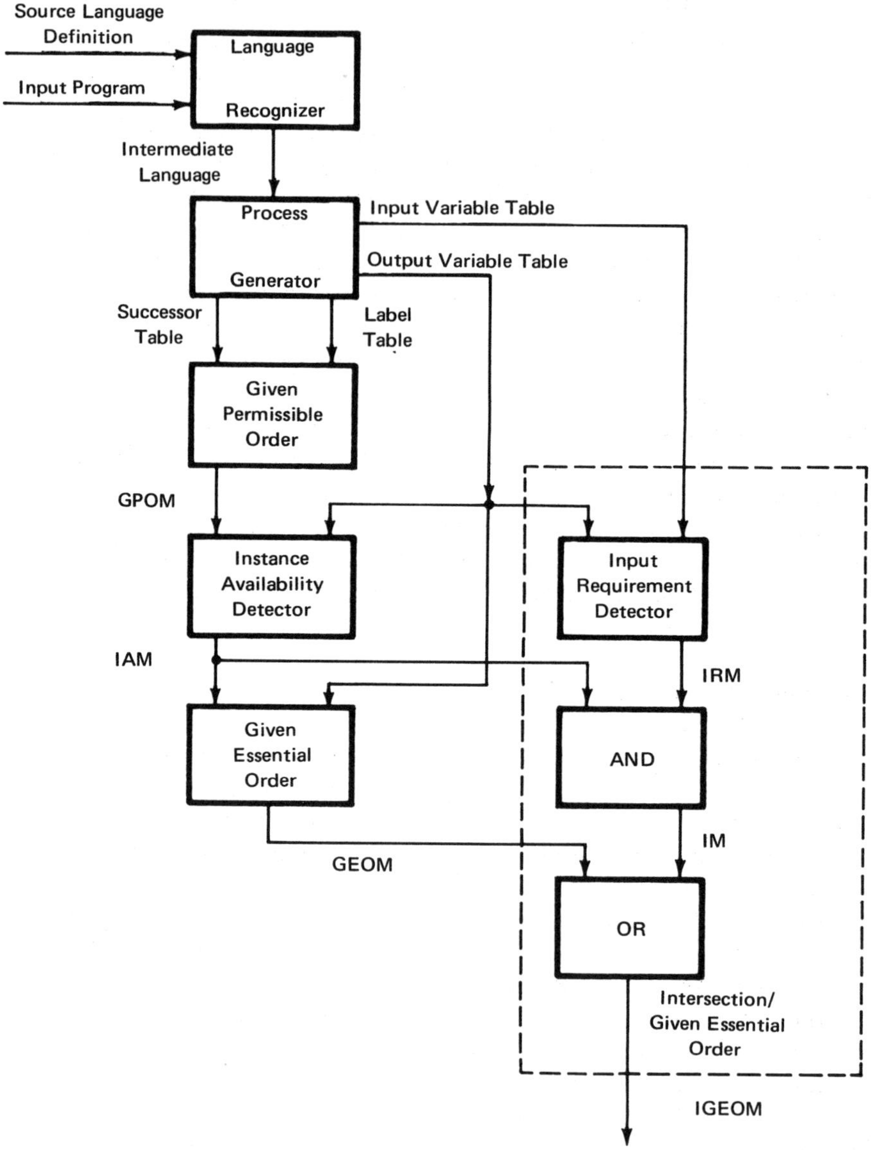

Fig. 1. Block Diagram of Detection Program

applied is also required to define the processes. The level of analysis described here is implicitly specified at the statement level. Each unconditional statement in the input program is considered as a process, and each conditional statement is considered as three processes:

1) a process that evaluates the condition
2) a process for the false alternative
3) a process for the true alternative.

Each process is represented in the following set of tables. The output variable table contains for each process its output set. The input variable table contains for each process its input set. The successor table contains for each process either a single unconditional successor process or a pair of true and false conditional successor processes. The successor processes are identified either by a process number or a label. The label table relates each label to the process number of its associated process.

After the operation of the language recognizer/process generator on the entire input program, the formation of the processes is complete. The parallelism analysis is begun using the information in the input, output, successor, and label tables.

Permissible Order

The function of the permissible order is to transform the given permissible ordering information from table form (successor and label

GPOM	= given permissible order matrix
IAM	= instance availability matrix
GEOM	= given essential order matrix
IRM	= input requirement matrix
IM	= intersection matrix
IGEOM	= intersection/given essential order matrix
MASK	= temporary column vector
$M(i, k)$	= element at row i column k of matrix M
$M(*, k)$	= column vector k of M
$M(i, *)$	= row vector i of M
$M(*, *)$	= matrix M
$\phi(i)$	= set of output variables of process i
$I(k)$	= set of input variables of process k
N	= number of processes
v	= variable in $\phi(*)$ or $I(*)$
$\langle i, v \rangle$	= variable v of process i
TRU	= pseudo-output of head process of conditional true branch
FALS	= pseudo-output of head process of conditional false branch

Fig. 2. Symbols Used in Algorithm Description

tables) into Boolean matrix form. In addition to this given permissible order matrix (GPOM), a list of backward arcs is formed to identify loops. The GPOM is initially empty (all false or zeros).

The method of operation is to sequence through the successor table (referencing the label table when necessary), and at each step examine the process number (i) and its successor process number (j). If $i < j$ then insert a "1" into the given permissible order matrix at position (i,j). If $i \geq j$, then this represents a backward arc, and the pair of process numbers are placed on the list of backward arcs. The process numbers assigned to the given permissible order are such that the Boolean matrix is upper triangular except for backward arcs.

Instance Availability Detector

The function of the instance availability detector is to use the process outputs (from the output variable table) and the given permissible order matrix (GPOM) to determine all instances of variables which could be available when each process is executed.

The resulting information is represented in an instance availability matrix (IAM) that contains one column for each process (indexed by the process number) and one row for each output instance (indexed by the process number and the variable name). The information in this matrix may be interpreted as follows. The instance of variable v produced by process P is assumed to index row i. If the entry at row i and column k of the matrix is true, "1," then in the serial execution of the program the value of variable v produced by process P could be available when process number k is executed. If the entry is false, the value of variable v produced by process P cannot be available to process k.

The method of operation of the instance availability detector is shown in Fig. 3. The instance availability matrix (IAM) is initially empty (all false). Each process P_k is considered, one at a time, in increasing order of process numbers. A column vector of instances of variables which could be available to direct successors of P_k after the execution of P_k is determined. Since the processes are being considered in the order of increasing process numbers, all predecessors of P_k will have been considered before P_k. Thus, column k of the IAM will contain all instances of variables which could be available to P_k (whether in the input set of P_k or not). The instances of variables available from P_k is formed as all instances of variables available to P_k, with

```
IAM (*, *) ← 0

for k ← 1 step 1 until N - 1 Do
    MASK (*) ← IAM (*, k)
    for each v ∈ φ (k) Do
        MASK (<k, v>) ← true
        if v ≠ TRU and v ≠ FALS then
            for i ← 1 step 1 until k - 1 Do
                if v ∈ φ (i) then
                    MASK (<i, v>) ← false
        for i ← k + 1 step 1 until N Do
            if GPOM (k, i) then
                IAM (*, i) ← IAM (*, i) or MASK (*)
```

Fig. 3. Instance Availability Matrix Algorithm

the exception of those for which another instance of the variable unconditionally appears in the output set of P_k. The output set of P_k is also included in the instances of variables available from P_k. This column vector from P_k is then logically ORed into columns of the IAM corresponding to each direct successor (as indicated in GPOM) of P_k. Repetition of this sequence for each process completes the IAM.

Given Essential Order

The function of the given essential order is the generation of the given essential order matrix (GEOM) from the instance availability matrix (IAM) and the output variable table. The given essential order matrix (GEOM) reflects the essential ordering resulting from processes being in one but not both branches of a conditional statement.

The method of operation is shown in Fig. 4. Recall that each conditional has been assigned three process numbers: let them be P_{i-1}, P_i and P_{i+1}. P_{i-1} computes the condition and has the other two as direct successors. P_i has a pseudo-output variable FALS and is a successor to the false alternative of the condition. P_{i+1} has a pseudo-output variable TRU and is a successor to the true alternative of the condition.

For any process P_k:

1) if the variable FALS output from P_i is available to process P_k and the variable TRU output from P_{i+1} is not available to P_k then P_i is essentially ordered before P_k.
2) If the variable TRU output from P_{i+1} is available to process P_k and the variable FALS output from P_i is not available to P_k, then P_{i+1} is essentially ordered before P_k.
3) P_{i-1} is essentially ordered before P_i and before P_{i+1}.

These three steps are performed for each conditional in the program and the resulting information is represented in GEOM.

Intersection/Given Essential Order

The function of the intersection/given essential order is to determine the essential order in a program from the instance availability matrix (IAM), the given essential order matrix (GEOM), and the input and output variable tables. The result is the intersection/given essential order matrix (IGEOM).

```
for i ← 1 step 1 until N Do
    if FALS ∈ φ (i) then
        GEOM (i - 1, i) ← true
        GEOM (i - 1, i + 1) ← true
        for k ← i + 2 step 1 until N Do
            if IAM (<i, FALS>, k) and not IAM (<i + 1, TRU>, k) then
                GEOM (i, k) ← true
            if IAM (<i + 1, TRU>, k) and not IAM (<i, FALS>, k) then
                GEOM (i + 1, k) ← true
```

Fig. 4. Given Essential Order Matrix Algorithm

```
IGEOM (*, *) ← GEOM (*, *)
IRM (*, *) ← 0
IM (*, *) ← 0
for k ← 2 step 1 until N Do
    for each v ∈ I (k) Do
        for i ← k - 1 step - 1 until 1 Do
            if v ∈ φ (i) then
                IRM (<i, v>, k) ← true
                if IAM (<i, v>, k) = true then
                    IM (<i, v>, k) ← true
                    IGEOM (i, k) ← true
```

Fig. 5. Intersection/Given Essential Order Algorithm

The intersection/given essential order algorithm is shown in Fig. 5 and functions as follows:

1) Compute directly from the process descriptions an input requirement matrix (IRM). The IRM has the same row and column dimensions and designations as the IAM. The IRM is determined by examining the input variables required for each process P_k. For each input variable v required by process P_k, the output sets of all possible predecessor processes $P_i (i < k)$ are examined. If variable v is in the output instance set of P_i (i.e. $v \in \phi(i)$), then true (1) is entered in IRM at row $\langle i,v \rangle$, column k.
2) Determine the entire intersection matrix (IM) as the Boolean component-by-component ANDing of the IAM and the IRM.
3) Determine the intersection/given essential order relation (IGEOM) by Boolean component-by-component ORing the given essential order matrix (GEOM) and the modified intersection matrix (IM). The intersection matrix (IM) as computed has row indices of the form (i,v) instead of process number i. ORing all rows of IM (i, v) with the same i forms the single row of IGEOM representing process i.

The IGEOM shows (1) the process partial order, (2) the data paths between processes, and (3) the extent of different branches of conditionals. This last point is of value in machine implementation since when a conditional is executed it is desired to quickly determine which set of processes need not be executed.

EXPLOITATION

Since the time required to control the sequencing of events should be small compared with the time for executing the events, hardware techniques (as opposed to software methods) are required to efficiently exploit the parallelism at the instruction level. For example, in conventional machines it is generally desirable to have a hardware instruction counter rather than to have to store and retrieve the "next instruction address" to and from memory.

A hardware mechanism has been designed to control the sequencing of partially ordered instructions in a parallel processing system. This mechanism permits initiation of instructions as soon as all predecessor instructions have been executed. The mechanism is asynchronous in the sense that there is no inessential fixed order of instruction initiation and the sequencing control is independent of variable instruction duration. The order control information used by the mechanism is represented in Boolean matrix form as generated by the detection algorithm. This matrix (IGEOM) contains the order control (precedence) information between every pair of processes, the information for passing data between processes, and the information necessary to quickly identify the set of processes in one but not both branches of a conditional.

This mechanism may be used in a sequencing unit of a parallel processing system to control the sequencing of instructions and to allow the exploitation of parallelism. When a segment of program is to be executed, its set of instructions and order control information (IGEOM) is sent to the sequencing unit. The order control information (IGEOM) is transferred to a set of flip-flops arranged in array form. This flip-flop array (also called IGEOM) and the associated E, A, IP, and I Boolean vectors are shown in Fig. 6 and in more detail in Fig. 7.

The E (executed), A (allowable), IP (in progress), and I (initiable) vectors are each of length n, where n is the number of instructions in the segment. Each position of these vectors is associated with an instruction and has the following interpretation:

1) E_i = 1/0: Instruction i has/has not been executed.
2) A_i = 1/0: All/some predecessor instructions of instruction i have/have not been executed.
3) IP_i = 1/0: Instruction i is/is not in progress (initiated but not completed).
4) I_i = 1/0: Instruction i may/may not be initiated.

Within the IGEOM matrix, a bit set (to 1) at row i, column j means that process i must precede process j in execution. In addition to

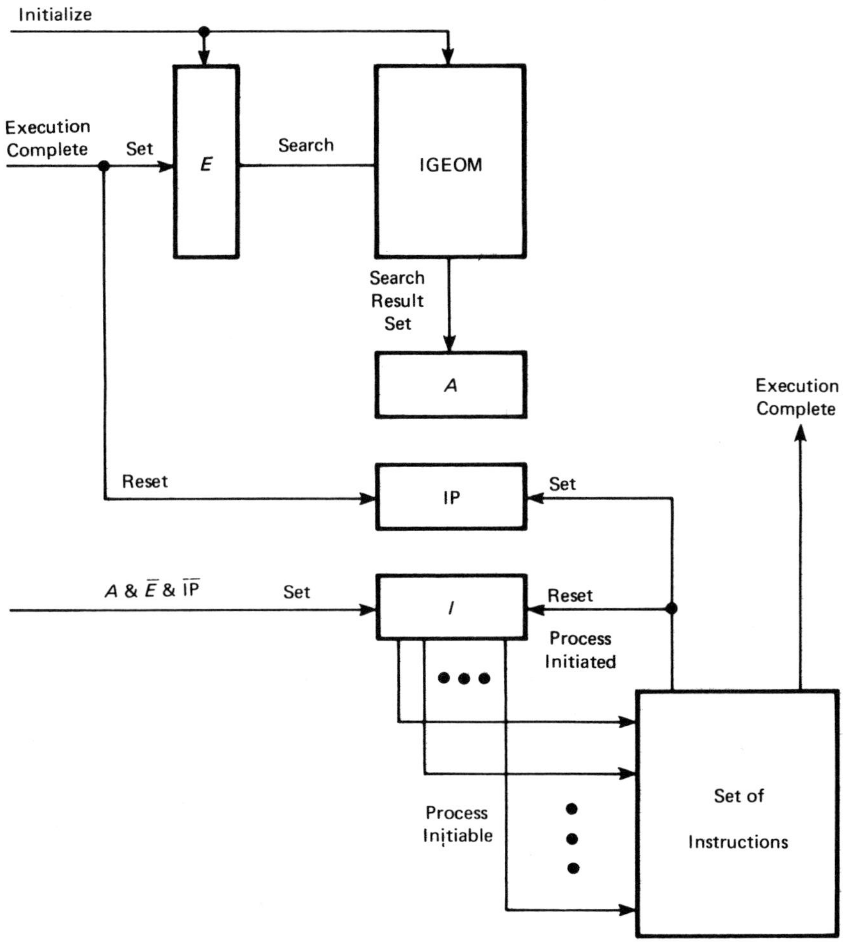

Fig. 6. Sequencing Unit, Block Diagram

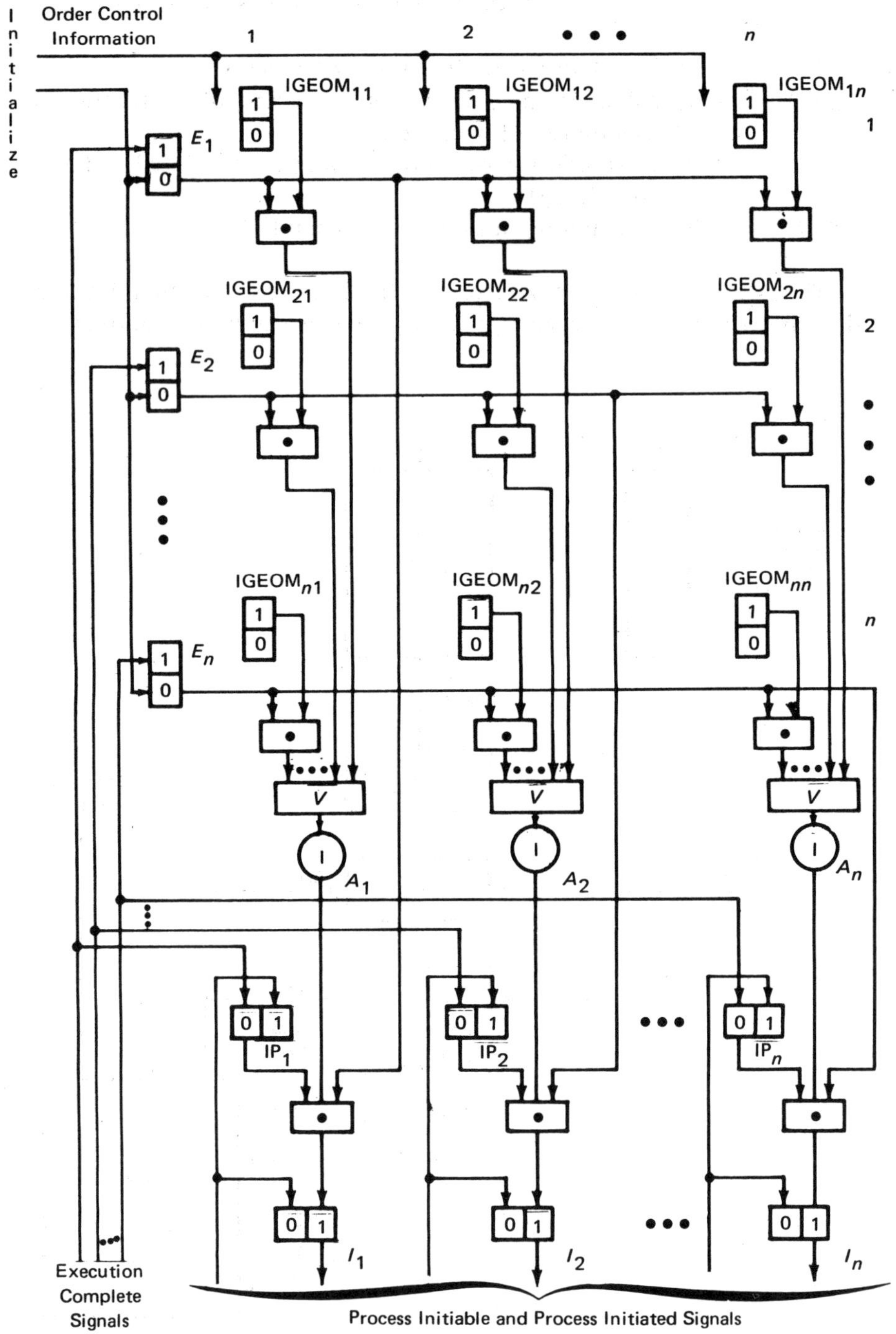

Fig. 7. Sequencing Mechanism

providing the precedence information for sequence control, the IGEOM matrix also provides information for passing of operand values. A bit set at position (i,j) also means that the output of instruction i can be used as input to instruction j.

When a set of instructions and the associated IGEOM matrix information is sent to a sequencing unit, the E, IP, and I vectors are set to "0." To begin processing, there must be a (nonnull) set of first processes (associated column of IGEOM is all zeros). When an instruction has been executed, the instruction number (let us say j) is used to set the appropriate bit in E (E_j) and to reset the associated bit in IP (IP_j). Then the following two operations are performed simultaneously:

1) The jth row of IGEOM is searched for ones. For each column which has a one in the jth row there is an instruction that requires the result value just computed. This value is passed simultaneously to the operand positions of all identified instructions in the instruction set.
2) The "new" E vector is used to search the IGEOM matrix column by column (all columns simultaneously). Any column which has a zero everywhere that E has a zero indicates an instruction which is allowable in the sense that all of its predecessors have been executed. For all columns satisfying this search, the associated positions in vector A are a one. All instructions whose position in A is a one and bit in E is a zero (not executed) and bit in IP is zero (not in progress) are initiable and hence their I bit is set (i.e. $A\&\overline{E}\&\overline{IP} \rightarrow I$). A more detailed algorithm for this step is now stated which includes the case of conditionals.

The following procedure will produce the next set of processes that are ready for initiation of execution (initiable):

1) When execution of process j is completed, set the associated bit position E_j in the executed vector E and reset the corresponding bit position IP_j in the in progress vector IP.
2) If every position of the E vector is set (all processes have been executed), then exit the procedure; otherwise,
3) If the process just executed is a conditional, then all processes in the branch not chosen are made to appear to have been executed as follows; otherwise, go to step (4).
   a) Construct a temporary search vector with search criteria "zero" in the process position associated with the chosen

branch of the conditional, and "one" in the process position of the branch not chosen.

b) Search for a match in the rows of IGEOM corresponding to these two positions, masking all other positions. This produces in vector A, ones in the positions corresponding to processes that are exclusively in the branch not chosen.

c) OR vector A (from step [3b]) with the vector E and store into E. This causes the processes in the not chosen branch to appear to have been executed.

d) In addition, set in E the two positions following the conditional. If j is the conditional, then set E_{j+1} and E_{j+2}, the pseudo-processes representing alternative path heads.

4) Using the vector E, search the IGEOM matrix column by column (all columns simultaneously) for columns whose row positions are zero everywhere that vector E is zero. A process whose column satisfies the search criterion is considered to be allowed since all of its predecessor processes have been executed. This search produces a vector A of allowed processes. The output A is a vector indicating the set of processes that are not waiting for any other process to be completed. However, this also includes all processes already executed and all processes presently in progress. Therefore:

5) For each process position j, if E_j is zero (j is not executed) and IP_j is zero (j is not in progress) and A_j is one (j is allowable), then set I_j. The resulting vector I now indicates the set of processes that are ready for initiation of execution (initiable).

6) The vector I produced by step (5) may then be used to select from the list of all processes the subset ready for initiation of execution.

7) When process j is actually initiated, IP_j is set to one (now in progress) and I_j is reset to zero (not to be reinitiated). An example of this procedure will be shown in the next section.

EXAMPLE

This section illustrates the function of the parallelism detection program and the partial order sequencing mechanism. The input program described here is a conditional example constructed for explication. The functions performed by the various sections of the parallelism detection program (See Fig. 1.) are illustrated by arrays (program output). The arrays are actually Boolean (i.e. contents are either 0 or 1); however, in the program output, the 0's are suppressed and the 1's are translated to the units digit of the associated column for expository purposes.

The flow chart of the program and the flow diagram illustrating the results (process descriptions) of the language recognizer and process generator sections are shown in Fig. 8. The following program outputs are shown in Fig. 9:

1) (*a*)–the given permissible order matrix (GPOM)
2) (*b*)–the instance availability matrix (IAM)
3) (*c*)–the given essential order matrix (GEOM)
4) (*d*)–the input requirement matrix (IRM)
5) (*e*)–the intersection matrix (IM)
6) (*f*)–the intersection/given essential order matrix (IGEOM)

This example is also used to demonstrate the operation of the sequencing mechanism. The intersection/given essential order matrix (IGEOM) developed from the parallelism detection program, is shown in Fig. 9(*f*) and is used in the sequencing mechanism.

The function of the sequencing mechanism is described using "snapshots" of the *E*,*A*,IP, and *I* vectors showing each step of the operation of the mechanism (Fig. 10). Recall that the *E* vector represents the executed processes, the *A* vector represents the allowable processes, the IP vector represents the processes in progress of being executed, and the

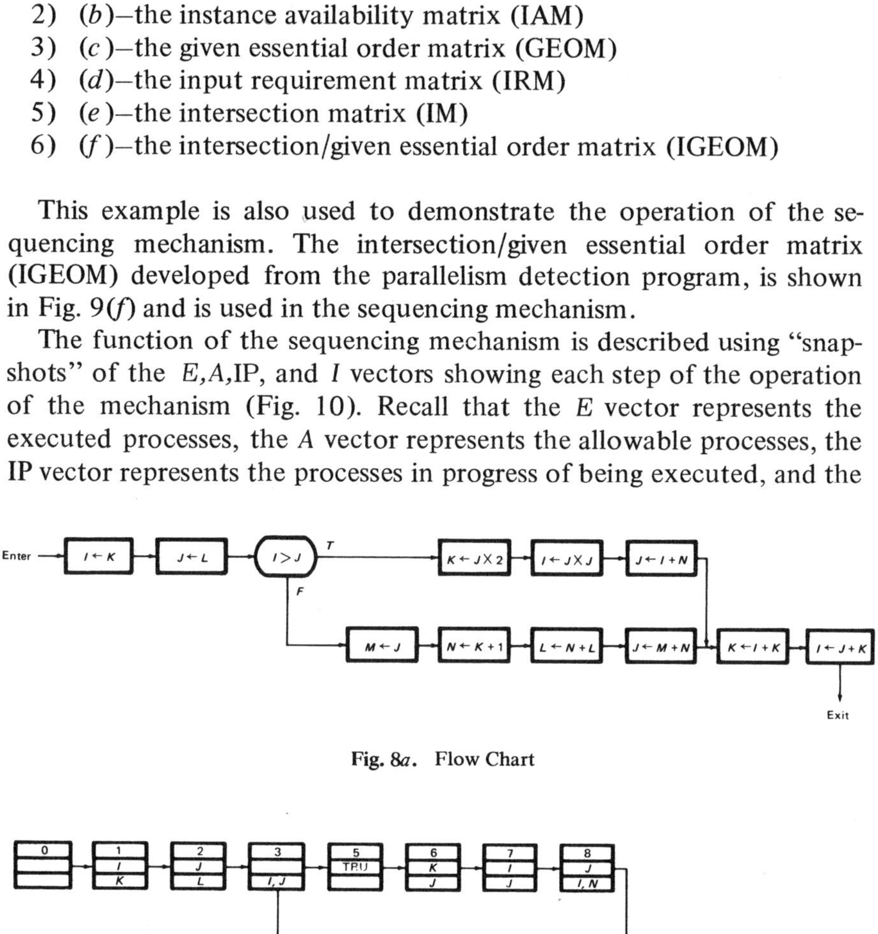

Fig. 8*a*. Flow Chart

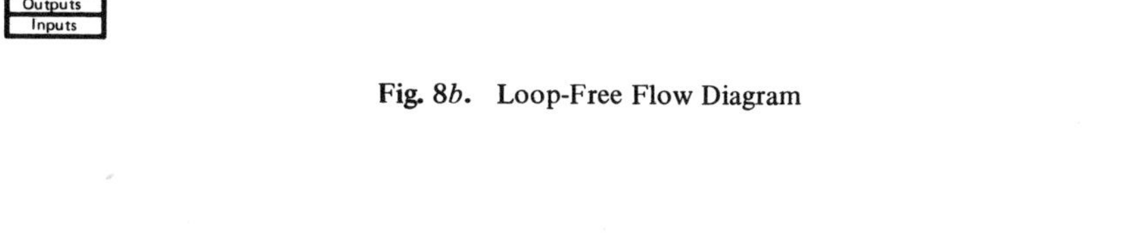

Fig. 8*b*. Loop-Free Flow Diagram

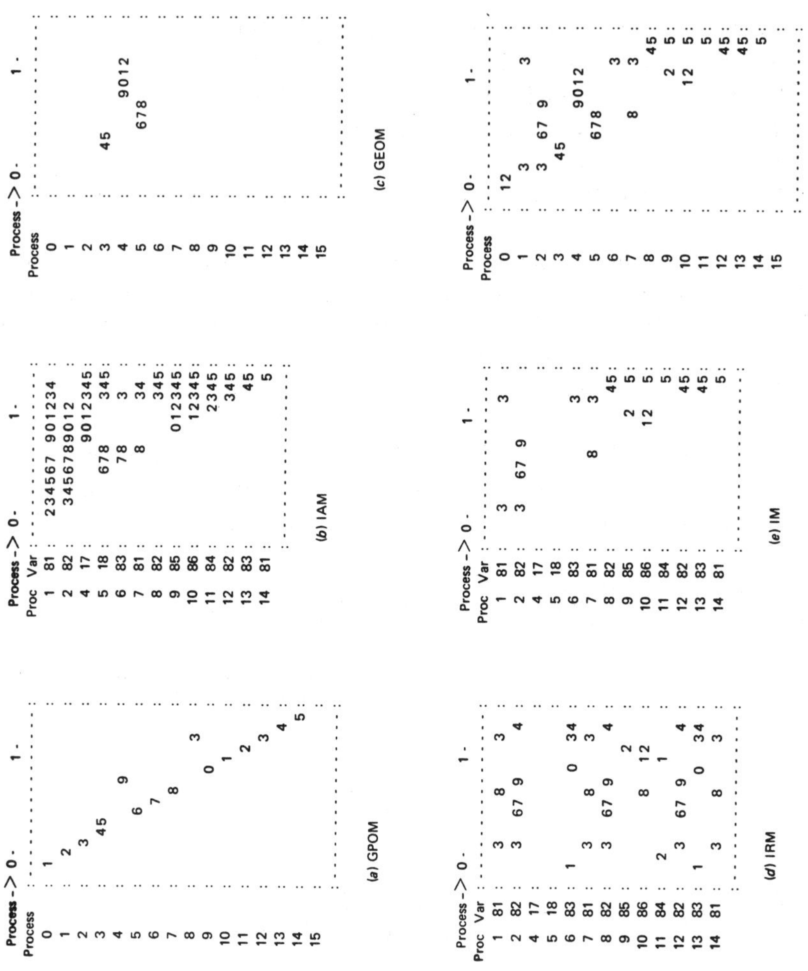

Fig. 9. Detection Program Outputs for Example

State		0 1 2 3 4 5 6 7 8 9 10 11 12 13 14 15	Algorithm Step	
0	E			Initial State
	A	0		
	IP			
	I			
1	E		4, 5	Search: result—process 0 initiable
	A	0		Set I_0
	IP			
	I	0		
2	E		6, 7	Initiate process 0
	A	0		Set IP_0, Reset I_0
	IP	0		
	I			
3	E	0	1, 2, 3	Process 0 execution complete
	A	0 1 2		Set E_0. Reset IP_0
	IP			
	I			
4	E	0	4, 5	Search: result—processes 1, 2 initiable
	A	0 1 2		Set I_1, I_2
	IP			
	I	1 2		
5	E	0	6, 7	Initiate processes 1, 2
	A	0 1 2		Set IP_1, IP_2. Reset I_1, I_2
	IP	1 2		
	I			
6	E	0 1	1, 2, 3	Process 1 execution complete
	A	0 1 2		Set E_1. Reset IP_1
	IP	2		
	I			
7	E	0 1	4, 5	Search: result—no process initiable
	A	0 1 2		
	IP	2		
	I			
8	E	0 1 2	1, 2, 3	Process 2 execution complete
	A	0 1 2 3		Set E_2. Reset IP_2
	IP			
	I			
9	E	0 1 2	4, 5	Search: result—process 3 initiable
	A	0 1 2 3		Set I_3
	IP			
	I	3		

Fig. 10. Example Showing Operation of Sequencing Mechanism

State		0 1 2 3 4 5 6 7 8 9 10 11 12 13 14 15	Algorithm Step	
10	E	0 1 2	6, 7	Initiate process 3
	A	0 1 2 3		Set IP_3. Reset I_3
	IP	3		
	I			
11	E	0 1 2 3	1, 2, 3	Process 3 execution complete
	A	0 1 2 3		Set E_3/Reset IP_3
	IP			Since process 3 is conditional,
	I			a special case
			3a	Assume false branch is chosen—then temporary search vector is: E_4 (chosen) = 0, E_5 (not chosen) = 1
12	E	5	3b	Search: result is those processes exclusively in true (not chosen) branch are indicated in A (A_6, A_7, A_8 are true)*/The positions 4 and 5 of the temporary search vector are shown in E
	A	6 7 8		
	IP			
	I			
13	E	0 1 2 3 6 7 8	3c	A or $E \rightarrow E$ (using E from State 11)
	A			
	IP			
	I			
14	E	0 1 2 3 4 5 6 7 8	3d	Set E_4 and E_5 since they are the two processes after the conditional
	A			
	IP			
	I			
15	E	0 1 2 3 4 5 6 7 8	4, 5	Resume normal operation
	A	0 1 2 3 4 5 6 7 8 9 10 13		Search: result—processes 9, 10 and 13 initiable**
	IP			
	I	9 10 13		Set I_9, I_{10}, I_{13}
16	E	0 1 2 3 4 5 6 7 8	6, 7	Initiate processes 9, 10, and 13
	A	0 1 2 3 4 5 6 7 8 9 10 13		Set IP_9, IP_{10}, IP_{13}
	IP	9 10 13		Reset I_9, I_{10}, I_{13}
	I			
17	E	0 1 2 3 4 5 6 7 8 9	1, 2, 3	Process 9 execution complete
	A	0 1 2 3 4 5 6 7 8 9 10 13		Set E_9/Reset IP_9
	IP	10 13		
	I			

*Note: If true branch were chosen, temporary search vector would be: $E_4 = 1$, $E_5 = 0$. Search result: those processes exclusively in false branch would be indicated in A (A_9, A_{10}, A_{11}, A_{12}).

**Process 13 although being outside both branches of the conditional, is initiable as soon as the condition is determined to be false.

Fig. 10. *(continued)*

State		0 1 2 3 4 5 6 7 8 9 10 11 12 13 14 15	Algorithm Step	
18	*E*	0 1 2 3 4 5 6 7 8 9	4, 5	Search: result—no process can be initiated
	A	0 1 2 3 4 5 6 7 8 9 10 13		
	IP	10 13		
	I			
19	*E*	0 1 2 3 4 5 6 7 8 9 10 13	1, 2, 3	Processes 10 and 13 execution complete at same time
	A	0 1 2 3 4 5 6 7 8 9 10 11 12 13		Set E_{10}, E_{13}/Reset IP_{10}, IP_{13}
	IP			
	I			
20	*E*	0 1 2 3 4 5 6 7 8 9 10 13	4, 5	Search: result—processes 11 and 12 are initiable
	A	0 1 2 3 4 5 6 7 8 9 10 11 12 13		Set I_{11}, I_{12}
	IP			
	I	11 12		
21	*E*	0 1 2 3 4 5 6 7 8 9 10 13	6, 7	Initiate processes 11 and 12
	A	0 1 2 3 4 5 6 7 8 9 10 11 12 13		Set IP_{11}, IP_{12}
	IP	11 12		
	I			
22	*E*	0 1 2 3 4 5 6 7 8 9 10 11 12 13	1, 2, 3	Process 11 and 12 execution completed at same time
	A	0 1 2 3 4 5 6 7 8 9 10 11 12 13 14		Set E_{11}, E_{12}/Reset IP_{11}, IP_{12}
	IP			
	I			
23	*E*	0 1 2 3 4 5 6 7 8 9 10 11 12 13	4, 5	Search: result—process 14 initiable
	A	0 1 2 3 4 5 6 7 8 9 10 11 12 13 14		Set I_{14}
	IP			
	I	14		
24	*E*	0 1 2 3 4 5 6 7 8 9 10 11 12 13	6, 7	Initiate process 14
	A	0 1 2 3 4 5 6 7 8 9 10 11 12 13 14		Set IP_{14}/Reset I_{14}
	IP	14		
	I			
25	*E*	0 1 2 3 4 5 6 7 8 9 10 11 12 13 14	1, 2, 3	Process 14 execution completed
	A	0 1 2 3 4 5 6 7 8 9 10 11 12 13 14 15		Set E_{14}/Reset IP_{14}
	IP			
	I			
26	*E*	0 1 2 3 4 5 6 7 8 9 10 11 12 13 14	4, 5	Search: result—process 15 initiable
	A	0 1 2 3 4 5 6 7 8 9 10 11 12 13 14 15		Set I_{15}
	IP			
	I	15		
27	*E*	0 1 2 3 4 5 6 7 8 9 10 11 12 13 14	6, 7	Initiate process 15
	A	0 1 2 3 4 5 6 7 8 9 10 11 12 13 14 15		Set IP_{15}, Reset I_{15}
	IP	15		
	I			

Fig. 10. *(continued)*

State		0 1 2 3 4 5 6 7 8 9 10 11 12 13 14 15	Algorithm Step	
28	E	0 1 2 3 4 5 6 7 8 9 10 11 12 13 14 15	1, 2	Process 15 execution complete
	A	0 1 2 3 4 5 6 7 8 9 10 11 12 13 14 15		Set E_{15}/Reset IP_{15}
	IP			E = all ones/segment execution complete
	I			

Fig. 10. *(continued)*

I vector represents the initiable processes. The E vector (since it is the only column vector) is transposed for ease of representation. Associated with each vector quadruple are numbers that refer to the steps of the procedure which yield the result. Any assumption made concerning the sequence of events is stated at the point of assumption.

CONCLUSIONS

The primary goal of this investigation was to develop techniques for exposing and exploiting parallelism at the instruction level. At this level, it was contended that (1) the parallelism be exposed by a detection program rather than programmer specification, and (2) the exploitation be accomplished using special hardware rather than software methods. A detection algorithm was developed and a program was written to demonstrate the feasibility of the algorithm. A hardware mechanism was designed to control the sequencing of the partially ordered instructions using the ordering information generated by the detection program.

The evaluation of new techniques to be included in computing systems involves the comparison of the potential gains resulting from their implementation balanced against the additional costs for the implementation.

The potential gains were mentioned as motivations for the study of parallelism and include faster execution times and more efficient use of system resources (especially memories). Of course, these gains are a function of the existence of parallelism at this level, which is an area of study not covered in this investigation.

The additional costs include:

1) increased time for program compilation because of inclusion of the algorithm for automatic detection
2) increased memory required for storage of order control information

3) extra logic and possible increased time for sequencing of the partially ordered events.

The increased compilation time is estimated from the times to execute the parallelism detection program, although much of the execution time of the program is to perform functions already existing in compilers (e.g. language recognizer). The estimated increase in execution time for compilation with automatic detection of parallelism is 10 per cent to 50 per cent. Of course, the time required for detection (and in fact for compilation) becomes negligible for those sections of code that are executed repeatedly such as production programs or code within loop bodies.

The amount of memory required for storage of order control information depends on the number of processes per segment. If there are n processes per segment, then an n^2(or $n[n-1]/2$ if upper triangular is used) Boolean array is needed. Note that only one copy of the partial order information among the processes in a loop body is necessary since the partial order is the same for all iterations. The percentage increase in storage necessary for the partial order information depends upon (1) the number of bits required to describe a process, and (2) the number of processes (program size) relative to the data storage requirements.

The extra logic necessary for the sequencing mechanism depends upon the number of processes per segment. For 16 processes in a segment, the sequencing mechanism requires approximately 500 gates. Since only a few connections are needed to the package, and the logic is of a regular structure, the device is well suited to fabrication in LSI.

The time required for determining the "next set of processes to be initiated" using the suggested device is one clock time. This time may be completely overlapped with instruction execution time just as the instruction counter incrementing is in conventional machines. Thus, no extra time need be spent to control the sequencing of the partially ordered processes.

The gains resulting from the implementation of the techniques developed depends on the degree of parallelism that exists in problems and problem descriptions. The determination of statistics on the degree of parallelism requires extensive analysis of a large number and variety of programs. A greater payoff will be in the obvious areas of real-time problems, production programs, and loops within programs.

A more detailed description of this work, its relation with the work of other investigators, and a bibliography is contained in the suggested readings.

Readings

1. H. W. Bingham and E. W. Reigel, "Parallelism Exposure and Exploitation in Digital Computing Systems," *ECOM*, 02463-F, June, 1969.
2. E. W. Reigel, "Parallelism Exposure and Exploitation in Digital Computing Systems," (Ph.D. dissertation, University of Pennsylvania, Philadelphia, 1969; [also, TR-69-4, Burroughs Corporation, 1969]).